Deepen Your Mind

Deepen Your Mind

前言

巨量資料發展至今,早已不是一個新興詞語,巨量資料的應用已經無處不在。在巨量資料時代,我們面臨的不僅是巨量的資料,更重要的是巨量資料所帶來的資料的擷取、儲存、處理等各方面的問題。為了更快速、更全面地展示巨量資料的實作應用,本書以一個資料倉儲專案為切入點,帶領讀者一步步揭開巨量資料的面紗。

資料倉儲專案是學習巨量資料的重要基礎。本書以資料倉儲的架設為主線,從架設之初的架構選型、資料服務的整體策劃到資料的流向,資料的擷取、儲存和計算,循序漸進,一步步地展開,進行細緻剖析。在對資料傳輸過程的說明中,穿插了資料倉儲的相關理論知識及巨量資料關鍵架構元件的說明,務必讓讀者對巨量資料有更深刻的了解,更加全面地了解巨量資料生態系統。

本書共 9 章,包含巨量資料與資料倉儲概論、專案需求描述、專案部署的環境準備、使用者行為資料獲取模組、業務資料獲取模組、資料倉儲架設模組、資料視覺化模組、即席查詢模組、中繼資料管理模組。

木專案採用主流的資料倉儲建模方式(確定業務過程、宣告粒度、確定維度、確實事實),覆蓋目前主流架構——擷取,Flume/Kafka/Sqoop;儲存,MySQL/Hadoop/HBase;計算,Hive/Tez;查詢,Presto/Druid/Kylin;視覺化,Superset;任務排程,Azkaban;中繼資料管理,Atlas;指令稿,Shell。整套專案包含業務指標近 100 個、Shell 指令稿 40 多個、使用者行為原始表 11 張,業務原始表 24 張、資料倉儲總表近 100 張。閱讀本書要求讀者具有一定的程式設計基礎,至少掌握一種程式語言(如 Java)及 SQL 查詢語言。

作者

本書書附資源,請至本公司官網 https://deepmind.com.tw,尋找對應書號下載。

目錄

01 巨量資料與資料倉儲概論

02 專案需求描述

05 業務資料獲取模組

06 資料倉儲架設模組

07 資料視覺化模組

08 即席查詢模組

09 中繼資料管理模組

巨量資料與資料倉儲概論

知其然知其所以然,在正式開始學習之前,本章先為讀者解答一些基本的概念問題。

- 什麼是巨量資料?
- 巨量資料牛熊圈的主要組成是什麼?
- 巨量資料應用在哪些企業?
- 什麼是資料倉儲?
- 資料倉儲可以用來做什麼?

本書的學習需要讀者具備一定的基礎,本章會列出說明。同時,對學習後讀者可以收穫的成果進行簡單的介紹。

1.1 巨量資料概論

1.1.1 什麼是巨量資料

讀者首先需要了解什麼是資料?什麼是巨量資料?

資料在我們的生活中無處不在,清晨起床,用手機開啟新聞資訊,此時就產生了資料;早上上班乘坐捷運,刷悠遊卡進站,又產生了資料;開啟購物網站,下單購買商品,還會產生資料……生活在當今這個高度資訊化的世界,一切行為幾乎都可以用資料來描述,這種情況發生在每個人的身上。每時每刻都有上億筆資料產生,這些巨量資料流進了那些提供網際網路服務的公司,儲存在他們的系統中。如果不對其加以利用,則這些資料只是拖慢系統的沉重負擔,但如果善於採擷,則這些資料就是蘊藏極大價值的寶藏!

那麼巨量資料究竟是什麼?國際頂級權威諮詢機構麥肯錫列出定義,巨量資料指的是所涉及的資料集規模已經超過了傳統資料庫軟體取得、儲存、管理和分析的能力。這是一個被故意設計成主觀性的定義,並且是一個關於多大的資料集才能被認為是巨量資料的可變定義,即並不定義大於一個特定資料的 TB 才叫巨量資料。因為隨著技術的不斷發展,符合巨量資料標準的資料集容量也會增長;並且定義隨不同產業也有變化,這依賴於在一個特定企業通常何種軟體和資料集有多大。因此,巨量資料在今天不同產業中的範圍可以從幾十 TB 到幾 PB。隨著資料量越來越大,巨量資料的儲存和分析計算面臨的挑戰也越來越大。

1.1.2 巨量資料生態圈簡介

在巨量資料高速發展的幾年中,已經形成了一個完備多樣的巨量資料生態圈,如圖 1-1 所示。從圖 1-1 中可以看出,巨量資料生態圈分為 7 層,這 7 層如果進一步概括,可以歸納為資料獲取層、資料計算層和資料應用層 3 層結構。

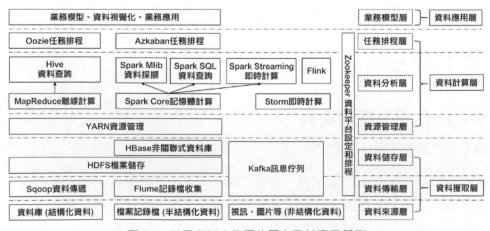

圖 1-1 巨量資料生態圈的層次及其應用舉例

1. 資料獲取層

資料獲取層是整個巨量資料平台的源頭，是整個巨量資料系統的基礎。目前許多公司的業務平台每天都會產生巨量的記錄檔資料，收集記錄檔資料供離線和線上的分析系統使用是記錄檔收集系統需要做的事情。除記錄檔資料外，巨量資料系統的資料來源還包含業務資料庫的結構化資料，以及視訊、圖片等非結構化資料。隨著巨量資料的重要性逐漸突顯，巨量資料擷取系統的合理架設就顯得尤為重要。

巨量資料擷取過程中的挑戰越來越多，主要來自以下幾個方面。

① 資料來源多種多樣。
② 資料量大且變化快。
③ 如何保障所擷取資料的可用性。
④ 如何避免擷取重複的資料。
⑤ 如何保障所擷取資料的品質。

針對這些挑戰，記錄檔收集系統需要具有高可用性、高可靠性、可擴允性等特徵。現在主流的資料傳輸層的工具有 Sqoop、Flume、DataX 等，透過多種工具的配合使用，可以滿足多種資料來源的擷取傳輸工作。同時資料傳輸層大部分的情況下還需要對資料進行初步的清洗、過濾、整理、格式化等一系列轉換操作，使資料轉為適合查詢的格式。資料獲取完成後，需要選用合適的資料儲存系統，考慮到資料儲存的可用性及後續計算的便利性，通常選用分散式檔案系統，如 HDFS 和 HBase 等。

2. 資料計算層

巨量資料僅被擷取到資料儲存系統是遠遠不夠的，只有透過整合計算，資料中的潛在價值才可以被採擷出來。

資料計算層可以劃分為離線資料計算和即時資料計算。離線資料計算主要是指傳統的資料倉儲概念，資料計算可以以天為單位，還可以細分為小時或整理為以周和月為單位，主要以 $T+1$ 的模式進行，即每天凌晨處理上一天的資料。

隨著業務的發展，部分業務需求對即時性的要求逐漸加強，即時計算開始佔有較大的比例，即時計算的應用場景也越來越廣泛，舉例來說，電子商務即時交易資料更新、裝置即時執行狀態報告、活躍使用者區域分佈即時變化等。生活中比較常見的有地圖與位置服務應用即時分析路況、天氣應用即時分析天氣變化趨勢等。

巨量資料的計算需要使用的資源是極大的，大量的資料計算任務通常需要透過資源管理系統共用一個叢集的資源，YARN 便是資源管理系統的典型代表。透過資源管理系統可以加強叢集的使用率、降低運行維護成本。巨量資料的計算通常不是獨立的，一個計算任務的執行很大可能依賴於另一個計算任務的結果，使用任務排程系統可以極佳地處理任務之間的相依關係，實現任務的自動化執行。

無論何種資料計算，進行資料計算的前提是標準合理地規劃資料，架設標準統一的資料倉儲系統。透過架設合理的、全面的資料倉儲系統，儘量避開資料容錯和重複計算等問題，使資料的價值發揮到最大。為此，資料倉儲分層理念被逐漸豐富完善，目前應用比較廣泛的資料倉儲分層理念將資料倉儲分為 4 層，分別是原始資料層、明細資料層、整理資料層和應用資料層。透過資料倉儲不同層次之間的分工分類，使資料更加規範化，可以幫助使用者需求獲得更快實現，並且可以更加清楚明確地管理資料。

3. 資料應用層

當資料被整合計算完成之後，需要最後提供給使用者使用，這就是資料應用層。不同的資料平台針對其不同的資料需求有各自對應的資料應用層的規劃設計，資料的最後需求計算結果可以建置在不同的資料庫上，舉例來說，MySQL、HBase、Redis、Elasticsearch 等。透過這些資料庫，使用者可以很方便地存取最後的結果資料。

最後的結果資料由於針對的使用者不同，可能有不同層級的資料呼叫量，面臨著不同的挑戰。如何能更穩定地提供給使用者服務、滿足各種使用者複雜的資料業務需求、保障資料服務介面的高可用性等，都是資料應用層需要考慮的問題。

1.1.3 巨量資料應用場景

巨量資料無處不在，包含金融、汽車、網際網路、電信、物流、電影娛樂等在內的社會各行各業都已經融入了巨量資料的印跡。

金融業：巨量資料在金融企業的應用範圍很廣，比較典型的金融巨量資料的應用場景集中在使用者經營、資料風控、產品設計和決策支援等，舉例來說，巨量資料分析為客戶推薦合適的產品、預測未來理財產品的受歡迎程度。金融企業的資料大部分是結構化資料，儲存在傳統關聯式資料庫中，透過資料採擷可以分析出隱藏在交易資料中的極大商業價值。

汽車企業：巨量資料在汽車企業釋放出的極大價值引發越來越多汽車企業人士的關注，最基本的有利用巨量資料分析消費者的行為決定汽車行銷方向、分析使用者維保行為幫助二手車真實價值評估、智慧導航巨量資料為智慧化交通提供更多的空間和可能，此外利用巨量資料和物聯網技術製造的無人駕駛汽車，在不遠的未來將走入我們的日常生活中。

網際網路企業：巨量資料在網際網路企業的應用已經滲透到了各方面，幾乎客戶的所有行為都會在網際網路平台上留下痕跡，所以網際網路企業可以方便地取得大量的使用者行為資訊，透過分析使用者行為，可以制定更有針對性的服務策略。除利用巨量資料提升自己的業務以外，網際網路企業已經開始實現資料業務化，利用巨量資料發現新的商業價值。以阿里巴巴為例，它不僅在加強個性化推薦，開發「千人千面」這種針對消費者的巨量資料應用，並且還在嘗試利用巨量資料進行智慧客戶服務。

電信業：電信業巨量資料的發展仍處於探索階段，目前電信業者對巨量資料的應用主要表現在利用巨量資料進行基礎設施建設最佳化、網路營運管理最佳化及市場精準行銷，並利用巨量資料實現客戶線下分析，及時掌握客戶線下傾向，實施客戶挽留措施。

物流業：物流業透過巨量的物流資料，以及運輸、倉儲、包裝、運輸等環節中有關的資料資訊等，採擷出新的深層價值，進而加強運輸與配送效率、減少物流成本，更有效地滿足客戶服務要求。實際表現在利用巨量資料最佳化物流網路，車貨比對，庫存預測，供應鏈協作管理，進一步加強物流效率，降低物流成本。

電影娛樂：巨量資料在精準行銷中所起的作用尤其反映在電影領域，採擷精準人群的意義更加非同一般，因為不同類型的影片背後是截然不同的受眾群眾，針對不同的受眾，還可以利用巨量資料去採擷受眾的所思、所想、所好，為內容策劃甚至演員選角提供更多的資料支援。

巨量資料的價值，遠遠不止於此，巨量資料對各行各業的滲透，大幅推動了社會生產和生活的發展，未來必將產生重大而深遠的影響。

1.2 資料倉儲概論

1.2.1 什麼是資料倉儲

資料倉儲，英文名稱為 Data Warehouse，可簡寫為 DW 或 DWH。資料倉儲是為企業所有等級的決策制定過程，提供所有類型資料支援的資源集合。它出於分析性報告和決策支援目的而建立。

隨著技術的高速發展，經過多年的資料累積，各網際網路公司已儲存了巨量的原始資料和各種業務資料，所以資料倉儲技術是各網際網路公司目前需要注重發展的技術領域。資料倉儲是針對分析的整合化資料環境。透過對資料倉儲中的資料進行分析，可以幫助企業改進業務流程、控制成本、加強產品品質等。

1.2.2 資料倉儲能幹什麼

資料倉儲系統是一個資訊服務和管理平台，它從業務處理系統獲得資料，主要以星形模型和雪花模型組織資料，並為使用者從資料中取得資訊和知識提供各種方法。

按照功能結構劃分，資料倉儲系統至少應該包含資料取得（Data Acquisition）、資料儲存（Data Storage）和資料存取（Data Access）三個關鍵部分。

企業資料倉儲的建設，是以現有企業業務系統和大量業務資料的累積為基礎的。資料倉儲不是靜態的概念，只有把資訊及時交給需要這些資訊的使用者，幫助他們做出改善其業務經營的決策，資訊才能發揮作用、才有意義。而把資訊加以整理歸納和重組，並及時提供給對應的管理決策人員，是資料倉儲的根本任務。因此，從企業的角度看，資料倉儲的建設是一個工程。

1.2.3 資料倉儲的特點

1. 資料倉儲中的資料是針對主題的

與傳統資料庫針對應用進行資料組織的特點相對應，資料倉儲中的資料是針對主題進行資料組織的。什麼是主題呢？首先，主題是一個抽象的概念，是較高層次上企業資訊系統中的資料綜合、歸類並進行分析利用的抽象。在邏輯意義上，它對應企業中某　巨觀分析領域所涉及的分析物件。針對主題的資料組織方式，就是在較高層次上對分析物件的資料的完整、一致的描述，能完整、統一地刻畫各分析物件所涉及企業的各項資料，以及資料之間的聯繫。所謂較高層次是相對於針對應用的資料組織方式而言的，是指按照主題進行資料組織的方式具有更高的資料抽象等級。

2. 資料倉儲中的資料是整合的

資料倉儲中的資料是從原有的、分散的資料庫中取出來的，取出的資料可分為操作類型資料和分析類型資料兩大類，兩者之間差別甚大。第一，資料倉儲的每個主題所對應的來源資料在原有的各分散資料庫中有許多重複和不一致的地方，且來自不同連線系統的資料都和不同的應用邏輯綁定在一起；第二，資料倉儲中的資料不是從原有的資料庫系統中直接獲得的。因此，資料在進入資料倉儲之前，必然要經過統一與綜合，這一步是資料倉儲建設中最關鍵、最複雜的一步，所要完成的工作如下。

① 要統一來源資料中的所有矛盾之處，如欄位的名稱相同異義、異名同義、單位不統一、字組長度不一致等。

② 進行資料綜合和計算。資料倉儲中的資料綜合工作可以在從原有資料庫中取出資料時完成，但大多數是在資料倉儲內部完成的，即進入資料倉儲以後進行資料綜合。

3. 資料倉儲中的資料是不可更新的

資料倉儲中的資料主要供企業管理者決策分析使用，所涉及的資料操作主要是資料查詢，一般情況下並不進行修改操作。資料倉儲中的資料反映的是相當長的一段時間內歷史資料的內容，是不同時間的資料庫快照的集合，以及以這些快照為基礎進行統計、綜合和重組的匯出資料，而非連線處理的資料。資料庫中進行連線處理的資料經過整合輸入資料倉儲中，一旦資料倉儲儲存的資料已經超過資料倉儲的資料儲存期限，這些資料會被刪除。因為資料倉儲只能進行資料查詢操作，所以資料倉儲管理系統相比資料庫管理系統而言要簡單得多。資料庫管理系統中的許多技術困難，如完整性保護、平行處理控制等，在資料倉儲管理系統中幾乎可以忽略。但是在資料倉儲中要查詢的資料量通常很大，所以就對資料查詢提出了更高的要求，它要求採用各種複雜的索引技術，同時由於資料倉儲針對的是商業企業的高層管理者，他們會對資料查詢介面的人性化性和資料表示提出更高的要求。

4. 資料倉儲中的資料是隨時間不斷變化的

資料倉儲中的資料不可更新是針對應用來說的，也就是說，資料倉儲的使用者在進行資料分析和處理時是不進行資料更新操作的。但並不是說，在從資料整合輸入資料倉儲開始到最後被刪除的整個資料生存週期中，資料倉儲中的資料都是永遠不變的。

資料倉儲中的資料是隨時間不斷變化的，這是資料倉儲的第 4 個特點。這一特點表現在以下 3 個方面。

① 資料倉儲隨著時間的變化不斷增加新的資料內容。資料倉儲系統必須不斷捕捉 OLTP 資料庫中變化的資料，並追加到資料倉儲中，也就是要不

斷地產生 OLTP 資料庫的快照，經過統一整合後增加到資料倉儲中；但對於確實不再變化的資料庫快照，如果捕捉到新的變化資料，則只產生一個新的資料庫快照增加進去，而不會對原有的資料庫快照進行修改。

② 資料倉儲隨時間的變化不斷刪除舊的資料內容。資料倉儲中的資料也有儲存期限，一旦超過這一期限，就要被刪除。只是資料倉儲中的資料時限要遠遠長於操作型環境中的資料時限。在操作型環境中一般只儲存 60 ～ 90 天的資料，而在資料倉儲中則需要儲存較長時限的資料（如 5 ～ 10 年），以滿足 DSS 進行趨勢分析的要求。

③ 資料倉儲中包含了大量的綜合資料，其中很多資料與時間密切相關，如數據經常按照時間段進行綜合，或隔一定的時間進行抽樣等。這些資料要隨著時間的變化不斷地進行重新綜合。因此，資料倉儲的資料特徵都包含時間項，以標明資料的歷史時期。

1.3 學前導讀

在開始學習之前，希望讀者仔細閱讀以下內容，便於開啟巨量資料學習之門。

1.3.1 學習的基礎要求

在學習本書之前，讀者需要提前了解一些基礎知識，有助更加輕鬆、快速地掌握巨量資料的相關內容，在後續專案的架設過程中能更加得心應手，為深入學習巨量資料打下堅實的基礎。

首先，學習巨量資料技術，讀者一定要掌握一個操作巨量資料技術的利器，這個利器就是一種程式語言，如 Java、Scala、Python、R 等。本書以 Java 為基礎進行撰寫，所以學習本書需要讀者具備一定的 Java 基礎知識和 Java 程式設計經驗。

其次，讀者還需要掌握一些資料庫知識，如 MySQL、Oracle 等，並熟練使用 SQL，本書將出現大量的 SQL 操作。

最後，讀者還需要掌握一種作業系統技術，即在伺服器領域佔主導地位的 Linux，只要能夠熟練使用 Linux 的常用系統指令、檔案操作指令和一些基本的 Linux Shell 程式設計即可。巨量資料系統需要處理業務系統伺服器產生的巨量記錄檔資料資訊，這些資料通常儲存在伺服器端，各大網際網路公司常用的作業系統是在實際工作中安全性和穩定性很高的 Linux 或 UNIX。巨量資料生態圈的各架構元件也普遍執行在 Linux 上。

1.3.2 你將學到什麼

本書將帶領讀者完成一個完整的資料倉儲架設及需求實現專案，大致可以劃分為 3 部分：資料倉儲概論及專案需求描述、專案架構架設和專案需求實現。

在專案需求及架構說明部分，讀者可以更加了解一個資料倉儲專案的實際需求，以及根據需求如何完成架構選型的過程。

在專案架構架設部分，讀者將跟隨本書從作業系統開始，一步步架設自己的虛擬機器系統，了解各架構的基礎，完成各架構的基本設定，最後形成一個可以正常執行的巨量資料虛擬機器系統。

在專案需求實現部分，本書將從使用者行為資料獲取模組、業務資料獲取模組、資料倉儲架設模組、即席查詢模組、中繼資料管理模組 5 個方面對需求進行實現，讀者透過本部分的學習將了解一個完整的資料倉儲系統從資料來源到資料的最後展示是如何實現的，同時還能學到資料倉儲相關的理論知識，掌握 Hive、Sqoop、Flume 等記錄檔資料獲取工具的工作原理及應用方法。本部分對電子商務資料倉儲的常見實戰指標及困難實戰指標進行了透徹說明，實際指標包含每日、每週、每月活躍裝置明細，留存使用者比例，沉默使用者、回流使用者、流失使用者統計，最近連續 3 周活躍使用者統計，最近 7 天內連續 3 天活躍使用者統計等。

透過對資料倉儲系統的學習，讀者能夠對資料倉儲專案建立起清晰、明確的概念，系統、全面地掌握各項資料倉儲專案技術，輕鬆應對各種資料倉儲的難題。

1.4 本章歸納

本章主要為讀者說明了巨量資料的概念及巨量資料的實際應用場景，讓讀者
對資料倉儲建立了初步的認識，還介紹了資料倉儲的概念、作用及特點，並
為讀者明確了學習本書應該具備的技術基礎以及最後的學習成果目標，為讀
者之後的學習做好準備。

專案需求描述

隨著網際網路的迅速發展,電子商務變得越來越重要,線上購物已經成為廣大消費者主要的消費方式之一。滿足使用者需求,讓使用者可以方便購物、快樂購物,讓使用者可以輕易、方便地找到自己所需的商品,商品能夠更加精準地發送到需要的使用者眼前,如此,才可以實現一個電子商務系統的真正價值,做到利潤最大化。

電子商務系統在滿足大量使用者的存取和購物需求的同時,還會產生大量的使用者行為資料及業務互動資料。使用者行為資料是指使用者在使用系統的時候進行點擊、瀏覽、增加購物車、收藏等行為時產生的記錄檔資料,在這些記錄檔資料中儲存著使用者的 id、點擊時間、手機型號、通路來源等資訊。業務互動資料是指在系統執行過程中由後端程式記錄下來的業務資料。舉例來說,使用者的訂單表、支付情況表、訂單詳情表、使用者詳情表等,這些資料或存在於前端伺服器中,或存在於後端伺服器的傳統資料庫中,如果加以利用則能採擷出其中極大的潛在價值。

為了大幅地採擷資料中的潛在價值,我們需要架設資料倉儲系統。資料倉儲系統從資料的擷取流程開始,將不同來源的資料統一擷取進資料倉儲中,在資料倉儲中對資料進行合理的分析、分類、儲存和計算,這個資料倉儲將針對所有有資料分析需求的使用者,包含企業決策者、營運人員、資料分析師等,提供給使用者多樣的資料服務,解決使用者對資料方面的需求。

本章將重點介紹本書中電子商務資料項目的需求，主要包含專案的產品描述、功能架構設計、系統流程圖設計、各模組的功能業務描述，以及根據需求確定的硬體和軟體環境。

本專案主要針對想要了解更多巨量資料專案相關知識的巨量資料從業人員、有一定程式設計基礎想要對巨量資料有所了解的巨量資料技術的初學者，以及想要更多地了解巨量資料的開發流程以期能更進一步地應用巨量資料的使用者。

2.1 任務概述

專案的需求說明非常重要，不僅可以讓程式設計師了解產品的需求，知道開發的目標結果，也可以讓系統應用人員了解業務流程的設計是否完善。所以在專案需求說明書中必須有詳細的系統功能說明。

2.1.1 產品描述

本資料倉儲專案將資料獲取、資料同步匯入、資料分層架設、需求分層實現、指令稿實現、任務定時排程、中繼資料管理等功能集合，提供資料展示頁面，讓使用者透過本資料倉儲專案，將電子商務系統產生的使用者行為資料和業務互動資料及時同步到資料倉儲中，並對資料需求進行分析計算，個別需求可以透過 Web 頁面獲得展示。還對即席查詢引擎進行了探索連線，平行考慮了三種即席查詢引擎，提供給使用者即席查詢服務。

2.1.2 系統目標

資料倉儲系統需要實現的目標如下：

- 環境架設完整，技術選型合理，架構服務分配合理；
- 資訊流完整，包含資料產生、資料獲取、資料倉儲建模、資料即席查詢；
- 能應對巨量資料的分析查詢；
- 實現中繼資料管理。

2.1.3 系統功能結構

如圖 2-1 所示，該資料倉儲系統主要分為 4 個功能模組，分別是資料獲取、資料倉儲平台、資料視覺化和即席查詢。

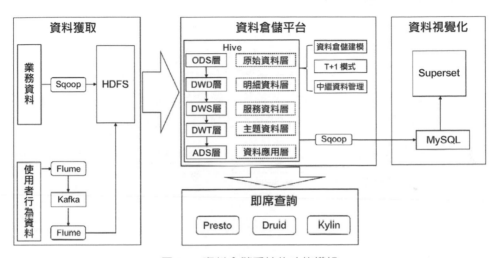

圖 2-1　資料倉儲系統的功能模組

資料獲取模組主要負責將電子商務系統前端的使用者行為資料以及業務互動資料獲取到巨量資料儲存系統中，所以資料獲取模組共分為兩大系統：使用者行為資料獲取系統和業務互動資料獲取系統。使用者行為資料主要以記錄檔的形式落盤（儲存在伺服器磁碟中，下同），採用 Flume 作為資料獲取架構對資料進行即時監控擷取；業務互動資料主要儲存在 MySQL 中，採用 Sqoop$T+1$ 形式的擷取。

資料獲取模組負責將原始資料獲取到資料倉儲中，合理建表，並針對資料進行清洗、逸出、分類、重組、合併、拆分、統計等，將資料合理分層，相當大地減少資料重複計算的情況。在針對固定長期需求進行資料倉儲的合理建設的同時，還應考慮使用者的即席查詢需求，需對外提供即席查詢介面。一方面是為了讓使用者能夠更高效率地採擷和使用資料；另一方面是為了讓平台管理人員能夠更加有效地做好系統的維護管理工作，對資料倉儲的中繼資料資訊建立管理。

資料視覺化主要負責將最後需求結果資料匯入 MySQL 中，供資料使用者使用或對資料進行 Web 頁面展示。

2.1.4 系統流程圖

資料倉儲系統主要流程如圖 2-2 所示。前端埋點（指資料獲取的技術方式，下同）使用者行為資料經生產層 Flume Agent、Kafka、消費層 Flume Agent 落盤到 HDFS 中，業務互動資料經 Sqoop 擷取到 HDFS 中，HDFS 中的資料經過 Hive 的相關操作，將資料進行分析轉換，形成合理分層，最後獲得需求結果資料，將資料匯出 MySQL 中，實現資料視覺化，並提供即席查詢服務。

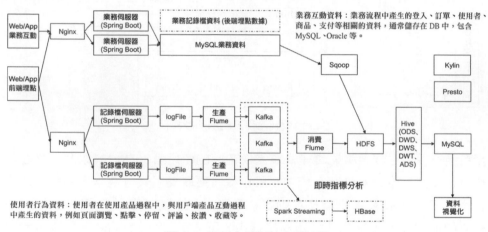

圖 2-2 資料倉儲系統主要流程

2.2 業務描述

2.2.1 擷取模組業務描述

1. 資料產生模組之使用者行為資料基本格式

使用者執行的一些操作會產生使用者行為資料發送到伺服器，資料分為公共欄位和業務欄位。

■ 公共欄位：基本所有 Android 手機包含的欄位。
■ 業務欄位：前端埋點上報的欄位，有實際的業務類型。

以下範例表示業務欄位的上傳。

```
{
"ap":"xxxxx",                           // 專案資料來源
"cm": {                                 // 公共欄位
    "mid": "",                          // (String) 裝置唯一標識
        "uid": "",                      // (String) 使用者標識
        "vc": "1",                      // (String) versionCode，程式版本編號
        "vn": "1.0",                    // (String) versionName，程式版本名
        "l": "zh",                      // (String) 系統語言
        "sr": "",                       // (String) 通路號，應用從哪個通路來
        "os": "7.1.1",                  // (String) Android版本
        "ar": "CN",                     // (String) 區域
        "md": "BBB100-1",               // (String) 手機型號
        "ba": "blackberry",             // (String) 手機品牌
        "sv": "V2.2.1",                 // (String) sdkVersion
        "g": "",                        // (String) gmail
        "hw": "1620x1080",              // (String) heightXwidth，螢幕長寬
        "t": "1506047606608",           // (String) 用戶端記錄檔產生的時間
        "nw": "WIFI",                   // (String) 網路模式
        "ln": 0,                        // (double) lng，經度
        "la": 0                         // (double) lat，緯度
    },
"et": [                                 //事件
        {
            "ett": "1506047605364", // 用戶端事件產生的時間
            "en": "display",            // 事件名稱
            "kv":                       // 事件結果，以key-value的形式自行定義
                "goodsid": "236",
                "action": "1",
                "extend1": "1",
                "place": "2",
                "category": "75"
            }
        }
    ]
}
```

範例記錄檔（伺服器時間戳記 | 記錄檔）如下。

```
1540934156385|{
    "ap": "gmall",
    "cm": {
        "uid": "1234",
        "vc": "2",
        "vn": "1.0",
        "l": "EN",
        "sr": "",
        "os": "7.1.1",
        "ar": "CN",
        "md": "BBB100-1",
        "ba": "blackberry",
        "sv": "V2.2.1",
        "g": "abc@gmail.com",
        "hw": "1620x1080",
        "t": "1506047606608",
        "nw": "WIFI",
        "ln": 0,
        "la": 0
    },
    "et": [
        {
            "ett": "1506047605364",  // 用戶端事件產生的時間
            "en": "display",          // 事件名稱
            "kv": {                    // 事件結果，以key-value的形式自行定義
                "goodsid": "236",
                "action": "1",
                "extend1": "1",
                "place": "2",
                "category": "75"
            }
        },{
        "ett": "1552352626835",
        "en": "active_background",
        "kv": {
          "active_source": "1"
        }
```

```
                }
            ]
        }
    }
```

2. 資料產生模組之事件記錄檔資料

（1）商品列表頁載入過程的事件名稱為 loading，產生的記錄檔資料的實際欄位名稱及欄位描述如表 2-1 所示。

表 2-1　商品列表頁載入過程 loading

欄位名稱	欄位描述
action	動作：開始載入 =1，載入成功 =2，載入失敗 =3
loading_time	載入時長：計算從下拉開始到介面傳回資料的時間（開始載入上報 0，載入成功或載入失敗上報實際時間）
loading_way	載入類型：讀取快取 =1，從介面拉新資料 =2（載入成功才會上報載入類型）
extend1	擴充欄位
extend2	擴充欄位
type	載入方式：自動載入 =1，使用者下拉頁面載入 =2，點擊底部按鈕載入 =3
type1	載入失敗碼：將載入失敗狀態碼報回來（報空為載入成功，沒有失敗）

（2）商品點擊的事件名稱為 display，產生的記錄檔資料的實際欄位名稱及欄位描述如表 2-2 所示。

表 2-2　商品點擊 display

欄位名稱	欄位描述
action	動作：曝光商品 =1，點擊商品 =2
goodsid	商品 id（伺服器端下發的商品 id）
place	順序（第幾件商品，第一件為 0，第二件為 1，依此類推）
extend1	曝光類型：第一次曝光 =1，重複曝光 =2
category	品項 id（伺服器端定義的品項 id）

（3）商品詳情展示的事件名稱為 newsdetail，產生的記錄檔資料的實際欄位名稱及欄位描述如表 2-3 所示。

表 2-3 商品詳情展示 newsdetail

欄位名稱	欄位描述
entry	頁面入口來源：應用首頁 =1，push=2，詳情頁相關推薦 =3
action	動作：開始載入 =1，載入成功 =2，載入失敗 =3，退出頁面 =4
goodsid	商品 id（伺服器端下發的商品 id）
show_style	商品樣式：無圖 =0，一張大圖 =1，兩張大圖 =2，三張小圖 =3，一張小圖 =4，一張大圖兩張小圖 =5
news_staytime	頁面停留時長：從商品開始載入到使用者關閉頁面所用的時間。若中途又跳躍到了其他頁面，則暫停計時，待回到詳情頁時恢復計時；若中途跳躍的時間超過 10 分鐘，則本次計時作廢，不上報本次數據；若未載入成功就退出，則報空
loading_time	載入時長：從頁面開始載入到介面傳回資料的時間（開始載入報 0，載入成功或載入失敗上報實際時間）
type1	載入失敗碼：將載入失敗狀態碼報回來（報空為載入成功）
category	品項 id（伺服器端定義的品項 id）

（4）廣告點擊的事件名稱為 ad，產生的記錄檔資料的實際欄位名稱及欄位描述如表 2-4 所示。

表 2-4 廣告點擊 ad

欄位名稱	欄位描述
entry	入口：商品列表頁 =1，應用首頁 =2，商品詳情頁 =3
action	動作：請求廣告 =1，取得快取廣告 =2，廣告位展示 =3，廣告展示 =4，廣告點擊 =5
content	狀態：成功 =1，失敗 =2
detail	失敗碼（沒有則上報空）
source	廣告來源：AdMob=1，Facebook=2，ADX（百度）=3，VK（俄羅斯）=4
behavior	使用者行為：主動取得廣告 =1，被動取得廣告 =2
newstype	Type：圖文 =1，圖集 =2，段子 =3，GIF=4，視訊 =5，調查 =6，純文字 =7，視訊 + 圖文 =8，GIF+ 圖文 =9，其他 =0
show_style	內容樣式：無圖（純文字）=6，一張大圖 =1，三張小圖 + 文 =4，一張小圖 =2，一張大圖兩張小圖 + 文 =3，圖集 + 文 =5，一張大圖 + 文 =11，GIF（大圖）+ 文 =12，視訊（大圖）+ 文 = 13。來自詳情頁相關推薦的商品，上報樣式都為 0（因為都是左文右圖）

（5）訊息通知的事件名稱為 notification，產生的記錄檔資料的實際欄位名稱及欄位描述如表 2-5 所示。

表 2-5 訊息通知 notification

欄位名稱	欄位描述
action	動作：通知產生 =1，通知出現 =2，通知點擊 =3，常駐通知展示（不重複上報，一天之內只報一次）=4
type	通知 id：預警通知 =1，天氣預報（早 =2，晚 =3），常駐 =4
ap_time	用戶端出現時間
content	備用欄位

（6）使用者後台活躍的事件名稱為 active_background，產生的記錄檔資料的實際欄位名稱及欄位描述如表 2-6 所示。

表 2-6 使用者後台活躍 active_background

欄位名稱	欄位描述
active_source	upgrade=1，download（下載）=2，plugin_upgrade=3

（7）使用者評價的事件名稱為 comment，產生的記錄檔資料的實際欄位名稱及欄位描述如表 2-7 所示。

表 2-7 使用者評價 comment

欄位名稱	欄位描述	欄位類型	長度	允許空	預設值
comment_id	評價表	int	10,0		
userid	使用者 id	int	10,0	√	0
p_comment_id	父級評價 id（為 0 則是一級評價，不為 0 則是回覆）	int	10,0	√	
content	評價內容	string	1000	√	
add_time	建立時間	string		√	
other_id	評價的相關 id	int	10,0	√	
praise_count	按讚數量	int	10,0	√	0
reply_count	回覆數量	int	10,0	√	0

（8）使用者收藏的事件名稱為 favorites，產生的記錄檔資料的實際欄位名稱及欄位描述如表 2-8 所示。

表 2-8　使用者收藏 favorites

欄位名稱	欄位描述	欄位類型	長度	允許空	預設值
id	主鍵	int	10,0		
course_id	商品 id	int	10,0	√	0
userid	使用者 id	int	10,0	√	0
add_time	建立時間	string		√	

（9）使用者按讚的事件名稱為 praise，產生的記錄檔資料的實際欄位名稱及欄位描述如表 2-9 所示。

表 2-9　使用者按讚 praise

欄位名稱	欄位描述	欄位類型	長度	允許空	預設值
id	主鍵	int	10,0		
userid	使用者 id	int	10,0	√	
target_id	按讚物件的 id	int	10,0	√	
type	按讚類型：問答按讚 =1，問答評價按讚 =2，文章按讚 =3，評價按讚 =4	int	10,0	√	
add_time	建立時間	string		√	

（10）產生的錯誤記錄檔資料的實際欄位名稱及欄位描述如表 2-10 所示。

表 2-10　錯誤記錄檔

欄位名稱	欄位描述
errorBrief	錯誤摘要
errorDetail	錯誤詳情

3. 資料產生模組之啟動記錄資料

啟動記錄，產生的記錄檔資料的實際欄位名稱及欄位描述如表 2-11 所示。

表 2-11　啟動記錄

欄位名稱	欄位描述
entry	入口：push=1，widget=2，icon=3，notification=4，lockscreen_widget=5
open_ad_type	啟動廣告類型：啟動原生廣告 =1，啟動 APP 中廣告 =2
action	狀態：成功 =1，失敗 =2
loading_time	載入時長：從下拉開始到介面傳回資料的時間（開始載入報 0，載入成功或載入失敗上報實際時間）
detail	失敗碼（沒有則上報空）
extend1	失敗的 message（沒有則上報空）
en	記錄檔類型：en=start

4. 資料獲取模組之生產資料

透過執行生產記錄檔的 jar 套件來模擬資料產生的過程，獲得記錄檔資料。

5. 資料獲取模組

資料獲取模組主要擷取並落盤到伺服器資料夾中的記錄檔資料，需要監控多個記錄檔產生資料夾並能夠做到中斷點續傳，實現資料消費 "at least once" 語義，以及能夠根據擷取到的記錄檔內容對記錄檔進行分類擷取落盤，發往不同的 Kafka topic。Kafka 作為一個訊息中介軟體造成記錄檔緩衝作用，避免同時發生的大量讀 / 寫入請求造成 HDFS 效能下降，能對 Kafka 的記錄檔生產擷取過程進行即時監控，避免消費層 Flume 在落盤 HDFS 過程中產生大量小資料檔案，而降低 HDFS 執行效能，並對落盤資料採取適當壓縮措施，儘量節省儲存空間，降低網路 I/O。

業務資料獲取要求按照業務資料庫表結構在資料倉儲中同步建表，並且根據業務資料庫表性質指定對應的同步策略，進行合理的關係建模和維度建模。

2.2.2 資料倉儲需求業務描述

1. 資料分層建模

資料倉儲被分為 5 層，描述如下。

- ODS（Operation Data Store）層：原始資料層，儲存原始資料，直接載入原始記錄檔、資料，資料保持原貌不做處理。
- DWD（Data Warehouse Detail）層：明細資料層，結構和粒度與 ODS 層保持一致，對 ODS 層的資料進行清洗（去除空值、無效資料、超過極限範圍的資料）。
- DWS（Data Warehouse Service）層：服務資料層，以 DWD 層的資料為基礎，進行輕度整理。一般聚集到以使用者當日、裝置當日、商家當日、商品當日等的粒度。在這層通常會以某一個維度為線索，組成跨主題的寬表，舉例來說，由一個使用者當日的簽到數、收藏數、評價數、抽獎數、訂閱數、按讚數、瀏覽商品數、增加購物車數、下單數、支付數、退款數及點擊廣告數組成的寬表。
- DWT（Data Warehouse Topic）層：主題資料層，按照主題對 DWS 層資料進行進一步聚合，建置每個主題的全量寬表。
- ADS（Application Data Store）層：資料應用層，也有人把這層稱為 APP 層、DAL 層、DM 層等。針對實際的資料需求，以 DWD 層、DWS 層和 DWT 層的資料為基礎，組成各種統計報表，統計結果最後同步到關聯式資料庫，如 MySQL，以供 BI 或應用系統查詢使用。

讀者需要按照命名標準合理建表。

2. 需求實現

電子商務業務發展日益成熟，但是如果缺少精細化營運的意識和資料驅動的經驗，那麼發展將陷入瓶頸。作為電子商務資料分析的重要工具——資料倉儲的作用就是為營運人員和決策團隊提供關鍵指標的分析資料。電子商務平台的資料分析主要關注五大關鍵資料指標，包含活躍使用者量、轉化、留存、複購、GMV（指成交金額），以及三大關鍵想法：商品營運、使用者營運和產品營運。圍繞這一原則，本專案中要求實現的主要需求如下。

- 當日、當周、當月的活躍裝置數；
- 每日新增裝置數；
- 沉默裝置數；
- 本周回流裝置數；
- 流失裝置數；
- 流存率；
- 最近連續三周活躍裝置數；
- 最近七天內連續三天活躍裝置數；
- 每日活躍會員數；
- 每日新增會員數；
- 每日新增付費會員數；
- 每日總付費會員數；
- 總會員數；
- 會員活躍率；
- 會員付費率；
- 會員新鮮度；
- 使用者行為漏斗分析；
- 商品銷量排名；
- 商品收藏排名；
- 商品加入購物車排名；
- 商品退款率排名；
- 商品差評率排名；
- 每日下單資訊統計；
- 每日支付資訊統計；
- 複購率。

要求將全部需求實現的結果資料儲存在 ADS 層，並且完成可用於工作排程的指令稿，實現任務自動排程。

2.2.3 資料視覺化業務描述

在 MySQL 中根據 ADS 層的結果資料建立對應的表，使用 Sqoop 工具定時將結果資料匯出到 MySQL 中，並使用資料視覺化工具對資料進行展示。

2.3 系統執行環境

2.3.1 硬體環境

在實際生產環境中，我們需要進行伺服器的選型，伺服器是選擇物理機還是雲主機呢？

1. 機器成本考慮

物理機，以 128GB 記憶體、20 核物理 CPU、40 執行緒、8TB HDD 和 2TB SSD 的戴爾品牌機為例，單台報價約 20 萬元，並且還需要考慮託管伺服器的費用，一般物理機壽命為 5 年左右。

雲主機，以阿里雲為例，與上述物理機的設定相似，每年的費用約 25 萬元。

2. 運行維護成本考慮

物理機需要由專業運行維護人員進行維護，雲主機的運行維護工作由服務提供方完成，運行維護工作相對輕鬆。

實際上，伺服器的選型除了參考上述條件，還應該根據資料量來確定叢集規模。

在本專案中，讀者可在個人電腦上架設測試叢集，建議將電腦設定為 16GB 記憶體、8 核物理 CPU、i7 處理器、1TB SSD。測試伺服器規劃如表 2-12 所示。

表 2-12　測試伺服器規劃

服務名稱	子服務	節點伺服器 hadoop102	節點伺服器 hadoop103	節點伺服器 hadoop104
HDFS	NameNode	√		
	DataNode	√	√	√
	SecondaryNameNode			√
YARN	NodeManager	√	√	√
	ResourceManager		√	
Zookeeper	Zookeeper Server	√	√	√
Flume（擷取記錄檔）	Flume	√	√	
Kafka	Kafka	√	√	√
Flume（消費 Kafka）	Flume			√
Hive	Hive	√		
MySQL	MySQL	√	√	√

服務名稱	子服務	節點伺服器 hadoop102	節點伺服器 hadoop103	節點伺服器 hadoop104
Keepalived	Keepalived		√	√
Sqoop	Sqoop	√		
Superset	Superset	√		
Presto	Coordinator	√		
	Worker		√	√
Azkaban	AzkabanWebServer	√		
	AzkabanExecutorServer	√		
Druid	Druid	√	√	√
HBase	HRegionServer	√	√	√
Kylin	Kylin	√		
Solr	Solr	√	√	√
Atlas	Atlas	√		
服務數總計		18	12	12

2.3.2 軟體環境

1. 技術選型

資料獲取運輸方面，在本專案中主要完成三個方面的需求：將伺服器中的記錄檔資料即時擷取到巨量資料儲存系統中，以防止資料遺失及資料堵塞；將業務資料庫中的資料獲取到資料倉儲中；同時將需求計算結果匯出到關聯式資料庫方便進行展示。為此我們選用了 Flume、Kafka 和 Sqoop。

Flume 是一個高可用、高可靠、分散式的巨量資料收集系統，可從多種來源資料系統擷取、聚集和移動大量的資料聯集中儲存。Flume 提供了豐富多樣的元件供使用者使用，不同的元件可以自由組合，組合方式以使用者設定的設定檔為基礎，非常靈活，可以滿足各種資料獲取傳輸需求。

Kafka 是一個提供容錯儲存、高即時性的分散式訊息佇列平台。我們可以將它用在應用和處理系統間高即時性和高可用性的流式資料儲存中，也可以即時地為流式應用傳送和回饋流式資料。

Sqoop 用於在關聯式資料庫（RDBMS）和 HDFS 之間傳輸資料，啟用了一個 MapReduce 任務來執行資料獲取任務，傳輸大量結構化或半結構化資料的過程是完全自動化的。其主要透過 JDBC 和關聯式資料庫進行互動，理論上支援 JDBC 的 Database 都可以使用 Sqoop 和 HDFS 進行資料互動。

資料儲存方面，在本專案中主要完成對巨量原始資料及轉化後各層資料倉儲中的資料的儲存和對最後結果資料的儲存。對巨量原始資料的儲存，我們選用了 HDFS。HDFS 是 Hadoop 的分散式檔案系統，適合應用於大規模的資料集上，將大規模的資料集以分散式檔案的方式儲存於叢集中的各台節點伺服器上，加強檔案儲存的可用性。對最後結果資料的儲存，由於資料體量比較小，且為了方便存取，我們選用了 MySQL。

資料計算方面，我們選用設定了 Tez 執行引擎的 Hive。Hive 是以 Hadoop 為基礎的資料倉儲工具，可以將結構化的資料檔案對映為一張資料庫表，並提供 SQL 查詢功能，將 SQL 敘述轉化為 MapReduce 任務進行執行，可以説在 Hadoop 之上提供了資料查詢的功能，主要解決非關係類型資料的查詢問題。Tez 執行引擎可以將多個有依賴的作業轉為一個作業，這樣可以減少中間計算過程產生的資料的落盤次數，進一步大幅提升作業的計算效能。

即席查詢模組，我們對目前比較流行的三種即席查詢都進行了探索實驗，分別是 Presto、Druid 和 Kylin。三種即席查詢各有千秋，Presto 基於記憶體計算，Druid 是優秀的時序資料處理引擎，Kylin 基於預 Cube 建立計算。

面對巨量資料的處理，對中繼資料的管理會隨著資料體量的增大而顯得尤為重要。為尋求資料治理的開放原始碼解決方案，Hortonworks 公司聯合其他廠商與使用者於 2015 年發起資料治理倡議，包含資料分類、集中策略引擎、資料血緣、安全、生命週期管理等方面。Apache Atlas 專案就是這個倡議的結果，社區夥伴持續地為該專案提供新的功能和特性。該專案用於管理共用中繼資料、資料分級、稽核、安全性、資料保護等方面。

歸納如下。

- 資料獲取與傳輸：Flume、Kafka、Sqoop。
- 資料儲存：MySQL、HDFS。
- 資料計算：Hive、Tez。
- 任務排程：Azkaban。
- 即席查詢：Presto、Druid、Kylin。
- 中繼資料管理：Atlas。

2. 架構選型

架構版本的選型要求滿足資料倉儲平台的幾大核心需求：子功能不設侷限、國內外資料及社區儘量豐富、元件服務的成熟度和流行度較高。待選擇版本如下。

- Apache：運行維護過程煩瑣，元件間的相容性需要自己研究（本次選用）。
- CDH：不開放原始碼，不用擔心元件相容問題。
- IIDP：開放原始碼，但沒有 CDH 穩定，使用較少。

筆者經過考量決定選擇 Apache 原生版本巨量資料架構，一方面可以自由訂製所需功能元件；另一方面 CDH 和 HDP 版本架構體量較大，對伺服器設定要求相對較高。本專案中用到的元件較少，Apache 原生版本即可滿足需要。

筆者經過對版本相容性的研究，確定的版本選型如表 2-13 所示。

表 2-13 版本選型

產品	版本
Hadoop	2.7.2
Flume	1.7.0
Kafka	0.11.0.2
Hive	2.3.1
Sqoop	1.4.6
MySQL	5.6.24

產品	版本
Azkaban	2.5.0
Java	1.8
Zookeeper	3.4.10
Presto	0.196
Druid	2.7.10
HBase	1.3.1
Kylin	2.5.1
Solr	5.2.1
Atlas	0.8.4

2.4　本章歸納

本章主要對本書的專案需求進行了介紹，首先介紹了本專案即將架設的資料倉儲產品需要實現的系統目標、系統功能結構和系統流程圖；然後對各主要功能模組進行了重點描述，並對每個模組的重點需求進行了介紹；最後根據專案的整體需求對系統執行的硬體環境和軟體環境進行了設定選型。

專案部署的環境準備

透過上一章的分析，我們已經明確了將要使用的架構類型和實現方式，本章將根據上一章的需求分析，架設一個完整的專案開發環境，即使讀者的電腦中已經具備這些環境，也建議瀏覽一遍本章內容，因為其對後續開發過程中程式和命令列的了解很有幫助。

3.1 Linux 環境準備

3.1.1 VMware 安裝

本節介紹的虛擬機器軟體是 VMware，VMware 可以讓使用者在一台電腦上同時執行多個作業系統，還可以像 Windows 應用程式一樣來回切換。使用者可以如同操作真實安裝的系統一樣操作虛擬機器系統，甚至可以在一台電腦上將幾個虛擬機器系統連接為一個區域網或連接到網際網路。

在虛擬機器系統中，每台虛擬產生的電腦都被稱為「虛擬機器」，而用來儲存所有虛擬機器的電腦則被稱為「宿主機」。使用 VMware 虛擬機器軟體安裝虛擬機器可以減少因安裝新系統導致的資料遺失問題，還可以讓使用者方便地體驗各種系統，以進行學習和測試。

VMware 支援多種平台，可以安裝在 Windows、Linux 等作業系統上，初學者大多使用 Windows，可下載 VMware Workstation for Windows 版本。VMware

的安裝非常簡單，與其他 Windows 軟體類似，本書不進行詳細說明。值得一提的是，在安裝過程中安裝的類型包含典型安裝或自訂安裝，筆者建議初學者選擇「典型」安裝。

VMware 安裝完成啟動後，即可進行 Linux 的安裝部署。

推薦使用版本：VMware Workstation Pro 或 VMware Workstation Player。其中，Player 版本供個人使用者使用，非商業用途，是免費的，其他的 VMware 版本在此不進行過多介紹。

3.1.2　CentOS 安裝

在安裝 CentOS 之前，使用者需要檢查本機 BIOS 是否支援虛擬化，開機後進入 BIOS 介面，不同電腦進入 BIOS 介面的操作有所不同，然後進入 Security 下的 Virtualization，選擇 Enable 即可。

啟動 VMware，進入主介面，依次進行新虛擬機器的設定，然後選擇設定類型，如圖 3-1 所示。

圖 3-1　選擇設定類型

點擊「下一步」按鈕,進入「安裝客戶端作業系統」介面,選擇「稍後安裝作業系統」選項,如圖 3-2 所示。

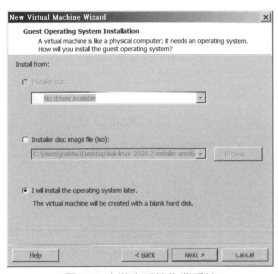

圖 3-2 安裝客戶端作業系統

點擊「下一步」按鈕,進入「選擇客戶端作業系統」介面,選擇 "Linux" 選項,然後在「版本」下拉清單中選擇要安裝的對應的 Linux 版本,此處選擇 "CentOS" 選項,如圖 3-3 所示。

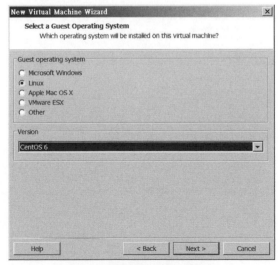

圖 3-3 選擇客戶端作業系統

點擊「下一步」，進入「命名虛擬機器」介面，給虛擬機器起一個名字，如
"CentOS 6.3" 或 "PlayBoy"，然後點擊「瀏覽」按鈕，選擇虛擬機器系統安裝
檔案的儲存位置，如圖 3-4 所示。

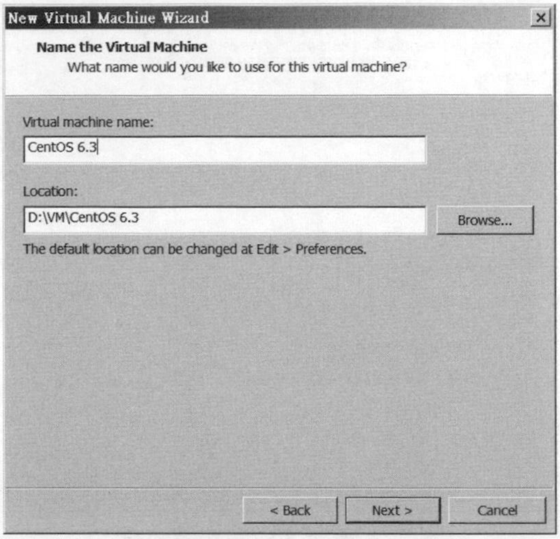

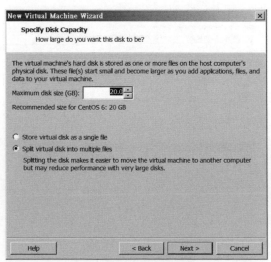

圖 3-4 命名虛擬機器

點擊「下一步」按鈕，進入「指定磁碟容量」介面。預設虛擬的最大磁碟大小
為 20GB（虛擬出來的磁碟會以檔案形式儲存在虛擬機器系統安裝目錄中），
如圖 3-5 所示。

圖 3-5 指定磁碟容量

點擊「下一步」按鈕,進入「已準備好建立虛擬機器」介面,確認虛擬機器設定,若無須改動,則點擊「完成」按鈕,開始產生虛擬機器,如圖 3-6 所示。

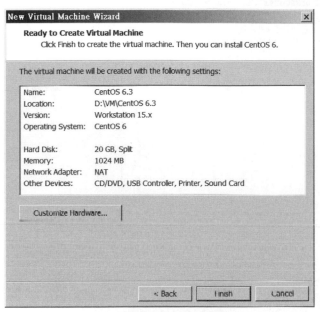

圖 3-6 準備建立虛擬機器

我們可以略做調整,點擊「自訂硬體」按鈕,開啟「硬體」對話方塊。為使虛擬機器中的系統執行速度快一點,我們可以選擇「記憶體」選項來調整虛擬機器記憶體大小,建議調整為 4GB,但是虛擬機器記憶體不要超過宿主機記憶體的一半。CentOS 6.x 最少需要 628MB 的記憶體,否則會開啟簡易安裝過程,如圖 3-7 所示。

選擇「新 CD/DVD(IDE)」選項,可以進行光碟設定。如果選擇「使用物理驅動器」選項,則虛擬機器會使用宿主機的物理光碟,如果選擇「使用 ISO 映射檔案」選項,則可以直接載入 ISO 映射檔案,點擊「瀏覽」按鈕找到 ISO 映射檔案的位置即可,如圖 3-8 所示。

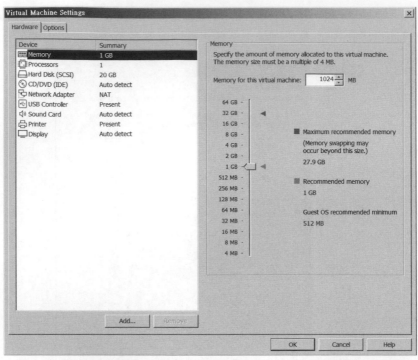

圖 3-7 硬體調整

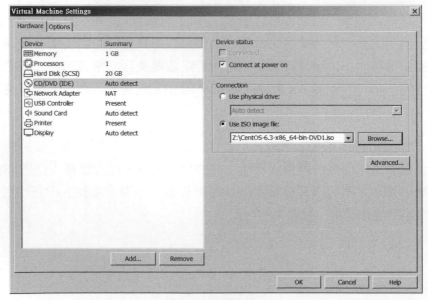

圖 3-8 光碟設定

點擊「關閉」按鈕即可。如果還想調整虛擬機器的硬體規格,則可以選擇「虛擬機器」下拉式功能表中的「設定」指令,重新進入「硬體」對話方塊,如圖 3-9 所示。

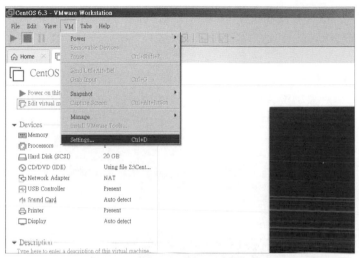

圖 3-9 調整虛擬機器的硬體規格

選擇「電源」→「開啟此虛擬機器電源」選項,開啟虛擬機器,就能看到 CentOS6 的安裝歡迎介面了,如圖 3-10 所示。

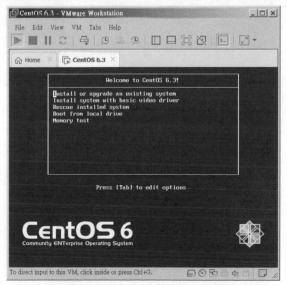

圖 3-10 CentOS6 安裝歡迎頁面

選擇 "Install or upgrade an existing system" 選項，安裝一個全新的系統。進入
安裝環境後，精靈首先會詢問是否檢測安裝媒體的完整性，如圖 3-11 所示。
這是為了避免因為安裝來源不正確，造成無法順利安裝而產生損失，一般情
況下，如果下載過程中沒有出現問題，則無須檢測（檢測時間較久），直接點
擊 "Skip" 按鈕跳過即可。

> ★ **注意**：在虛擬機器和宿主機之間，滑鼠是不能同時有作用的，如果從宿主
> 機進入虛擬機器，則需要把滑鼠指標移入虛擬機器；如果從虛擬機器傳回宿主
> 機，則按 Ctrl+Alt 組合鍵退出。

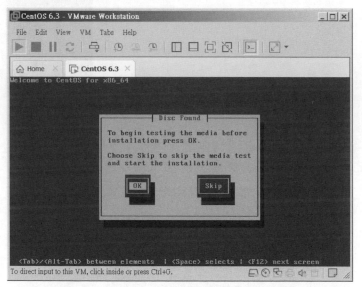

圖 3-11 檢測安裝媒體

進入 CentOS 6.3 歡迎介面，點擊 "Next" 按鈕，進入選擇安裝系統的預設語言
介面，可以根據需要自行選擇，舉例來說，選擇「中文」。選擇完成後，點擊
"Next" 按鈕，進入鍵盤設定介面，選擇預設的美國式鍵盤。

點擊「下一步」按鈕，進入存放裝置選擇介面，選擇「基本存放裝置」選
項，會出現存放裝置警告。警告安裝操作會導致存放裝置中的資料遺失，然
後點擊「是，忽略所有資料」按鈕，如圖 3-12 所示。

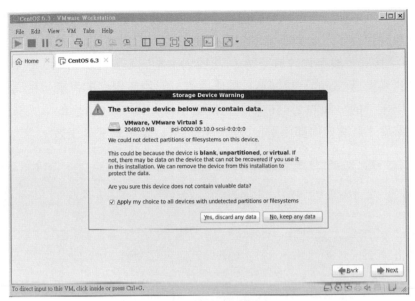

圖 3-12 存放裝置警告

點擊「下一步」按鈕，進入主機名稱設定介面，預設主機名稱是 "localhost.
localdomain"，可以自行更改，如圖 3-13 所示。在此介面中還可以設定網路，使
用者也可以在安裝完成後執行 setup 或 ifconfig 指令進行網路設定，這裡略過。

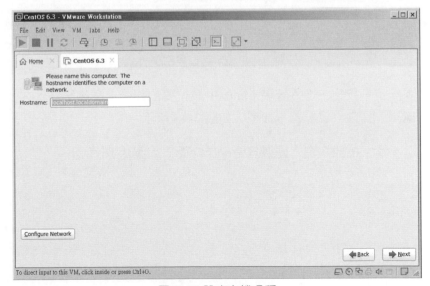

圖 3-13 設定主機名稱

點擊「下一步」按鈕,進入時區選擇介面,如果住在中國,則選擇「亞洲 / 上海」選項就可以了,建議不選取「系統時脈使用 UTC 時間」核取方塊。點擊「下一步」按鈕,設定管理員密碼(「根密碼」指的是管理員密碼,在 Linux 中管理員的名稱為 "root",翻譯為「根使用者」)。用於學習的系統,密碼設定簡單是可以接受的,如 "123456",但可能會出現如圖 3-14 所示的「脆弱密碼」提示,點擊「無論如何都使用」按鈕,依然可以讓脆弱密碼生效。

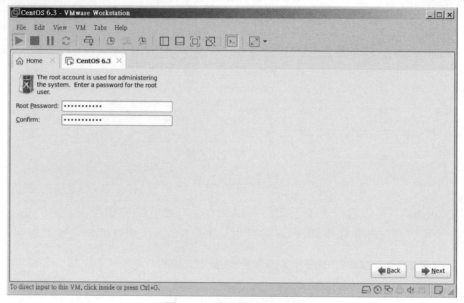

圖 3-14 設定管理員密碼

點擊「下一步」按鈕,進入安裝 Linux 中最重要的部分:硬碟分區。在此,筆者推薦選擇「建立自訂配置」類型,如圖 3-15 所示。

點擊「下一步」按鈕,進入硬碟分區操作介面,如圖 3-16 所示。

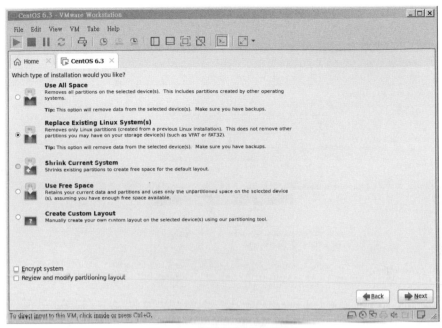

圖 3-15 選擇硬碟分區類型

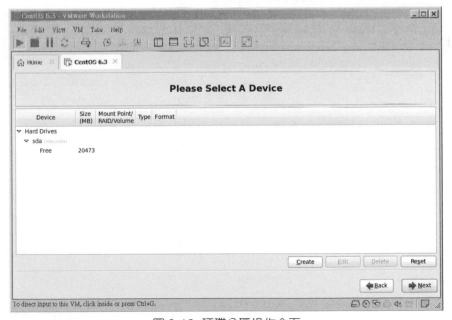

圖 3-16 硬碟分區操作介面

點擊「建立」按鈕,產生分區,如圖 3-17 所示。

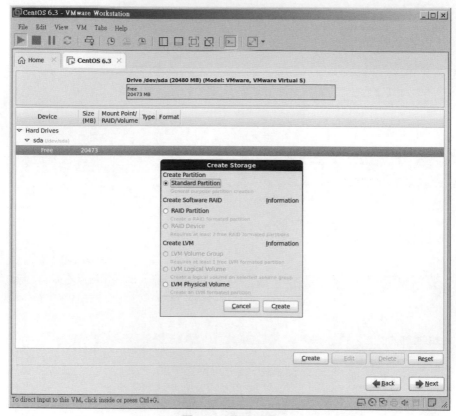

圖 3-17　產生分區

點擊「建立」按鈕,進入「增加分區」介面,如圖 3-18、圖 3-19 和圖 3-20 所示。在此介面,我們可以建立 /boot 分區、/ 分區、/home 分區、swap 分區等。

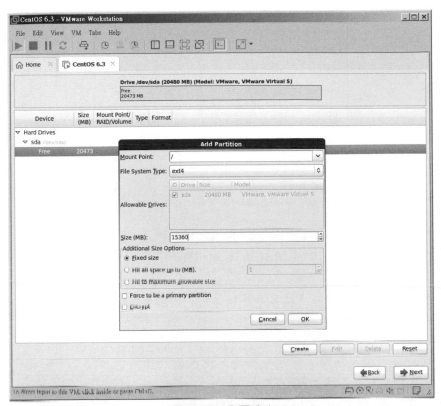

圖 3-18 / 分區建立

圖 3-19 swap 分區建立

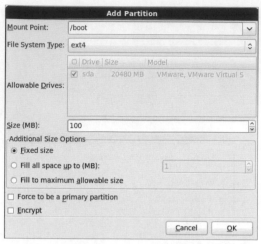

圖 3-20 /boot 分區建立

★ **注意**：swap 分區是在「檔案系統類型」下拉清單中選擇的，而非在「掛載點」下拉清單中選擇的。

分區建立完成後，點擊「確定」按鈕，出現格式化警告，點擊「格式化」按鈕，進入啟動載入程式安裝介面，如圖 3-21 所示。

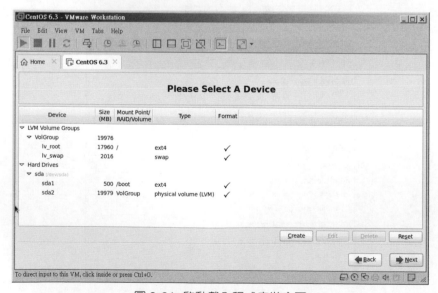

圖 3-21 啟動載入程式安裝介面

點擊「下一步」按鈕，在出現的介面中選擇 "Desktop" 選項，並選擇「現在自訂」選項，訂製系統軟體，如圖 3-22 所示。

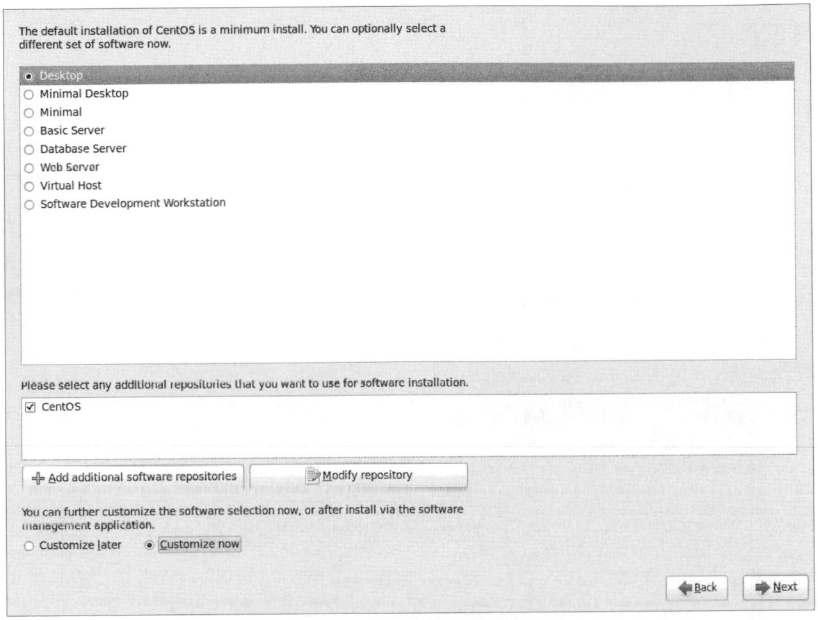

圖 3-22　訂製系統軟體

點擊「下一步」按鈕，進入系統服務自訂選擇介面，建議基本系統部分選擇「相容程式庫」和「基本」，應用程式選擇「網際網路瀏覽器」，桌面除「KDE 桌面」外全部選取，語言支援選擇「中文支援」，其餘部分全部不選取。

完成設定後，開始安裝 CentOS，會等待一段時間，螢幕顯示目前安裝的軟體套件及其簡介、預估剩餘時間以及安裝的進度。安裝完成後，點擊「重新啟動」按鈕，重新啟動後就可以進入登入介面了。還記得 Linux 的根使用者是 root 嗎？還記得安裝時輸入的 root 密碼嗎？輸入正確的使用者名稱和密碼就可以登入系統了。

3.1.3　遠端終端機安裝

大多數伺服器的日常管理操作，都是透過遠端系統管理工具進行的。常見的遠端系統管理方法包含如 VNC 的圖形遠端系統管理、如 Webmin 的以瀏覽器為基礎的遠端系統管理，不過常用的還是命令列操作。在 Linux 中遠端系統管理使用的是 SSH 協定，本節先介紹兩個遠端系統管理工具的使用方法。

1. PuTTY

PuTTY 是一個完全免費的 Windows 遠端系統管理用戶端工具，體積小，操作簡單，是免安裝軟體，無須安裝，下載後即可使用。對經常到客戶公司提供技術支援和維護的使用者，相當方便，只要隨身帶一個隨身碟，即可隨處登入。

下載 PuTTY 後雙擊 putty.exe，出現如圖 3-23 所示的「PuTTY 設定」對話方塊。

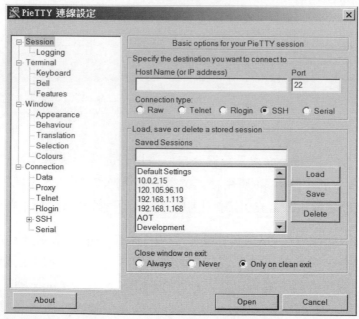

圖 3-23「PuTTY 設定」對話方塊

在「主機名稱（或 IP 位址）」文字標籤中輸入遠端登入主機的 IP 位址，如 192.168.44.8，「通訊埠」根據使用的協定有所區別（選擇不同的「連接類型」選項，通訊埠會自動變化，建議選擇 "SSH" 選項）。在「儲存的階段」文字標籤中輸入一個名稱，點擊「儲存」按鈕即可把本次的連接設定儲存起來。設定完成後點擊「開啟」按鈕，即可出現如圖 3-24 所示的操作介面。

圖 3-24　PuTTY 操作介面

圖 3-25　「快速連接」對話方塊

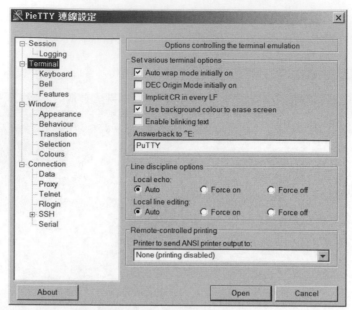

圖 3-26 PuTTY 模擬設定

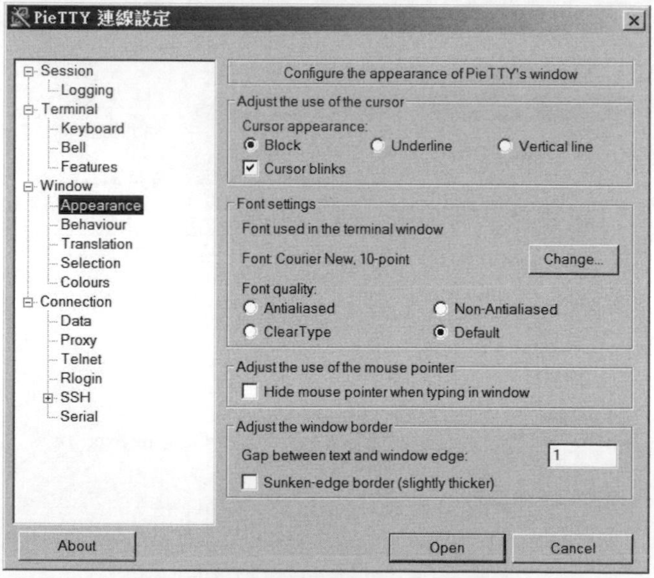

圖 3-27 PuTTY 視窗和文字外觀設定

至此，我們就架設好了初步的學習實驗環境。

3.2 Linux 環境設定

3.2.1 網路設定

對安裝好的 VMware 進行網路設定，方便虛擬機器連接網路，本次設定建議選擇 NAT（網路位址編譯）模式，需要宿主機的 Windows 和虛擬機器的 Linux 能夠進行網路連接，同時虛擬機器的 Linux 可以透過宿主機的 Windows 進入網際網路。

選擇「編輯」→「虛擬網路編輯器」指令，如圖 3-28 所示，對虛擬機器進行網路設定。

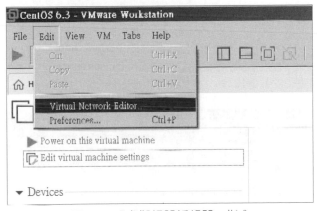

圖 3-28「虛擬網路編輯器」指令

在開啟的「虛擬網路編輯器」對話方塊中，選擇 NAT 模式，並修改虛擬機器的子網 IP 位址，如圖 3-29 所示。

點擊「NAT 設定」按鈕，在開啟的「NAT 設定」對話方塊中，檢視閘道設定，如圖 3-30 所示。

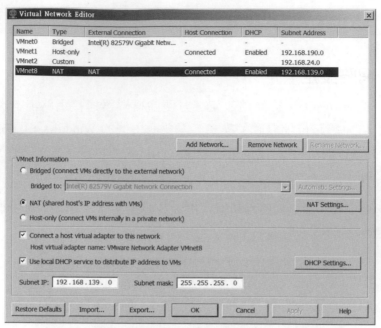

圖 3-29 選擇 NAT 模式並修改虛擬機器的子網 IP 位址

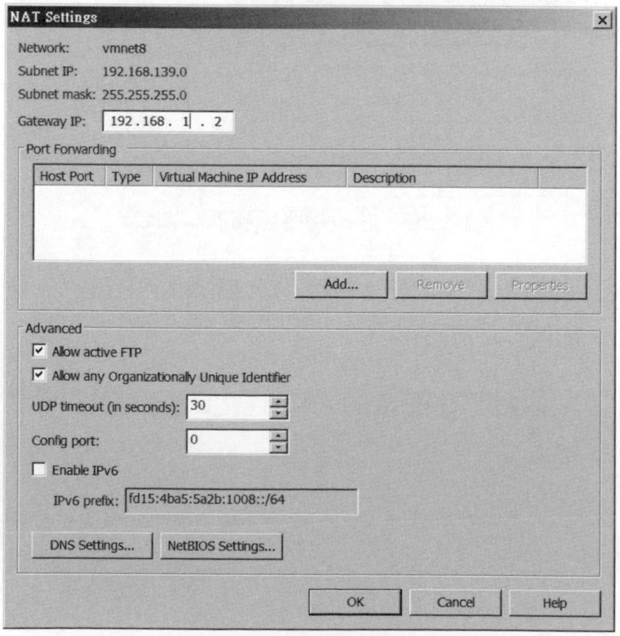

圖 3-30 檢視閘道設定

檢視 Windows 環境中的 vmnet8 網路設定，如圖 3-31 所示，檢視路徑為「主控台」→「網路和 Internet」→「網路連接」。

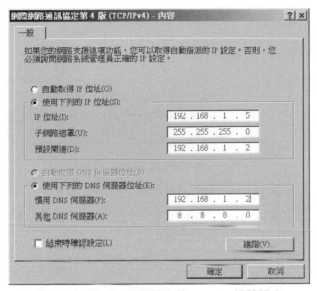

圖 3-31 Windows 環境中的 vmnet8 網路設定

3.2.2 網路 IP 位址設定

修改網路 IP 位址為靜態 IP 位址，避免 IP 位址經常變化，進一步方便節點伺服器間的互相通訊。

```
[root@hadoop100 桌面]#vim /etc/sysconfig/network-scripts/ifcfg-eth0
```

以下粗體的項必須修改，有值的按照下面的值修改，沒有該項的則需要增加。

```
DEVICE=eth0                                  #介面名稱（裝置，網路卡）
HWADDR=00:0C:2x:6x:0x:xx                      #物理IP位址
TYPE=Ethernet                                #網路類型（通常是Ethernet）
UUID=926a57ba-92c6-4231-bacb-f27e5e6a9f44    #隨機id
#系統啟動的時候網路介面是否有效（yes/no）
ONBOOT=yes
#IP位址的設定方法[none（啟動時不使用協定）|static（靜態設定IP位址）|bootp
（BOOTP協定）|
```

```
#dhcp（DHCP協定）]
BOOTPROTO=static
#IP位址
IPADDR=192.168.1.101
#閘道
GATEWAY=192.168.1.2
#域名解析器
DNS1=192.168.1.2
```

修改 IP 位址後的結果如圖 3-32 所示，執行 ":wq" 指令，儲存退出。

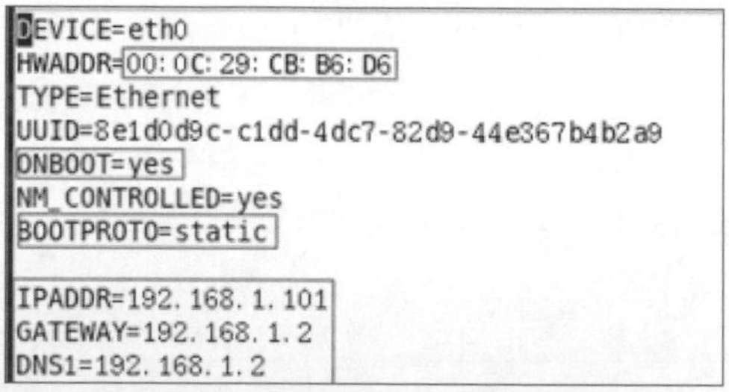

```
DEVICE=eth0
HWADDR=00:0C:29:CB:B6:D6
TYPE=Ethernet
UUID=8e1d0d9c-c1dd-4dc7-82d9-44e367b4b2a9
ONBOOT=yes
NM_CONTROLLED=yes
BOOTPROTO=static

IPADDR=192.168.1.101
GATEWAY=192.168.1.2
DNS1=192.168.1.2
```

圖 3-32 修改 IP 位址後的結果

執行 service network restart 指令，重新啟動網路服務，如圖 3-33 所示。

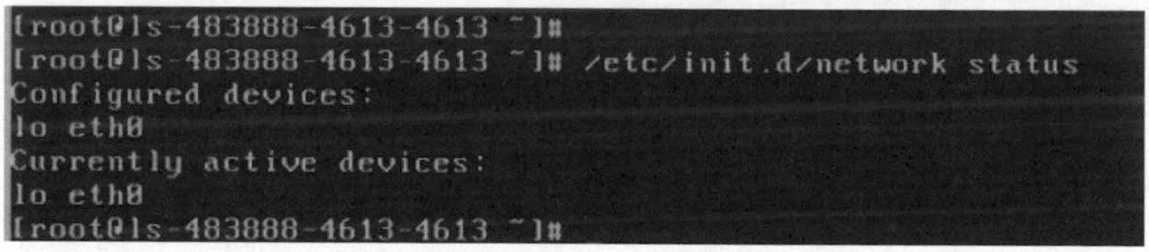

```
[root@ls-483888-4613-4613 ~]#
[root@ls-483888-4613-4613 ~]# /etc/init.d/network status
Configured devices:
lo eth0
Currently active devices:
lo eth0
[root@ls-483888-4613-4613 ~]#
```

圖 3-33 重新啟動網路服務

如果顯示出錯，則執行 "reboot" 指令，重新啟動虛擬機器。

3.2.3 主機名稱設定

修改主機名稱為一系列有規律的主機名稱，並修改 hosts 檔案增加我們需要的主機名稱和 IP 位址對映，以便方便管理且方便節點伺服器間透過主機名稱進行通訊。

1. 修改 **Linux** 的主機對映檔案（**hosts** 檔案）

（1）進入 Linux 檢視本機的主機名稱。執行 hostname 指令進行檢視。

```
[root@hadoop100 桌面]# hostname
hadoop100
```

（2）如果感覺此主機名稱不合適，則可以進行修改。編輯成功 /etc/sysconfig/network 檔案進行修改。

```
[root@hadoop100 桌面]# vim /etc/sysconfig/network
NETWORKING=yes
NETWORKING_IPV6=no
HOSTNAME= hadoop100
```

> ★ **注意**：主機名稱不要有 "_"（底線）。

（3）開啟 /etc/sysconfig/network 檔案後，可以看到主機名稱，在此處可以完成對主機名稱的修改，本例不做修改，仍為 hadoop100。

（4）儲存並退出。

（5）開啟 /etc/hosts 檔案。

```
[root@hadoop100 桌面]# vim /etc/hosts
```

增加以下內容。

```
192.168.1.100 hadoop100
192.168.1.101 hadoop101
192.168.1.102 hadoop102
192.168.1.103 hadoop103
192.168.1.104 hadoop104
192.168.1.105 hadoop105
192.168.1.106 hadoop106
192.168.1.107 hadoop107
192.168.1.108 hadoop108
```

（6）重新啟動裝置，檢視主機名稱，可以看到已經修改成功。

2. 修改 **Windows** 的主機對映檔案（**hosts** 檔案）

（1）進入 C:\Windows\System32\drivers\etc 路徑。

（2）複製 hosts 檔案到桌面上。

（3）開啟桌面上的 hosts 檔案並增加以下內容。

```
192.168.1.100 hadoop100
192.168.1.101 hadoop101
192.168.1.102 hadoop102
192.168.1.103 hadoop103
192.168.1.104 hadoop104
192.168.1.105 hadoop105
192.168.1.106 hadoop106
192.168.1.107 hadoop107
192.168.1.108 hadoop108
```

（4）用桌面上的 hosts 檔案覆蓋 C:\Windows\System32\drivers\etc 路徑中的 hosts 檔案。

3.2.4 防火牆設定

為了使 Windows 或其他系統可以存取 Linux 虛擬機器內的服務，我們有時候需要關閉虛擬機器的防火牆服務，以下是常見的防火牆啟動 / 關閉指令。

1. 臨時關閉防火牆

（1）檢視防火牆狀態。

```
[root@hadoop100桌面]   # service iptables status
```

（2）臨時關閉防火牆。

```
[root@hadoop100桌面]   # service iptables stop
```

2. 開機啟動時關閉防火牆

（1）檢視開機啟動時防火牆狀態。

```
[root@hadoop100桌面]   #chkconfig iptables --list
```

（2）設定開機時關閉防火牆。

```
[root@hadoop100桌面]   #chkconfig iptables off
```

3.2.5 一般使用者設定

root 使用者具有太大的操作許可權，而在實際操作中又需要對使用者有所限制，所以我們需要建立一般使用者。

（1）建立 atguigu 使用者。

（2）設定 atguigu 使用者具有 root 許可權，接下來的所有操作都將在一般使用者身份下完成。

① 增加 atguigu 使用者，並設定密碼。

```
[root@hadoop100 ~]#useradd atguigu
[root@hadoop100 ~]#passwd atguigu
```

② 修改設定檔。

```
[root@hadoop100 ~]#vim /etc/sudoers
```

修改 /etc/sudoers 檔案，找到第 91 行，在 root 下面增加一行。

```
## Allow root to run any commands anywhere
root        ALL=(ALL)      ALL
atguigu     ALL=(ALL)      ALL
```

或設定成執行 sudo 指令時，不需要輸入密碼。

```
## Allow root to run any commands anywhere
root        ALL=(ALL)      ALL
atguigu     ALL=(ALL)      NOPASSWD:ALL
```

修改完畢後，使用者使用 atguigu 帳號或執行 sudo 指令進行登入，即可獲得 root 操作許可權。

3.3 Hadoop 環境架設

在架設完 Linux 環境之後，我們正式開始架設 Hadoop 分散式叢集環境。

3.3.1 虛擬機器環境準備

1. 複製虛擬機器

關閉要被複製的虛擬機器，按右鍵虛擬機器名稱，在出現的快顯功能表中選擇「管理」→「複製」指令，如圖 3-34 所示。

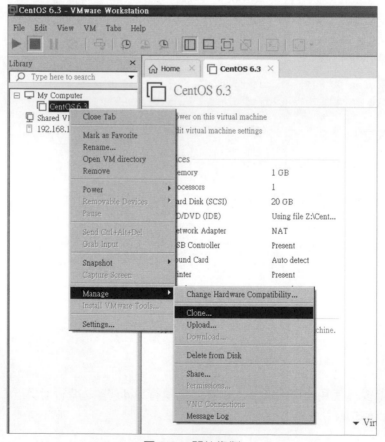

圖 3-34 開始複製

在歡迎介面點擊「下一步」按鈕,開啟「複製虛擬機器精靈」對話方塊,選擇「虛擬機器中的目前狀態」選項,複製虛擬機器,如圖 3-35 所示。

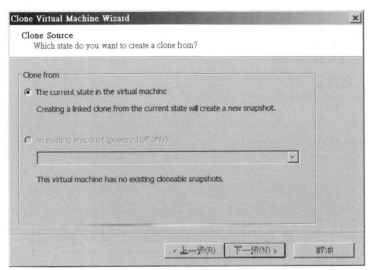

圖 3-35 複製虛擬機器

設定「複製方法」為「建立完整複製」,如圖 3-36 所示。

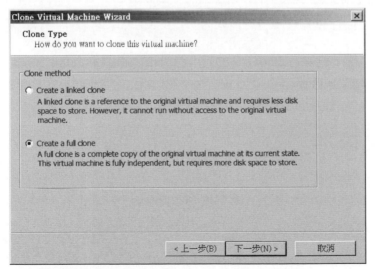

圖 3-36 設定「複製方法」為「建立完整複製」

設定複製的「虛擬機器名稱」和「位置」，如圖 3-37 所示。

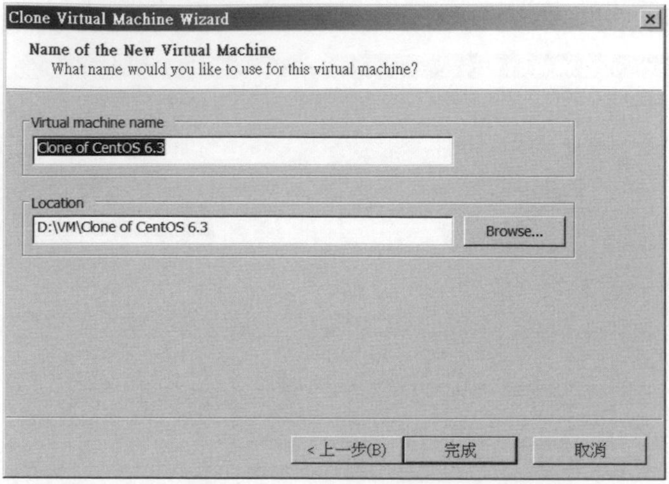

圖 3-37 設定複製的「虛擬機器名稱」和「位置」

點擊「完成」按鈕，開始複製，需要等待一段時間，複製完成後，點擊「關閉」按鈕。

修改複製後的虛擬機器的 IP 位址。

```
[root@hadoop110 /]#vim /etc/udev/rules.d/70-persistent-net.rules
```

進入以下頁面，刪除 "eth0" 所在的行，將 "eth1" 修改為 "eth0"，同時複製物理 IP 位址，如圖 3-38 所示。

圖 3-38 修改網路卡

修改 eth0 網路卡中的物理 IP 位址。

```
[root@hadoop110 /]#vim /etc/sysconfig/network-scripts/ifcfg-eth0
```

把複製的物理 IP 位址進行更新。

```
HWADDR=00:0c:29:34:c4:3f        #物理IP位址
```

修改為想要設定的 IP 位址。

```
IPADDR=192.168.1.102            #IP位址
```

按照 3.2.3 節中主機名稱的設定方法修改主機名稱。

重新啟動伺服器,按照上述操作分別複製 3 台虛擬機器,命名為 hadoop102、hadoop103、hadoop104,主機名稱和 IP 位址分別與 3.2.3 節中的 hosts 檔案設定一一對應。

2. 建立安裝目錄

(1)在 /opt 目錄下建立 module、software 資料夾。

```
[atguigu@hadoop102 opt]$ sudo mkdir module
[atguigu@hadoop102 opt]$ sudo mkdir software
```

(2)修改 module、software 資料夾的所有者。

```
[atguigu@hadoop102 opt]$ sudo chown atguigu:atguigu module/ software/
[atguigu@hadoop102 opt]$ ll
總用量 8
drwxr-xr-x. 2 atguigu atguigu 4096 1月   17 14:37 module
drwxr-xr-x. 2 atguigu atguigu 4096 1月   17 14:38 software
```

之後,所有的軟體安裝操作將在 module 和 software 資料夾中進行。

3. 設定三台虛擬機器免密登入

為什麼需要設定免密登入呢?這與 Hadoop 分散式叢集的架構有關。我們架設的 Hadoop 分散式叢集是「主從架構」,設定了節點伺服器間免密登入之後,就可以方便地透過主節點伺服器啟動從節點伺服器,而不用手動輸入使用者名稱和密碼。

第一步：設定 SSH。

（1）基本語法：假設要以使用者名稱 user 登入遠端主機 host，只需要輸入 ssh user@host，如 ssh atguigu@192.168.1.100，若本機使用者名稱與遠端使用者名稱一致，登入時則可以省略使用者名稱，如 ssh host。

（2）SSH 連接時出現 "Host key verification failed" 的錯誤訊息，直接輸入 yes 即可。

```
[atguigu@hadoop102 opt] $ ssh 192.168.1.103
The authenticity of host '192.168.1.103 (192.168.1.103)' can't be
established.
RSA key fingerprint is cf:1e:de:d7:d0:4c:2d:98:60:b4:fd:ae:b1:2d:ad:06.
Are you sure you want to continue connecting (yes/no)?
Host key verification failed.
```

第二步：無金鑰設定。

（1）免密登入原理如圖 3-39 所示。

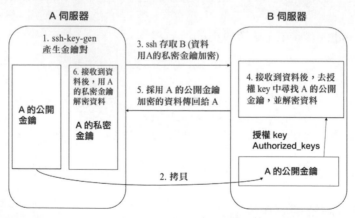

圖 3-39 免密登入原理

（2）產生公開金鑰和私密金鑰。

```
[atguigu@hadoop102 .ssh]$ ssh-key-gen -t rsa
```

連續按三次 Enter 鍵，就會產生兩個檔案：id_rsa（私密金鑰）、id_rsa.pub（公開金鑰）。

（3）將公開金鑰複製到要免密登入的目標伺服器上。

```
[atguigu@hadoop102 .ssh]$ ssh-copy-id hadoop102
[atguigu@hadoop102 .ssh]$ ssh-copy-id hadoop103
[atguigu@hadoop102 .ssh]$ ssh-copy-id hadoop104
```

.ssh 資料夾下的檔案功能解釋如下。

- known_hosts：記錄 SSH 存取過電腦的公開金鑰。
- id_rsa：產生的私密金鑰。
- id_rsa.pub：產生的公開金鑰。
- authorized_keys：儲存授權過的免密登入伺服器公開金鑰。

4. 設定時間同步

為什麼要設定節點伺服器間的時間同步呢？

即將架設的 Hadoop 分散式叢集需要解決兩個問題：資料的儲存和資料的計算。

Hadoop 對大型檔案的儲存採用分段的方法，將檔案切分成多塊，以塊為單位，分發到各台節點伺服器上進行儲存。當這個大型檔案再次被存取到的時候，需要從 3 台節點伺服器上分別拿出資料，然後進行計算。由於電腦之間的通訊和資料的傳輸一般是以時間為約定條件的，如果 3 台節點伺服器的時間不一致，就會導致在讀取區塊資料的時候出現時間延遲，可能會導致存取檔案時間過長，甚至失敗，所以設定節點伺服器間的時間同步非常重要。

第一步：設定時間伺服器（必須是 root 使用者）。

（1）檢查電腦中是否安裝了 ntp。

```
[root@hadoop102 桌面]# rpm -qa|grep ntp
ntp-4.2.6p5-10.el6.centos.x86_64
fontpackages-filesystem-1.41-1.1.el6.noarch
ntpdate-4.2.6p5-10.el6.centos.x86_64
```

（2）修改 ntp 設定檔。

```
[root@hadoop102 桌面]# vim /etc/ntp.conf
```

修改內容如下。

① 修改 1（設定本機網路上的主機不受限制），將以下設定前的 # 刪除，解開此行註釋。

```
#restrict 192.168.1.0 mask 255.255.255.0 nomodify notrap
```

② 修改 2（設定為不採用公共的伺服器）。

```
server 0.centos.pool.ntp.org iburst
server 1.centos.pool.ntp.org iburst
server 2.centos.pool.ntp.org iburst
server 3.centos.pool.ntp.org iburst
```

將上述內容修改為：

```
#server 0.centos.pool.ntp.org iburst
#server 1.centos.pool.ntp.org iburst
#server 2.centos.pool.ntp.org iburst
#server 3.centos.pool.ntp.org iburst
```

③ 修改 3（增加一個預設的內部時脈資料，使用它為區域網使用者提供服務）。

```
server 127.127.1.0
fudge 127.127.1.0 stratum 10
```

（3）修改 /etc/sysconfig/ntpd 檔案。

```
[root@hadoop102 桌面]# vim /etc/sysconfig/ntpd
```

增加以下內容（讓硬體時間與系統時間一起同步）。

```
SYNC_HWCLOCK=yes
```

重新啟動 ntpd 檔案。

```
[root@hadoop102 桌面]# service ntpd status
ntpd 已停
[root@hadoop102 桌面]# service ntpd start
正在啟動 ntpd:                                        [確定]
```

執行：

```
[root@hadoop102 桌面]# chkconfig ntpd on
```

第二步：設定其他伺服器（必須是 root 使用者）。
設定其他伺服器 10 分鐘與時間伺服器同步一次。

```
[root@hadoop103 hadoop-2.7.2]# crontab -e
```

撰寫指令稿。

```
*/10 * * * * /usr/sbin/ntpdate hadoop102
```

修改 hadoop103 的節點伺服器時間，使其與另外兩台節點伺服器時間不同步。

```
[root@hadoop103 hadoop]# date -s "2017-9-11 11:11:11"
```

10 分鐘後檢視該伺服器是否與時間伺服器同步。

```
[root@hadoop103 hadoop]# date
```

5. 撰寫叢集分發指令稿

叢集間資料的複製通用的兩個指令是 scp 和 rsync，其中，rsync 指令可以只對差異檔案進行更新，非常方便，但是使用時需要操作者頻繁輸入各種指令參數，為了能夠更方便地使用該指令，我們撰寫一個叢集分發指令稿，主要實現目前叢集間的資料分發。

第一步：指令稿需求分析。循環複製檔案到所有節點伺服器的相同目錄下。
（1）原始複製。

```
rsync -rv  /opt/module root@hadoop103:/opt/
```

（2）期望指令稿效果。

```
xsync path/filename     #要同步的檔案路徑或檔案名稱
```

（3）在 /home/atguigu/bin 目錄下儲存的指令稿，atguigu 使用者可以在系統任何地方直接執行。

第二步：指令稿實現。

（1）在 /home/atguigu 目錄下建立 bin 目錄，並在 bin 目錄下使用 vim 指令建立檔案 xsync，檔案內容如下。

```
[atguigu@hadoop102 ~]$ mkdir bin
[atguigu@hadoop102 ~]$ cd bin/
[atguigu@hadoop102 bin]$ touch xsync
[atguigu@hadoop102 bin]$ vim xsync
#!/bin/bash
#取得輸入參數個數，如果沒有參數，則直接退出
pcount=$#
if((pcount==0)); then
echo no args;
exit;
fi

#取得檔案名稱
p1=$1
fname=`basename $p1`
echo fname=$fname

#取得上級目錄到絕對路徑
pdir=`cd -P $(dirname $p1); pwd`
echo pdir=$pdir

#取得目前使用者名稱
user=`whoami`

#循環
for((host=103; host<105; host++)); do
        echo -------------------- hadoop$host -----------------
        rsync -rvl $pdir/$fname $user@hadoop$host:$pdir
done
```

（2）修改指令稿 xsync，使其具有執行許可權。

```
[atguigu@hadoop102 bin]$ chmod 777 xsync
```

（3）呼叫指令稿的形式：xsync 檔案名稱。

```
[atguigu@hadoop102 bin]$ xsync /home/atguigu/bin
```

3.3.2 JDK 安裝

JDK 是 Java 的開發工具箱，是整個 Java 的核心，包含 Java 執行環境、Java 工具和 Java 基礎類別庫，JDK 是學習巨量資料的基礎工具之一。即將架設的 Hadoop 分散式叢集的安裝程式就是用 Java 開發的，所有 Hadoop 分散式叢集 想要正常執行，必須安裝 JDK。

（1）在 3 台虛擬機器上分別移除現有的 JDK。

① 檢查電腦中是否已安裝 Java 軟體。

```
[atguigu@hadoop102 opt]$ rpm -qa | grep java
```

② 如果安裝的版本低於 1.7，則移除該 JDK。

```
[atguigu@hadoop102 opt]$ sudo rpm -e 實際軟體套件名
```

（2）將 JDK 匯入 opt 目錄下的 software 資料夾中。

① 在 Linux 下的 opt 目錄中檢視軟體套件是否進入成功。

```
[atguigu@hadoop102 opt]$ cd software/
[atguigu@hadoop102 software]$ ls
hadoop-2.7.2.tar.gz  jdk-8u144-linux-x64.tar.gz
```

② 解壓 JDK 到 /opt/module 目錄下，tar 指令用來解壓 .tar 或 .tar.gz 格式的壓縮檔，透過 -z 選項指定解壓 .tar.gz 格式的壓縮檔。-f 選項用於指定解壓檔案，-x 選項用於指定解壓縮操作，-v 選項用於顯示解壓過程，-C 選項用於指定解壓路徑。

```
[atguigu@hadoop102 software]$ tar -zxvf jdk-8u144-linux-x64.tar.gz -C
/opt/module/
```

（3）設定 JDK 環境變數，方便使用到 JDK 的程式能正常呼叫 JDK。

① 先取得 JDK 路徑。

```
[atgui@hadoop102 jdk1.8.0_144]$ pwd
/opt/module/jdk1.8.0_144
```

② 開啟 /etc/profile 檔案，需要注意的是，/etc/profile 檔案屬於 root 使用者，
需要使用 sudo vim 指令才可以對它進行編輯。

```
[atguigu@hadoop102 software]$ sudo vim /etc/profile
```

在 profile 檔案尾端增加 JDK 路徑，增加的內容如下。

```
#JAVA_HOME
export JAVA_HOME=/opt/module/jdk1.8.0_144
export PATH=$PATH:$JAVA_HOME/bin
```

儲存後退出。

```
:wq
```

③ 修改 /etc/profile 檔案後，需要執行 source 指令使修改後的檔案生效。

```
[atguigu@hadoop102 jdk1.8.0_144]$ source /etc/profile
```

（4）透過執行 java -version 指令，測試 JDK 是否安裝成功。

```
[atguigu@hadoop102 jdk1.8.0_144]# java -version
java version "1.8.0_144"
```

重新啟動（如果執行 java -version 指令可以正常檢視 Java 版本，說明 JDK 安
裝成功，則不用重新啟動）。

```
[atguigu@hadoop102 jdk1.8.0_144]$ sync
[atguigu@hadoop102 jdk1.8.0_144]$ sudo reboot
```

（5）分發 JDK 給所有節點伺服器。

```
[atguigu@hadoop102 jdk1.8.0_144]$ xsync /opt/module/jdk1.8.0_144
```

（6）分發環境變數。

```
[atguigu@hadoop102 jdk1.8.0_144]$ xsync /etc/profile
```

（7）執行 source 指令，使環境變數在每台虛擬機器上生效。

```
[atguigu@hadoop103 jdk1.8.0_144]$ source /etc/profile
[atguigu@hadoop104 jdk1.8.0_144]$ source /etc/profile
```

3.3.3 Hadoop 安裝

在架設 Hadoop 分散式叢集時，每個節點伺服器上的 Hadoop 設定大致相同，所以只需要在 hadoop102 節點伺服器上操作，設定完成之後同步到另外兩個節點伺服器上即可。

（1）將 Hadoop 的安裝套件 hadoop-2.7.2.tar.gz 匯入 opt 目錄下的 software 資料夾中，該資料夾被指定用來儲存各軟體的安裝套件。

① 進入 Hadoop 安裝套件路徑。

```
[atguigu@hadoop102 ~]$ cd /opt/software/
```

② 解壓安裝套件到 /opt/module 檔案中。

```
[atguigu@hadoop102 software]$ tar -zxvf hadoop-2.7.2.tar.gz -C /opt/
module/
```

③ 檢視是否解壓成功。

```
[atguigu@hadoop102 software]$ ls /opt/module/
hadoop-2.7.2
```

（2）將 Hadoop 增加到環境變數，可以直接使用 Hadoop 的相關指令操作，而不用指定 Hadoop 的目錄。

① 取得 Hadoop 安裝路徑。

```
[atguigu@ hadoop102 hadoop-2.7.2]$ pwd
/opt/module/hadoop-2.7.2
```

② 開啟 /etc/profile 檔案。

```
[atguigu@ hadoop102 hadoop-2.7.2]$ sudo vim /etc/profile
```

在 profile 檔案尾端增加 Hadoop 路徑，增加的內容如下。

```
##HADOOP_HOME
export HADOOP_HOME=/opt/module/hadoop-2.7.2
export PATH=$PATH:$HADOOP_HOME/bin
```

```
export PATH=$PATH:$HADOOP_HOME/sbin
```

③ 儲存後退出。

```
:wq
```

④ 執行 source 指令，使修改後的檔案生效。

```
[atguigu@ hadoop102 hadoop-2.7.2]$ source /etc/profile
```

（3）測試是否安裝成功。

```
[atguigu@hadoop102 ~]$ hadoop version
Hadoop 2.7.2
```

（4）重新啟動（如果 hadoop 指令可以用，則不用重新啟動）。

```
[atguigu@ hadoop101 hadoop-2.7.2]$ sync
[atguigu@ hadoop101 hadoop-2.7.2]$ sudo reboot
```

（5）分發 Hadoop 給所有節點伺服器。

```
[atguigu@hadoop100 hadoop-2.7.2]$ xsync /opt/module/hadoop-2.7.2
```

（6）分發環境變數。

```
[atguigu@hadoop100 hadoop-2.7.2]$ xsync /etc/profile
```

（7）執行 source 指令，使環境變數在每台虛擬機器上生效。

```
[atguigu@hadoop103 hadoop-2.7.2]$ source /etc/profile
[atguigu@hadoop104 hadoop-2.7.2]$ source /etc/profile
```

3.3.4 Hadoop 分散式叢集部署

Hadoop 的執行模式包含本機式、虛擬分散式及完全分散式三種模式。本次主要架設實際生產環境中比較常用的完全分散式模式，架設完全分散式模式之前需要對叢集部署進行提前規劃，不要將過多的服務集中到一台節點伺服器上。我們將負責管理工作的 NameNode 和 ResourceManager 分別部署在兩台節點伺服器上，另一台節點伺服器上部署 SecondaryNameNode，所有節點伺服器均承擔 DataNode 和 NodeManager 角色,並且 DataNode 和 NodeManager 通常儲存在同一台節點伺服器上,所有角色儘量做到均衡分配。

(1)叢集部署規劃如表 3-1 所示。

表 3-1 叢集部署規劃

	hadoop102	hadoop103	hadoop104
HDFS	NameNode DataNode	DataNode	SecondaryNameNode DataNode
YARN	NodeManager	ResourceManager NodeManager	NodeManager

(2)對叢集角色的分配主要依靠設定檔,設定叢集檔案的細節如下。

① 核心設定檔為 core-site.xml,該設定檔屬於 Hadoop 的全域設定檔,我們主要對分散式檔案系統 NameNode 的入口位址和分散式檔案系統中資料落地到伺服器本機磁碟的位置進行設定,程式如下。

```
[atguigu@hadoop102 hadoop]$ vim core-site.xml
<!-- 指定HDFS中NameNode的位址 -->
<property>
  <name>fs.defaultFS</name>
<!-- 其中,hdfs為協定名稱,hadoop102為NameNode的節點伺服器主機名稱,9000
為通訊埠-->
  <value>hdfs://hadoop102:9000</value>
</property>

<!-- 指定Hadoop執行時期產生的檔案的儲存目錄,該目錄需要單獨建立 -->
<property>
  <name>hadoop.tmp.dir</name>
  <value>/opt/module/hadoop-2.7.2/data/tmp</value>
</property>
```

② Hadoop 的環境設定檔為 hadoop-env.sh,在這個設定檔中我們主要需要指定 JDK 的路徑 JAVA_HOME,避免程式執行中出現 JAVA_HOME 找不到的異常。

```
[atguigu@hadoop102 hadoop]$ vim hadoop-env.sh
export JAVA_HOME=/opt/module/jdk1.8.0_144
```

③ HDFS 的設定檔為 hdfs-site.xml，在這個設定檔中我們主要對 HDFS 檔案系統的屬性進行設定。

```
[atguigu@hadoop102 hadoop]$ vim hdfs-site.xml
<!-- 指定HDFS儲存內容的備份個數 -->
<!-- Hadoop透過使用檔案的容錯來確保檔案儲存的可用性,由於有3個DataNode,
所以我們可以將備份數量設定為3 -->
<property>
  <name>dfs.replication</name>
  <value>3</value>
</property>
<!-- 設定Hadoop分散式叢集的SecondaryNameNode -->
<!-- SecondaryNameNode主要作為NameNode的輔助,通訊埠為50090 -->
<property>
        <name>dfs.namenode.secondary.http-address</name>
        <value>hadoop104:50090</value>
    </property>
```

④ YARN 的環境設定檔為 yarn-env.sh，同樣指定 JDK 的路徑 JAVA_HOME。

```
[atguigu@hadoop102 hadoop]$ vim yarn-env.sh
export JAVA_HOME=/opt/module/jdk1.8.0_144
```

⑤ 關於 YARN 的設定檔 yarn-site.xml，主要設定以下兩個參數。

```
[atguigu@hadoop102 hadoop]$ vim yarn-site.xml
<!-- reducer取得資料的方式 -->
<!-- yarn.nodemanager.aux-services是NodeManager上執行的附屬服務,其值需
要設定成mapreduce_shuffle才可以執行MapReduce程式 -->

 <property>
   <name>yarn.nodemanager.aux-services</name>
   <value>mapreduce_shuffle</value>
 </property>

 <!-- 指定YARN的ResourceManager的位址 -->
 <property>
  <name>yarn.resourcemanager.hostname</name>
  <value>hadoop103</value>
 </property>
```

⑥ MapReduce 的環境設定檔為 mapred-env.sh，同樣指定 JDK 的路徑 JAVA_
HOME。

```
[atguigu@hadoop102 hadoop]$ vim mapred-env.sh
export JAVA_HOME=/opt/module/jdk1.8.0_144
```

⑦ 關於 MapReduce 的設定檔 mapred-site.xml，主要設定一個參數，指明
MapReduce 的執行架構為 YARN。

```
[atguigu@hadoop102 hadoop]$ cp mapred-site.xml.template mapred-site.xml
[atguigu@hadoop102 hadoop]$ vim mapred-site.xml
<!-- MapReduce計算架構的資源交給YARN來管理 -->
 <property>
  <name>mapreduce.framework.name</name>
  <value>yarn</value>
 </property>
```

⑧ 主節點伺服器 NameNode 和 ResourceManager 的角色在設定檔中已經進行
了設定，還需指定從節點伺服器的角色，設定檔 slaves 就是用來設定 Hadoop
分散式叢集中各台從節點伺服器的角色。如下所示，對 slaves 檔案進行修改，
將 3 台節點伺服器全部指定為從節點伺服器，啟動 DataNode 和 NodeManager
處理程序。

```
/opt/module/hadoop-2.7.2/etc/hadoop/slaves
[atguigu@hadoop102 hadoop]$ vim slaves
hadoop102
hadoop103
hadoop104
```

⑨ 在叢集上分發設定好的 Hadoop 設定檔，這樣 3 台節點伺服器都可享有相同
的 Hadoop 的設定，接下來即可透過不同的處理程序啟動指令了。

```
[atguigu@hadoop102 hadoop]$ xsync /opt/module/hadoop-2.7.2/
```

⑩ 檢視檔案分發情況。

```
[atguigu@hadoop103 hadoop]$ cat /opt/module/hadoop-2.7.2/etc/hadoop/
core-site.xml
```

（3）建立資料目錄。

根據在 core-site.xml 檔案中設定的分散式檔案系統最後落地到各資料節點上的本機磁碟位置資訊 /opt/module/hadoop-2.7.2/data/tmp，自行建立該目錄。

```
[atguigu@hadoop102 hadoop-2.7.2]$ mkdir /opt/module/hadoop-2.7.2/data/tmp
[atguigu@hadoop103 hadoop-2.7.2]$ mkdir /opt/module/hadoop-2.7.2/data/tmp
[atguigu@hadoop104 hadoop-2.7.2]$ mkdir /opt/module/hadoop-2.7.2/data/tmp
```

（4）啟動 Hadoop 分散式叢集。

① 如果第一次啟動叢集，則需要格式化 NameNode。

```
[atguigu@hadoop102 hadoop-2.7.2]$ hadoop namenode -format
```

② 在設定了 NameNode 的節點伺服器後，透過執行 start-dfs.sh 指令啟動 HDFS，即可同時啟動所有的 DataNode 和 SecondaryNameNode。

```
[atguigu@hadoop102 hadoop-2.7.2]$ sbin/start-dfs.sh
[atguigu@hadoop102 hadoop-2.7.2]$ jps
4166 NameNode
4482 Jps
4263 DataNode
[atguigu@hadoop103 hadoop-2.7.2]$ jps
3218 DataNode
3288 Jps
[atguigu@hadoop104 hadoop-2.7.2]$ jps
3221 DataNode
3283 SecondaryNameNode
3364 Jps
```

③ 透過執行 start-yarn.sh 指令啟動 YARN，即可同時啟動 ResourceManager 和所有的 NodeManager。需要注意的是，NameNode 和 ResourceManager 如果不在同一台伺服器上，則不能在 NameNode 上啟動 YARN，應該在 ResourceManager 所在的伺服器上啟動 YARN。

```
[atguigu@hadoop103 hadoop-2.7.2]$ sbin/start-yarn.sh
```

透過執行 jps 指令可在各台節點伺服器上檢視處理程序啟動情況，若顯示以下內容，則表示啟動成功。

```
[atguigu@hadoop103 hadoop-2.7.2]$ sbin/start-yarn.sh
[atguigu@hadoop102 hadoop-2.7.2]$ jps
4166 NameNode
4482 Jps
4263 DataNode
4485 NodeManager
[atguigu@hadoop103 hadoop-2.7.2]$ jps
3218 DataNode
3288 Jps
3290 ResourceManager
3299 NodeManager
[atguigu@hadoop104 hadoop-2.7.2]$ jps
3221 DataNode
3283 SecondaryNameNode
3364 Jps
3389 NodeManager
```

（5）透過 Web UI 檢視叢集是否啟動成功。

① 在 Web 端輸入之前設定的 NameNode 的節點伺服器位址和通訊埠 50070，即可檢視 HDFS 檔案系統。舉例來說，在瀏覽器中輸入 http:// hadoop102:50070，可以檢查 NameNode 和 DataNode 是否正常。NameNode 的 Web 端如圖 3-40 所示。

圖 3-40 NameNode 的 Web 端

② 透過在 Web 端輸入 ResourceManager 的位址和通訊埠 8088，可以檢視
YARN 上任務的執行情況。舉例來說，在瀏覽器輸入 http://hadoop103:8088 ，
即可檢視本叢集 YARN 的執行情況。YARN 的 Web 端如圖 3-41 所示。

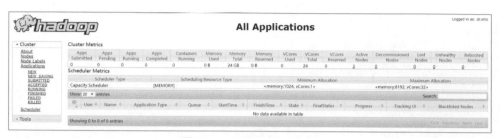

圖 3-41 YARN 的 Web 端

（6）執行 PI 實例，檢查叢集是否啟動成功。

在叢集任意節點伺服器上執行下面的指令，如果看到如圖 3-42 所示的執行結
果，則説明叢集啟動成功。

```
[atguigu@hadoop102 hadoop]$ cd /opt/module/hadoop-2.7.2/share/hadoop/
mapreduce/
[atguigu@hadoop102 mapreduce]$ hadoop jar hadoop-mapreduce-examples-
2.7.2.jar pi 10 10
```

圖 3-42 PI 實例執行結果

最後輸出為 Estimated value of Pi is 3.20000000000000000000。

3.3.5 設定 Hadoop 支援 LZO 壓縮

資料的壓縮對巨量資料的儲存非常重要，合理選用壓縮格式可以大幅縮小記憶體佔用空間，加強 I/O 傳輸效率，在 Flume 進行資料傳輸的過程中，我們需要將 Flume 落盤到 HDFS 檔案並儲存為壓縮格式。

HDFS 支援的壓縮格式很多，且各有優點，擷取到 HDFS 儲存的使用者行為記錄檔通常體量很大，落盤的單一檔案甚至會超過 HDFS 的檔案切片大小，當對這樣的大檔案進行處理時就會有關切片操作，支援切片同時壓縮率較高的壓縮格式為 LZO 壓縮，為此，需要設定 Hadoop 支援 LZO 壓縮，實際操作如下。

（1）先下載 LZO 的 jar 專案。

（2）下載後的檔案名稱是 hadoop-lzo-master，它是一個 .zip 格式的壓縮檔，先進行解壓，然後用 maven 進行編譯，產生 hadoop-lzo-0.4.20.jar。

（3）將編譯好後的 hadoop-lzo-0.4.20.jar 放入 hadoop-2.7.2/share/hadoop/common 中。

```
[atguigu@hadoop102 common]$ pwd
/opt/module/hadoop 2.7.2/share/hadoop/common
[atguigu@hadoop102 common]$ ls
hadoop-lzo-0.4.20.jar
```

（4）同步 hadoop-lzo-0.4.20.jar 到 hadoop103、hadoop104。

```
[atguigu@hadoop102 common]$ xsync hadoop-lzo-0.4.20.jar
```

（5）開啟 Hadoop 的設定檔 core-site.xml，增加支援 LZO 壓縮的設定，程式如下。

```
<property>
<name>io.compression.codecs</name>
```

```
<value>
org.apache.hadoop.io.compress.GzipCodec,
org.apache.hadoop.io.compress.DefaultCodec,
org.apache.hadoop.io.compress.BZip2Codec,
org.apache.hadoop.io.compress.SnappyCodec,
com.hadoop.compression.lzo.LzoCodec,
com.hadoop.compression.lzo.LzopCodec
</value>
</property>

<property>
    <name>io.compression.codec.lzo.class</name>
    <value>com.hadoop.compression.lzo.LzoCodec</value>
</property>
```

（6）同步 core-site.xml 檔案到 hadoop103、hadoop104。

```
[atguigu@hadoop102 hadoop]$ xsync core-site.xml
```

（7）啟動並檢視叢集。

```
[atguigu@hadoop102 hadoop-2.7.2]$ sbin/start-dfs.sh
[atguigu@hadoop103 hadoop-2.7.2]$ sbin/start-yarn.sh
```

① Web 和處理程序檢視。

- Web 檢視：http://hadoop102:50070。
- 處理程序檢視：執行 jps 指令，檢視各台節點伺服器的狀態。

② 當啟動發生錯誤時，採取以下措施。

- 檢視記錄檔：/opt/module/hadoop-2.7.2/logs。
- 如果進入安全模式，則可以透過執行 hdfs dfsadmin-safemode leave 指令強制離開安全模式。
- 停止所有處理程序，刪除 data 和 log 資料夾，然後執行 hdfs namenode-format 指令進行格式化（在叢集沒有重要資料的前提下）。

3.3.6 設定 Hadoop 支援 Snappy 壓縮

Hadoop 叢集本身不支援 Snappy 壓縮，若想使 Hadoop 支援 Snappy 壓縮，需要對 Hadoop 進行編譯，實際編譯步驟此處不再贅述。

（1）將編譯後支援 Snappy 壓縮的 Hadoop jar 套件解壓，將 lib/native 目錄下的所有檔案上傳到 hadoop102 的 /opt/module/hadoop-2.7.2/lib/native 目錄下，並分發到 hadoop103 和 hadoop104。

（2）重新啟動 Hadoop。

（3）檢查支援的壓縮方式。

參考程式如下。

```
[atguigu@hadoop102 native]$ hadoop checknative
hadoop:   true /opt/module/hadoop-2.7.2/lib/native/libhadoop.so
zlib:     true /lib64/libz.so.1
snappy:   true /opt/module/hadoop-2.7.2/lib/native/libsnappy.so.1
lz4:      true revision:99
bzip2:    false
```

3.4 本章歸納

本章主要對專案執行所需的環境進行了安裝和部署，從安裝虛擬機器和 CentOS 開始，到最後 JDK 和 Hadoop 的安裝，對每一步的安裝部署進行了詳細介紹。本章是整個專案的基礎，重點在於 Hadoop 叢集的架設和設定，讀者務必掌握。

使用者行為資料獲取模組

根據第 2 章中對擷取模組的整體分析,在本章中,我們將帶領讀者完成資料獲取模組的架設。

4.1 記錄檔產生

本專案需要讀者模仿前端記錄檔資料落盤過程自行產生模擬記錄檔資料,可以在虛擬機器的 /tmp/logs 目錄下產生每天的記錄檔資料。

1. 記錄檔啟動

(1)將取得的 jar 套件 log-collector-1.0-SNAPSHOT-jar-with-dependencies.jar 複製到 hadoop102 上,並同步到 hadoop103 的 /opt/module 目錄下。

```
[atguigu@hadoop102 module]$ xsync log-collector-1.0-SNAPSHOT-jar-with-
dependencies. jar
```

(2)在 hadoop102 上執行 jar 程式。

```
[atguigu@hadoop102 module]$ java -classpath log-collector-1.0-SNAPSHOT-
jar-with-dependencies.jar com.atguigu.appclient.AppMain >/opt/module/
test.log
```

（3）根據程式中設定檔的設定，系統會在虛擬機器的 /tmp/logs 目錄下產生記錄檔，產生的使用者行為記錄檔預設為伺服器的目前系統時間，若想產生不同時間的使用者行為記錄檔，則可以透過修改伺服器的時間來實現。

```
[atguigu@hadoop102 module]$ cd /tmp/logs/
[atguigu@hadoop102 logs]$ ls
app-2020-03-10.log
```

（4）在 /home/atguigu/bin 目錄下建立指令稿 dt.sh，用於統一修改伺服器的時間，以產生不同時間下的使用者行為記錄檔。

```
[atguigu@hadoop102 bin]$ vim dt.sh
```

（5）在指令稿中撰寫以下內容，分別在 3 台節點伺服器下使用 date 指令修改伺服器時間。

```
#!/bin/bash

for i in hadoop102 hadoop103 hadoop104
do
        echo "========== $i =========="
        ssh -t $i "sudo date -s $1"
done
```

（6）增加指令稿執行許可權。

```
[atguigu@hadoop102 bin]$ chmod 777 dt.sh
```

（7）啟動指令稿。

```
[atguigu@hadoop102 bin]$ dt.sh 2020-03-10
```

2. 叢集記錄檔產生啟動指令稿

將記錄檔產生的指令封裝成指令稿可以方便使用者呼叫執行，實際操作步驟如下。

（1）在 /home/atguigu/bin 目錄下建立指令稿 lg.sh。

```
[atguigu@hadoop102 bin]$ vim lg.sh
```

（2）指令稿想法：透過 i 變數在 hadoop102 和 hadoop103 節點伺服器間檢查，分別透過 ssh 指令進入兩台節點伺服器，執行 java 指令，執行記錄檔產生 jar 套件，在兩台節點伺服器的 /tmp/logs 目錄下產生模擬記錄檔。

在指令稿中撰寫以下內容。

```
#! /bin/bash

 for i in hadoop102 hadoop103
 do
  ssh $i "java -classpath /opt/module/log-collector-1.0-SNAPSHOT-jar-
with-dependencies.jar com.atguigu.appclient.AppMain $1 $2 >/opt/module/
test.log &"
 done
```

（3）增加指令稿執行許可權。

```
[atguigu@hadoop102 bin]$ chmod 777 lg.sh
```

（4）啟動指令稿。

```
[atguigu@hadoop102 module]$ lg.sh
```

（5）分別在 hadoop102 和 hadoop103 的 /tmp/logs 目錄下檢視產生的資料，判斷指令稿是否生效。

```
[atguigu@hadoop102 logs]$ ls
app-2020-03-10.log
[atguigu@hadoop103 logs]$ ls
app-2020-03-10.log
```

3. 叢集所有處理程序檢視指令稿

啟動叢集後，使用者需要透過 jps 指令檢視各台節點伺服器處理程序的啟動情況，操作起來比較麻煩，所以我們透過寫一個叢集所有處理程序檢視指令稿來實現使用一個指令稿檢視所有節點伺服器的所有處理程序的目的。

（1）在 /home/atguigu/bin 目錄下建立指令稿 xcall.sh。

```
[atguigu@hadoop102 bin]$ vim xcall.sh
```

（2）指令稿想法：透過 i 變數在 hadoop102、hadoop103 和 hadoop104 節點伺服器間檢查，分別透過 ssh 指令進入 3 台節點伺服器，執行傳入參數指定指令。

在指令稿中撰寫以下內容。

```
#! /bin/bash

for i in hadoop102 hadoop103 hadoop104
do
        echo --------- $i ----------
        ssh $i "$*"
done
```

（3）增加指令稿執行許可權。

```
[atguigu@hadoop102 bin]$ chmod 777 xcall.sh
```

（4）啟動指令稿。

```
[atguigu@hadoop102 bin]$ xcall.sh jps
```

4.2 擷取記錄檔的 Flume

如圖 4-1 所示，擷取記錄檔層 Flume 主要需要完成的任務為將記錄檔從落盤檔案中擷取出來，傳輸給訊息中介軟體 Kafka 叢集，這期間要保障資料不遺失，程式出現故障當機後可以快速重新啟動，對記錄檔進行初步分類，分別發往不同的 Kafka Topic，方便後續對記錄檔資料進行分別處理。

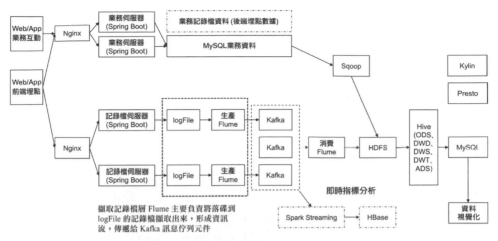

圖 4-1 擷取記錄檔層 Flume 的流向

4.2.1 Flume 元件

Flume 整體上是 Source-Channel-Sink 的三層架構，其中，Source 層完成對記錄檔的收集，將記錄檔封裝成 event 傳入 Channel 層中；Channel 層主要提供佇列的功能，對 Source 層中傳入的資料提供簡單的快取功能；Sink 層取出 Channel 層中的資料，將資料送入儲存檔案系統中，或對接其他的 Source 層。

Flume 以 Agent 為最小獨立執行單位，一個 Agent 就是一個 JVM，單一 Agent 由 Source、Sink 和 Channel 三大元件組成。

Flume 將資料表示為 event（事件），event 由一位組陣列的主體 body 和一個 key-value 結構的表頭 header 組成。其中，主體 body 中封裝了 Flume 傳送的資料，表頭 header 中容納的 key-value 資訊則是為了給資料增加標識，用於追蹤發送事件的優先順序和重要性，使用者可透過攔截器（Interceptor）進行修改。

Flume 的資料流由 event 貫穿始終，這些 event 由 Agent 外部的 Source 產生，當 Source 捕捉事件後會進行特定的格式化，然後 Source 會把事件推入 Channel 中，Channel 中的 event 會由 Sink 來拉取，Sink 拉取 event 後可以將 event 持久化或推向另一個 Source。

除此之外，Flume 還有一些使其應用更加靈活的元件：攔截器、Channel 選擇器（Selector）、Sink 組和 Sink 處理器。其功能如下。

- 攔截器可以部署在 Source 和 Channel 之間，用於對事件進行前置處理或過濾，Flume 內建了很多類型的攔截器，使用者也可以自訂自己的攔截器。
- Channel 選擇器可以決定 Source 接收的特定事件寫入哪些 Channel 元件中。
- Sink 組和 Sink 處理器可以幫助使用者實現負載平衡和容錯移轉。

4.2.2 Flume 安裝

在進行擷取記錄檔層的 Flume Agent 設定之前，我們首先需要安裝 Flume，Flume 需要安裝部署到每台節點伺服器上，實際安裝步驟如下。

（1）將 apache-flume-1.7.0-bin.tar.gz 上傳到 Linux 的 /opt/software 目錄下。

（2）解壓 apache-flume-1.7.0-bin.tar.gz 到 /opt/module/ 目錄下。

```
[atguigu@hadoop102 software]$ tar -zxf apache-flume-1.7.0-bin.tar.gz -C
/opt/module/
```

（3）修改 apache-flume-1.7.0-bin 的名稱為 flume。

```
[atguigu@hadoop102 module]$ mv apache-flume-1.7.0-bin flume
```

（4）將 flume/conf 目錄下的 flume-env.sh.template 檔案的名稱修改為 flume-env.sh，並設定 flume-env.sh 檔案，在設定檔中增加 JAVA_HOME 路徑，如下所示。

```
[atguigu@hadoop102 conf]$ mv flume-env.sh.template flume-env.sh
[atguigu@hadoop102 conf]$ vim flume-env.sh
export JAVA_HOME=/opt/module/jdk1.8.0_144
```

（5）將設定好的 Flume 分發到叢集中其他節點伺服器上。

4.2.3 擷取記錄檔 Flume 設定

1. Flume 設定分析

針對本專案，在撰寫 Flume Agent 設定檔之前，首先需要進行元件選型。

1）Source

本專案主要從一個即時寫入資料的資料夾中讀取資料，Source 可以選擇 Spooling Directory Source、Exec Source 和 Taildir Source。Taildir Source 相比 Exec Source、Spooling Directory Source 具有很多優勢。Taildir Source 可以實現中斷點續傳、多目錄監控設定。而在 Flume 1.6 以前需要使用者自訂 Source，記錄每次讀取檔案的位置，進一步實現中斷點續傳。Exec Source 可以即時搜集資料，但是在 Flume 不執行或 Shell 指令出錯的情況下，資料將遺失，進一步不能記錄資料讀取位置、實現中斷點續傳。Spooling Directory Source 可以實現目錄監控設定，但是不能即時擷取資料。

2）Channel

由於擷取記錄檔層 Flume 在讀取資料後主要將資料送往 Kafka 訊息佇列中，所以使用 Kafka Channel 是很好的選擇，同時選擇 Kafka Channel 可以不設定 Sink，加強了效率。

3）攔截器

本專案中主要部署兩個攔截器，一個用來過濾格式不正確的非法資料，這在實際生產環境中也是必不可少的，另一個用來分辨記錄檔類型，根據記錄檔類型給 event 增加 header 資訊，可以幫助 Channel 選擇器選擇記錄檔應該發往的 Channel。

4）Channel 選擇器

擷取記錄檔層 Flume 主要部署兩個 Kafka Channel，分別將資料發往不同的 Kafka Topic，兩個 Topic 儲存的資料不同，所以需要設定 Channel 選擇器決定記錄檔去向，並且設定選擇器類型為 multiplexing，在該模式下，會將 event 發送至特定的 Channel，而不會發送至所有 Channel，實現了記錄檔的分類分流。

2. Flume 的實際設定

在 /opt/module/flume/conf 目錄下建立 file-flume-kafka.conf 檔案。

```
[atguigu@hadoop102 conf]$ vim file-flume-kafka.conf
```

在檔案中設定以下內容。

```
#定義Agent必需的元件名稱，同時指定本設定檔的Agent名稱為a1
a1.sources=r1
a1.channels=c1 c2

#定義Source元件相關設定
#使用Taildir Source
a1.sources.r1.type = TAILDIR
#設定Taildir Source，儲存中斷點位置檔案的目錄
a1.sources.r1.positionFile = /opt/module/flume/test/log_position.json
#設定監控目錄組
a1.sources.r1.filegroups = f1
#設定目錄組下的目錄，可設定多個目錄
a1.sources.r1.filegroups.f1 = /tmp/logs/app.+

#設定Source發送資料的目標Channel
a1.sources.r1.channels = c1 c2

#攔截器
#設定攔截器名稱
a1.sources.r1.interceptors =  i1 i2
#設定攔截器名稱，需要寫明全類別名稱
a1.sources.r1.interceptors.i1.type = com.atguigu.flume.interceptor.
LogETLInterceptor$Builder
a1.sources.r1.interceptors.i2.type = com.atguigu.flume.interceptor.
LogTypeInterceptor$Builder

#設定Channel選擇器
#設定選擇器類型
a1.sources.r1.selector.type = multiplexing
#設定選擇器識別header中的key
a1.sources.r1.selector.header = topic
#設定不同的header資訊，發往不同的Channel
a1.sources.r1.selector.mapping.topic_start = c1
a1.sources.r1.selector.mapping.topic_event = c2
```

```
# configure channel設定Channel
#設定Channel類型為Kafka Channel
a1.channels.c1.type = org.apache.flume.channel.kafka.KafkaChannel
#設定Kafka叢集節點伺服器列表
a1.channels.c1.kafka.bootstrap.servers = hadoop102:9092,hadoop103:909
2,hadoop104:9092
#設定該Channel發往Kafka的Topic，該Topic需要在Kafka中提前建立
a1.channels.c1.kafka.topic = topic_start
#設定不將header資訊解析為event內容
a1.channels.c1.parseAsFlumeEvent = false
#設定該Kafka Channel所屬的消費者組名，為實現multiplexing類型的Channel選擇
器，應將2個Kafka Channel設定相同的消費者組
a1.channels.c1.kafka.consumer.group.id = flume-consumer

#設定同上
a1.channels.c2.type = org.apache.flume.channel.kafka.KafkaChannel
a1.channels.c2.kafka.bootstrap.servers = hadoop102:9092,hadoop103:
9092,hadoop104:9092
a1.channels.c2.kafka.topic = topic_event
a1.channels.c2.parseAsFlumeEvent = false
a1.channels.c2.kafka.consumer.group.id = flume-consumer
```

> ★ 注意：com.atguigu.flume.interceptor.LogETLInterceptor和com.atguigu.flume.interceptor.
> LogTypeInterceptor 是筆者自訂的攔截器的全類別名稱。讀者需要根據自己自
> 訂的攔截器進行對應修改。

4.2.4 Flume 的 ETL 攔截器和記錄檔類型區分攔截器

在本專案中自訂了兩個攔截器，分別是 ETL 攔截器、記錄檔類型區分攔截器。

ETL 是指將業務系統的資料經過取出、清洗轉換之後載入到資料倉儲的過
程，目的是將企業中分散、零亂、標準不統一的資料整合到一起，為企業的
管理者決策提供分析依據。在這裡可以簡單地了解為資料清洗。

ETL 攔截器主要用於過濾時間戳記非法和 JSON 資料不完整的記錄檔。

記錄檔類型區分攔截器主要用於將啟動記錄和事件記錄檔區分開，並增加 event 的 header 資訊，方便將不同類型的 event 發往 Kafka 的不同 Topic。

攔截器的定義步驟如下。

（1）建立 Maven 專案 flume-interceptor。

（2）建立套件名稱：com.atguigu.flume.interceptor。

（3）在 pom.xml 檔案中增加以下依賴。

```xml
<dependencies>
    <dependency>
        <groupId>org.apache.flume</groupId>
        <artifactId>flume-ng-core</artifactId>
        <version>1.7.0</version>
    </dependency>
</dependencies>

<build>
    <plugins>
        <plugin>
            <artifactId>maven-compiler-plugin</artifactId>
            <version>2.3.2</version>
            <configuration>
                <source>1.8</source>
                <target>1.8</target>
            </configuration>
        </plugin>
        <plugin>
            <artifactId>maven-assembly-plugin</artifactId>
            <configuration>
                <descriptorRefs>
                    <descriptorRef>jar-with-dependencies</descriptorRef>
                </descriptorRefs>
            </configuration>
            <executions>
                <execution>
                    <id>make-assembly</id>
                    <phase>package</phase>
```

```
                <goals>
                    <goal>single</goal>
                </goals>
            </execution>
        </executions>
    </plugin>
    </plugins>
</build>
```

（4）在 com.atguigu.flume.interceptor 套件中建立 LogETLInterceptor 類別名稱。
Flume 的 ETL 攔截器 LogETLInterceptor。

```java
package com.atguigu.flume.interceptor;

import org.apache.flume.Context;
import org.apache.flume.Event;
import org.apache.flume.interceptor.Interceptor;

import java.nio.charset.Charset;
import java.util.ArrayList;
import java.util.List;

public class LogETLInterceptor implements Interceptor {

    @Override
    public void initialize() {

    }

    @Override
    public Event intercept(Event event) {

        // 1 取得資料
        byte[] body = event.getBody();
        String log = new String(body, Charset.forName("UTF-8"));

        // 2 判斷資料是否合法
        if (log.contains("start")) {
            if (LogUtils.validateStart(log)){
```

```
                return event;
            }
        }else {
            if (LogUtils.validateEvent(log)){
                return event;
            }
        }

        // 3 傳回驗證結果
        return null;
}

@Override
public List<Event> intercept(List<Event> events) {

    ArrayList<Event> interceptors = new ArrayList<>();

    for (Event event : events) {
        Event intercept1 = intercept(event);

        if (intercept1 != null){
            interceptors.add(intercept1);
        }
    }

    return interceptors;
}

@Override
public void close() {

}

public static class Builder implements Interceptor.Builder{

    @Override
    public Interceptor build() {
        return new LogETLInterceptor();
    }
```

```
        @Override
        public void configure(Context context) {


        }
    }
}
```

（5）撰寫 Flume 記 錄 檔 過 濾 工 具 類 別 LogUtils，方 便 ETL 攔 截 器 LogETLInterceptor 呼叫。

```
package com.atguigu.flume.interceptor;
import org.apache.commons.lang.math.NumberUtils;

public class LogUtils {

    public static boolean validateEvent(String log) {
        // 伺服器時間 | JSON
        // 1549696569054 | {"cm":{"ln":"-89.2","sv":"V2.0.4","os":"8.2.0",
"g":"M67B4QYU@gmail.com","nw":"4G","l":"en","vc":"18","hw":"1080*1920",
"ar":"MX","uid":"u8678","t":"1549679122062","la":"-27.4","md":"sumsung-
12","vn":"1.1.3","ba":"Sumsung","sr":"Y"},"ap":"weather","et":[]}

        // 1 切割
        String[] logContents = log.split("\\|");

        // 2 驗證
        if(logContents.length != 2){
            return false;
        }

        //3 驗證伺服器時間
        if (logContents[0].length()!=13
|| !NumberUtils.isDigits(logContents[0])){
            return false;
        }

        // 4 驗證JSON
        if (!logContents[1].trim().startsWith("{")
```

```
|| !logContents[1].trim().endsWith("}")){
        return false;
    }


    return true;
}


public static boolean validateStart(String log) {

    if (log == null){
        return false;
    }


    // 驗證JSON
    if (!log.trim().startsWith("{") || !log.trim().endsWith("}")){
        return false;
    }


    return true;
}
}
```

（6）Flume 記錄檔類型區分攔截器 LogTypeInterceptor。

```
package com.atguigu.flume.interceptor;

import org.apache.flume.Context;
import org.apache.flume.Event;
import org.apache.flume.interceptor.Interceptor;

import java.nio.charset.Charset;
import java.util.ArrayList;
import java.util.List;
import java.util.Map;

public class LogTypeInterceptor implements Interceptor {
    @Override
    public void initialize() {
```

```
    }

    @Override
    public Event intercept(Event event) {

        // 區分記錄檔類型：body或header
        // 1 取得body資料
        byte[] body = event.getBody();
        String log = new String(body, Charset.forName("UTF-8"));

        // 2 取得header資料
        Map<String, String> headers = event.getHeaders();

        // 3 判斷資料類型並向header中設定值
        if (log.contains("start")) {
            headers.put("topic","topic_start");
        }else {
            headers.put("topic","topic_event");
        }

        return event;
    }

    @Override
    public List<Event> intercept(List<Event> events) {

        ArrayList<Event> interceptors = new ArrayList<>();

        for (Event event : events) {
            Event intercept1 = intercept(event);

            interceptors.add(intercept1);
        }

        return interceptors;
    }

    @Override
    public void close() {
```

```
    }

    public static class Builder implements  Interceptor.Builder{

        @Override
        public Interceptor build() {
            return new LogTypeInterceptor();
        }

        @Override
        public void configure(Context context) {

        }
    }
}
```

（7）包裝。

攔截器包裝之後，只需要單獨的壓縮檔，不需要將相依套件上傳。包裝之後
要放入 Flume 的 lib 目錄下，如圖 4-2 所示。

flume-interceptor-1.0-SNAPSHOT.jar	10KB	Executabl...	2018/11/2, 16:49
flume-interceptor-1.0-SNAPSHOT-jar-with-dependencies.jar	11.72MB	Executabl...	2018/11/2, 16:50

圖 4-2 攔截器壓縮檔

★ **注意：**為什麼不需要相依套件？因為相依套件在 Flume 的 lib 目錄下已經存在。

（8）需要先將打好的套件放入 hadoop102 的 /opt/module/flume/lib 目錄下。

```
[atguigu@hadoop102 lib]$ ls | grep interceptor
flume-interceptor-1.0-SNAPSHOT.jar
```

（9）分發 Flume 到 hadoop103 和 hadoop104。

```
[atguigu@hadoop102 module]$ xsync flume/
```

（10）執行 flume-ng agent 指令，將上述設定檔啟動，其中，--name 選項用於
指定本次指令執行的 Agent 名字，本設定檔中為 a1；--conf-file 選項用於指定
設定檔的儲存路徑。

```
[atguigu@hadoop102 flume]$ bin/flume-ng agent --name a1 --conf-file conf
/file-flume-kafka.conf
```

該 Flume Agent 的資料流向是 Kafka，由於我們還沒有安裝 Kafka，所以啟動後不能形成完整的資料流，若想看到資料的消費情況，讀者可以使用監控工具 Gangalia 進行檢視，此處不再贅述。

建議讀者後續學習了 Kafka 後再對該部分設定指令進行測試。

4.2.5 擷取記錄檔 Flume 啟動、停止指令稿

同記錄檔產生一樣，我們也將擷取記錄檔層 Flume 的啟動、停止指令封裝成指令稿，以方便後續呼叫執行。

（1）在 /home/atguigu/bin 目錄下建立指令稿 f1.sh。

```
[atguigu@hadoop102 bin]$ vim f1.sh
```

指令稿想法：透過比對輸入參數的值選擇是否啟動擷取程式，啟動擷取程式後，設定記錄檔不列印且程式在後台執行。

若停止程式，則透過管線符號切割等操作取得程式的編號，並透過 kill 指令停止程式。在指令稿中撰寫以下內容。

```
#! /bin/bash

case $1 in
"start"){
        for i in hadoop102 hadoop103
        do
                echo " --------啟動 $i 擷取Flume-------"
                ssh $i "source /etc/profile ; nohup /opt/module/flume/
bin/flume-ng agent --conf-file /opt/module/flume/conf/file-flume-kafka.
conf --name a1 -Dflume.root.logger-INFO,LOGFILE >/dev/null 2>&1 &"
        done
};;
"stop"){
        for i in hadoop102 hadoop103
```

```
        do
                echo " --------停止 $i 擷取Flume-------"
                ssh $i "ps -ef | grep file-flume-kafka | grep -v grep
|awk '{print \$2}' | xargs kill"
        done

};;
esac
```

指令稿説明如下。

說明 1：nohup 指令可以在使用者退出帳戶或關閉終端之後繼續執行對應的處理程序。nohup 指令就是不暫停的意思，不斷執行指令。

說明 2：/dev/null 代表 Linux 的空裝置檔案，所有往這個檔案裡面寫入的內容都會遺失，俗稱「黑洞」。企業在進行開發時，如果不想在主控台顯示大量的啟動過程記錄檔，就可以把記錄檔寫入「黑洞」，以減少磁碟儲存空間。

標準輸入 0：從鍵盤獲得輸入 /proc/self/fd/0。
標準輸出 1：輸出到主控台 /proc/self/fd/1。
錯誤輸出 2：輸出到主控台 /proc/self/fd/2。

說明 3：

① "ps -ef | grep file-flume-kafka" 用於取得 Flume 處理程序，檢視結果可以發現存在兩個處理程序 id，但是我們只想取得第一個處理程序 id 21319。

```
atguigu  21319      1 57 15:14 ?          00:00:03
......
atguigu  21428 11422  0 15:14 pts/1    00:00:00 grep file-flume-kafka
```

② "ps -ef | grep file-flume-kafka | grep -v grep" 用於過濾包含 grep 資訊的處理程序。

```
atguigu  21319      1 57 15:14 ?          00:00:03
......
```

③ "ps -ef | grep file-flume-kafka | grep -v grep |awk '{print \$2}'"，採用 awk，預設用空格分隔後，取第二個欄位，取得到 21319 處理程序 id。

④ "ps -ef | grep file-flume-kafka | grep -v grep |awk '{print \$2}' | xargs kill"，xargs 表示取得前一階段的執行結果，即 21319，作為下一個指令 kill 的輸入參數。實際執行的是 kill 21319。

（2）增加指令稿執行許可權。

```
[atguigu@hadoop102 bin]$ chmod 777 f1.sh
```

（3）f1 叢集啟動指令稿。

```
[atguigu@hadoop102 module]$ f1.sh start
```

（4）f1 叢集停止指令稿。

```
[atguigu@hadoop102 module]$ f1.sh stop
```

4.3 訊息佇列 Kafka

透過 Flume Agent 程式將記錄檔從落盤資料夾擷取出來之後，需要發送到 Kafka，Kafka 在這裡造成資料緩衝和負載平衡的作用，大幅減輕資料儲存系統的壓力。在向 Kafka 發送記錄檔之前，需要先安裝 Kafka，而在安裝 Kafka 之前需要先安裝 Zookeeper，為之提供分散式服務。本節主要帶領讀者完成 Zookeeper 和 Kafka 的安裝部署。

4.3.1 Zookeeper 安裝

Zookeeper 是一個能夠高效開發和維護分散式應用的協調服務，主要用於為分散式應用提供一致性服務，提供的功能包含維護設定資訊、名稱服務、分散式同步、組服務等。

Zookeeper 的安裝步驟如下。

1. 叢集規劃

在 hadoop102、hadoop103 和 hadoop104 三台節點伺服器上部署 Zookeeper。

2. 解壓安裝

（1）解壓 Zookeeper 安裝套件到 /opt/module/ 目錄下。

```
[atguigu@hadoop102 software]$ tar -zxvf zookeeper-3.4.10.tar.gz -C
/opt/module/
```

（2）在 /opt/module/zookeeper-3.4.10 目錄下建立 zkData 資料夾，用於儲存 Zookeeper 的相關資料。

```
mkdir -p zkData
```

3. 設定 zoo.cfg 檔案

（1）重新命名 /opt/module/zookeeper-3.4.10/conf 目錄下的 zoo_sample.cfg 為 zoo.cfg，我們可以對其中的設定進行自訂設定。

```
mv zoo_sample.cfg zoo.cfg
```

（2）實際設定，在設定檔中找到以下內容，將資料儲存目錄 dataDir 設定為上文中自行建立的 zkData 資料夾。

```
dataDir=/opt/module/zookeeper-3.4.10/zkData
```

增加以下設定，以下設定指出了 Zookeeper 叢集的 3 台節點伺服器資訊。

```
#######################cluster#########################
server.2=hadoop102:2888:3888
server.3=hadoop103:2888:3888
server.4=hadoop104:2888:3888
```

（3）設定參數解讀。

```
Server.A=B:C:D。
```

- A 是一個數字，表示第幾台伺服器；
- B 是這台伺服器的 IP 位址；

- C 是這台伺服器與叢集中的 Leader 伺服器交換資訊的通訊埠；
- D 表示當叢集中的 Leader 伺服器無法正常執行時期，需要一個通訊埠來重新進行選舉，選出一個新的 Leader 伺服器，而這個通訊埠就是用來執行選舉時伺服器相互通訊的通訊埠。

在叢集模式下設定一個檔案 myid，這個檔案在 dataDir 目錄下，其中有一個資料就是 A 的值，Zookeeper 啟動時讀取此檔案，並將裡面的資料與 zoo.cfg 檔案裡面的設定資訊進行比較，進一步判斷到底是哪台伺服器。

4. 叢集操作

（1）在 /opt/module/zookeeper-3.4.10/zkData 目錄下建立一個 myid 檔案，當叢集啟動時由 Zookeeper 讀取此檔案。

```
touch myid
```

> ✦ **注意**：當增加 myid 檔案時，一定要在 Linux 中建立，在文字編輯工具中建立有可能山現亂碼。

（2）編輯 myid 檔案。

```
vim myid
```

在檔案中增加與 Server 對應的編號，根據在 zoo.cfg 檔案中設定的 Server id 與節點伺服器的 IP 位址對應關係增加，如在 hadoop102 節點伺服器中增加 2。

（3）複製設定好的 Zookeeper 到其他伺服器上。

```
xsync /opt/module/zookeeper-3.4.10
```

分別修改 hadoop103 和 hadoop104 節點伺服器中 myid 檔案的內容為 3 和 4。

（4）在 3 台節點伺服器中分別啟動 Zookeeper。

```
[root@hadoop102 zookeeper-3.4.10]# bin/zkServer.sh start
[root@hadoop103 zookeeper-3.4.10]# bin/zkServer.sh start
[root@hadoop104 zookeeper-3.4.10]# bin/zkServer.sh start
```

（5）執行以下指令，在 3 台節點伺服器中檢視 Zookeeper 的服務狀態。

```
[root@hadoop102 zookeeper-3.4.10]# bin/zkServer.sh status
JMX enabled by default
Using config: /opt/module/zookeeper-3.4.10/bin/../conf/zoo.cfg
Mode: follower
[root@hadoop103 zookeeper-3.4.10]# bin/zkServer.sh status
JMX enabled by default
Using config: /opt/module/zookeeper-3.4.10/bin/../conf/zoo.cfg
Mode: leader
[root@hadoop104 zookeeper-3.4.5]# bin/zkServer.sh status
JMX enabled by default
Using config: /opt/module/zookeeper-3.4.10/bin/../conf/zoo.cfg
Mode: follower
```

4.3.2 Zookeeper 叢集啟動、停止指令稿

由於 Zookeeper 沒有提供多台伺服器同時啟動、停止的指令稿，使用單台節點伺服器執行伺服器啟動、停止指令顯然操作煩瑣，所以可將 Zookeeper 啟動、停止指令封裝成指令稿。實際操作步驟如下。

（1）在 hadoop102 的 /home/atguigu/bin 目錄下建立指令稿 zk.sh。

```
[atguigu@hadoop102 bin]$ vim zk.sh
```

指令稿想法：透過執行 ssh 指令，分別登入叢集節點伺服器，然後執行啟動、停止或檢視服務狀態的指令。在指令稿中撰寫以下內容。

```
#! /bin/bash

case $1 in
"start"){
 for i in hadoop102 hadoop103 hadoop104
 do
  ssh $i "source /etc/profile ; /opt/module/zookeeper-3.4.10/bin/
zkServer.sh start"
 done
};;
"stop"){
```

```
for i in hadoop102 hadoop103 hadoop104
do
  ssh $i "source /etc/profile ; /opt/module/zookeeper-3.4.10/bin/
zkServer.sh stop"
done
};;
"status"){
for i in hadoop102 hadoop103 hadoop104
do
  ssh $i "source /etc/profile ; /opt/module/zookeeper-3.4.10/bin/
zkServer.sh status"
done
};;
esac
```

（2）增加指令稿執行許可權。

```
[atguigu@hadoop102 bin]$ chmod 777 zk.sh
```

（3）Zookeeper 叢集啟動指令稿。

```
[atguigu@hadoop102 module]$ zk.sh start
```

（4）Zookeeper 叢集停止指令稿。

```
[atguigu@hadoop102 module]$ zk.sh stop
```

4.3.3 Kafka 安裝

Kafka 是一個優秀的分散式訊息佇列系統，透過將記錄檔訊息先發送至 Kafka，可以避開資料遺失的風險，增加資料處理的可擴充性，加強資料處理的靈活性和峰值處理能力，加強系統可用性，為訊息消費提供順序保障，並且可以控制最佳化資料流經系統的速度，解決訊息生產和訊息消費速度不一致的問題。

Kafka 叢集需要依賴 Zookeeper 提供服務來儲存一些中繼資料資訊，以保障系統可用性。在完成 Zookeeper 的安裝之後，就可以安裝 Kafka 了，實際安裝步驟如下。

（1）Kafka 叢集規劃如表 4-1 所示。

表 4-1　Kafka 叢集規劃

hadoop102	hadoop103	hadoop104
Zookeeper	Zookeeper	Zookeeper
Kafka	Kafka	Kafka

（2）下載安裝套件。
下載 Kafka 的安裝套件。

（3）解壓安裝套件。

```
[atguigu@hadoop102 software]$ tar -zxvf kafka_2.11-0.11.0.0.tgz -C /opt/
module/
```

（4）修改解壓後的檔案名稱。

```
[atguigu@hadoop102 module]$ mv kafka_2.11-0.11.0.0/ kafka
```

（5）在 /opt/module/kafka 目錄下建立 logs 資料夾，用於儲存 Kafka 執行過程中產生的記錄檔。

```
[atguigu@hadoop102 kafka]$ mkdir logs
```

（6）進入 Kafka 的設定目錄，開啟 server.properties，修改設定檔，Kafka 的設定檔都是以鍵值對的形式存在的，主要需要修改的內容如下。

```
[atguigu@hadoop102 kafka]$ cd config/
[atguigu@hadoop102 config]$ vim server.properties
```

找到對應的設定並按照以下內容進行修改。

```
#broker的全域唯一編號，不能重複
broker.id=0
#設定刪除Topic功能為true，即在Kafka中刪除Topic為真正刪除，而非標記刪除
delete.topic.enable=true
#處理網路請求的執行緒數量
num.network.threads=3
#用來處理磁碟I/O的執行緒數量
```

```
num.io.threads=8
#發送通訊端的緩衝區大小
socket.send.buffer.bytes=102400
#接收通訊端的緩衝區大小
socket.receive.buffer.bytes=102400
#請求通訊端的緩衝區大小
socket.request.max.bytes=104857600
#Kafka執行記錄檔儲存的路徑，設定為自行建立的logs資料夾
log.dirs=/opt/module/kafka/logs
#Topic在目前broker上的分區個數
num.partitions=1
#用來恢復和清理data下資料的執行緒數量
num.recovery.threads.per.data.dir=1
#資料檔案保留的最長時間，逾時則被刪除
log.retention.hours=168
#設定連接Zookeeper叢集的位址
zookeeper.connect=hadoop102:2181,hadoop103:2181,hadoop104:2181
```

（7）設定環境變數，將 Kafka 的安裝目錄設定到系統環境變數中，可以更加方便使用者執行 Kafka 的相關指令。在設定完環境變數後，需要執行 source 指令使環境變數生效。

```
[root@hadoop102 module]# vim /etc/profile
#KAFKA_HOME
export KAFKA_HOME=/opt/module/kafka
export PATH=$PATH:$KAFKA_HOME/bin
[root@hadoop102 module]# source /etc/profile
```

（8）安裝設定全部修改完成後，分發安裝套件和環境變數到叢集其他節點伺服器，並使環境變數生效。

```
[root@hadoop102 etc]# xsync profile
[atguigu@hadoop102 module]$ xsync kafka/
```

（9）修改 broker.id。

分別在 hadoop103 和 hadoop104 上修改設定檔 /opt/module/kafka/config/server.operties 中的 broker.id=1、broker.id=2。

> ★ 注意：broker.id 為識別 Kafka 叢集不同節點伺服器的標識，不可重複。

（10）啟動叢集。

依次在 hadoop102、hadoop103 和 hadoop104 上啟動 Kafka。

```
[atguigu@hadoop102 kafka]$ bin/kafka-server-start.sh config/server.properties &
[atguigu@hadoop103 kafka]$ bin/kafka-server-start.sh config/server.properties &
[atguigu@hadoop104 kafka]$ bin/kafka-server-start.sh config/server.properties &
```

（11）關閉叢集。

```
[atguigu@hadoop102 kafka]$ bin/kafka-server-stop.sh stop
[atguigu@hadoop103 kafka]$ bin/kafka-server-stop.sh stop
[atguigu@hadoop104 kafka]$ bin/kafka-server-stop.sh stop
```

4.3.4 Kafka 叢集啟動、停止指令稿

同 Zookeeper 一樣，將 Kafka 叢集的啟動、停止指令寫成指令稿，方便以後呼叫執行。

（1）在 /home/atguigu/bin 目錄下建立指令稿 kf.sh。

```
[atguigu@hadoop102 bin]$ vim kf.sh
```

在指令稿中撰寫以下內容。

```
#! /bin/bash

case $1 in
"start"){
        for i in hadoop102 hadoop103 hadoop104
        do
                echo " --------啟動 $i Kafka-------"

                ssh $i "source /etc/profile ; /opt/module/kafka/bin/
kafka-server-start.sh -daemon /opt/module/kafka/config/server.properties "
        done
};;
```

```
"stop"){
        for i in hadoop102 hadoop103 hadoop104
        do
                echo " --------停止 $i Kafka-------"
                ssh $i " source /etc/profile ; /opt/module/kafka/bin/
kafka-server-stop.sh stop"
        done
};;
esac
```

（2）增加指令稿執行許可權。

```
[atguigu@hadoop102 bin]$ chmod 777 kf.sh
```

（3）Kafka 叢集啟動指令稿。

```
[atguigu@hadoop102 module]$ kf.sh start
```

（4）Kafka 叢集停止指令稿。

```
[atguigu@hadoop102 module]$ kf.sh stop
```

4.3.5 Kafka Topic 相關操作

本節主要帶領讀者熟悉 Kafka 的常用命令列操作。在本專案中，學會使用命令列操作 Kafka 已經足夠，若想更加深入地了解 Kafka，體驗 Kafka 其餘的優秀特性。

（1）檢視 Kafka Topic 列表。

```
[atguigu@hadoop102 kafka]$ bin/kafka-topics.sh --zookeeper
hadoop102:2181 --list
```

（2）建立 Kafka Topic。
進入 /opt/module/kafka/ 目錄下，分別建立啟動記錄主題、事件記錄檔主題，建立的記錄檔主題名稱應該與擷取記錄檔層 Flume 的設定檔中的名稱相同。

① 建立啟動記錄主題。

```
[atguigu@hadoop102 kafka]$ bin/kafka-topics.sh --zookeeper
hadoop102:2181, hadoop103:2181, hadoop104:2181  --create
--replication-factor 1 --partitions 1 --topic topic_start
```

② 建立事件記錄檔主題。

```
[atguigu@hadoop102 kafka]$ bin/kafka-topics.sh --zookeeper
hadoop102:2181,hadoop103:2181,hadoop104:2181  --create
--replication-factor 1 --partitions 1 --topic topic_event
```

（3）刪除 Kafka Topic 指令。

若在建立主題時出現錯誤，則可以使用刪除主題指令對主題進行刪除。

① 刪除啟動記錄主題。

```
[atguigu@hadoop102 kafka]$ bin/kafka-topics.sh --delete --zookeeper
hadoop102:2181, hadoop103:2181,hadoop104:2181 --topic topic_start
```

② 刪除事件記錄檔主題。

```
[atguigu@hadoop102 kafka]$ bin/kafka-topics.sh --delete --zookeeper
hadoop102:2181,hadoop103:2181,hadoop104:2181 --topic topic_event
```

（4）Kafka 主控台生產訊息測試。

```
 [atguigu@hadoop102 kafka]$ bin/kafka-console-producer.sh \
--broker-list hadoop102:9092 --topic topic_start
>hello world
>atguigu  atguigu
```

（5）Kafka 主控台消費訊息測試。

```
[atguigu@hadoop102 kafka]$ bin/kafka-console-consumer.sh \
--bootstrap-server hadoop102:9092 --from-beginning --topic topic_start
```

其中，--from-beginning 表示將主題中以往所有的資料都讀取出來。使用者可根據業務場景選擇是否增加該設定。

（6）檢視 Kafka Topic 詳情。

```
[atguigu@hadoop102 kafka]$ bin/kafka-topics.sh --zookeeper hadoop102:2181 \
--describe --topic topic_start
```

（7）開啟 Kafka 主控台消費訊息測試，執行擷取記錄檔 Flume 啟動指令稿，檢視主控台是否有記錄檔列印，以驗證記錄檔是否擷取成功。

```
[atguigu@hadoop102 kafka]$ bin/kafka-console-consumer.sh \
--bootstrap-server hadoop102:9092 --from-beginning --topic topic_start
[atguigu@hadoop102 module]$ f1.sh start
```

4.4 消費 Kafka 記錄檔的 Flume

將記錄檔從擷取記錄檔層 Flume 發送到 Kafka 叢集後，接下來的工作需要將記錄檔資料進行落盤儲存，我們依然將這部分工作交給 Flume 完成，如圖 4-3 所示。

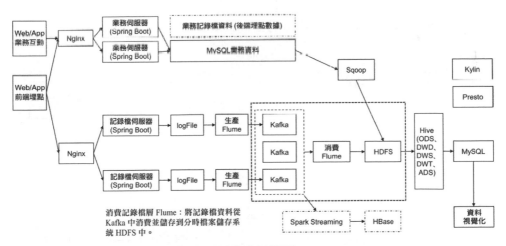

圖 4-3 消費記錄檔層 Flume

將消費記錄檔層 Flume Agent 程式部署在 hadoop104 上，實現 hadoop102、hadoop103 負責記錄檔的產生和擷取，hadoop104 負責記錄檔的消費儲存。在實際生產環境中應儘量做到將不同的任務部署在不同的節點伺服器上。消費記錄檔層 Flume 叢集規劃如表 4-2 所示。

表 4-2 消費記錄檔層 Flume 叢集規劃

	節點伺服器 hadoop102	節點伺服器 hadoop103	節點伺服器 hadoop104
Flume（消費 Kafka）			Flume

4.4.1 消費記錄檔 Flume 設定

1. Flume 設定分析

消費記錄檔層 Flume 主要從 Kafka 中讀取訊息，所以選用 Kafka Source。Channel 選用 File Channel，能大幅避免資料遺失。Sink 選用 HDFS Sink，可以將記錄檔直接落盤到 HDFS 中。

消費記錄檔層 Flume 設定分析如圖 4-4 所示，該層 Flume 需要從不同的 Kafka Topic 消費讀取訊息，再將記錄檔落盤到 HDFS 的不同目錄中，所以我們可以簡單架設兩個訊息通道，分別進行組裝。

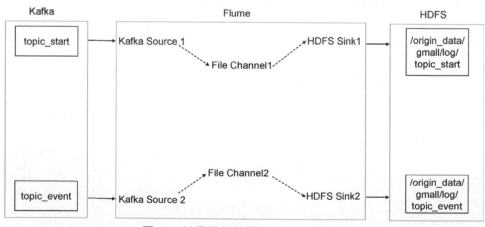

圖 4-4 消費記錄檔層 Flume 設定分析

2. Flume 實際設定

（1）在 hadoop104 的 /opt/module/flume/conf 目錄下建立 kafka-flume-hdfs.conf 檔案。

```
[atguigu@hadoop104 conf]$ vim kafka-flume-hdfs.conf
```

（2）在檔案中設定以下內容。

```
## Flume Agent元件宣告
a1.sources=r1 r2
a1.channels=c1 c2
a1.sinks=k1 k2

## Source1屬性設定
#設定Source類型為Kafka Source
a1.sources.r1.type = org.apache.flume.source.kafka.KafkaSource
#設定Katka Source每次從Kafka Topic中拉取的event個數
a1.sources.r1.batchSize = 5000
#設定拉取資料批次間隔為2000毫秒
a1.sources.r1.batchDurationMillis = 2000
#設定Kafka叢集位址
a1.sources.r1.kafka.bootstrap.servers =
hadoop102:9092,hadoop103:9092,hadoop104:9092
#設定Source 對接Kafka主題
a1.sources.r1.kafka.topics=topic_start

## source2屬性設定，與Source1設定類似，只是消費主題不同
a1.sources.r2.type = org.apache.flume.source.kafka.KafkaSource
a1.sources.r2.batchSize = 5000
a1.sources.r2.batchDurationMillis = 2000
a1.sources.r2.kafka.bootstrap.servers =
hadoop102:9092,hadoop103:9092,hadoop104:9092
a1.sources.r2.kafka.topics=topic_event

## Channel1屬性設定
#設定Channel類型為File Channel
a1.channels.c1.type = file
```

```
#設定儲存File Channel傳輸資料的中斷點資訊目錄
a1.channels.c1.checkpointDir = /opt/module/flume/checkpoint/behavior1
#設定File Channel傳輸資料的儲存位置
a1.channels.c1.dataDirs = /opt/module/flume/data/behavior1/
#設定File Channel的最大儲存容量
a1.channels.c1.maxFileSize = 2146435071
#設定File Channel最多儲存event的個數
a1.channels.c1.capacity = 1000000
#設定Channel滿時put交易的逾時
a1.channels.c1.keep-alive = 6


## Channel2屬性設定同Channel1，注意需要設定不同的目錄路徑
a1.channels.c2.type = file
a1.channels.c2.checkpointDir = /opt/module/flume/checkpoint/behavior2
a1.channels.c2.dataDirs = /opt/module/flume/data/behavior2/
a1.channels.c2.maxFileSize = 2146435071
a1.channels.c2.capacity = 1000000
a1.channels.c2.keep-alive = 6


## Sink1屬性設定
#設定Sink1類型為HDFS Sink
a1.sinks.k1.type = hdfs
#設定發到HDFS的儲存路徑
a1.sinks.k1.hdfs.path = /origin_data/gmall/log/topic_start/%Y-%m-%d
#設定HDFS落盤檔案的檔案名稱首
a1.sinks.k1.hdfs.filePrefix = logstart-


##Sink2 屬性設定同Sink1
a1.sinks.k2.type = hdfs
a1.sinks.k2.hdfs.path = /origin_data/gmall/log/topic_event/%Y-%m-%d
a1.sinks.k2.hdfs.filePrefix = logevent-


## 避免產生大量小檔案的相關屬性設定
a1.sinks.k1.hdfs.rollInterval = 10
a1.sinks.k1.hdfs.rollSize = 134217728
a1.sinks.k1.hdfs.rollCount = 0
```

```
a1.sinks.k2.hdfs.rollInterval = 10
a1.sinks.k2.hdfs.rollSize = 134217728
a1.sinks.k2.hdfs.rollCount = 0

## 控制輸出檔案是壓縮檔
a1.sinks.k1.hdfs.fileType = CompressedStream
a1.sinks.k2.hdfs.fileType = CompressedStream

a1.sinks.k1.hdfs.codeC = lzop
a1.sinks.k2.hdfs.codeC = lzop

## 拼裝
a1.sources.r1.channels = c1
a1.sinks.k1.channel= c1

a1.sources.r2.channels = c2
a1.sinks.k2.channel= c2
```

4.4.2 消費記錄檔 Flume 啟動、停止指令稿

將消費記錄檔層 Flume 程式的啟動、停止指令撰寫成指令稿，方便後續呼叫執行，指令稿包含啟動消費層 Flume 程式和根據 Flume 的任務編號停止其執行，與擷取記錄檔層 Flume 啟動、停止指令稿類似，撰寫步驟如下。

（1）在 /home/atguigu/bin 目錄下建立指令稿 f2.sh。

```
[atguigu@hadoop102 bin]$ vim f2.sh
```

在指令稿中撰寫以下內容。

```
#! /bin/bash

case $1 in
"start"){
        for i in hadoop104
        do
                echo " --------啟動 $i 消費flume-------"
```

```
              ssh $i " source /etc/profile ; nohup /opt/module/flume/
bin/flume-ng agent --conf-file /opt/module/flume/conf/kafka-flume-hdfs.
conf --name a1 -Dflume.root.logger=INFO,LOGFILE >/opt/module/flume/log.
txt   2>&1 &"
      done
};;
"stop"){
      for i in hadoop104
      do
              echo " --------停止 $i 消費flume------"
              ssh $i "ps -ef | grep kafka-flume-hdfs | grep -v grep
|awk '{print \$2}' | xargs kill"
      done

};;
esac
```

（2）增加指令稿執行許可權。

```
[atguigu@hadoop102 bin]$ chmod 777 f2.sh
```

（3）f2 叢集啟動指令稿。

```
[atguigu@hadoop102 module]$ f2.sh start
```

（4）f2 叢集停止指令稿。

```
[atguigu@hadoop102 module]$ f2.sh stop
```

4.5 擷取通道啟動、停止指令稿

在完成所有的擷取記錄檔落盤工作後，我們需要將本章有關的所有指令和指令稿統一封裝成擷取通道啟動、停止指令稿，否則一項一項開啟擷取通道的處理程序也是非常耗時的，撰寫步驟如下。

（1）在 /home/atguigu/bin 目錄下建立指令稿 cluster.sh。

```
[atguigu@hadoop102 bin]$ vim cluster.sh
```

在指令稿中撰寫以下內容。

```
#! /bin/bash

case $1 in
"start"){
 echo " -------- 啟動 叢集 -------"

 echo " -------- 啟動 hadoop叢集 -------"
 /opt/module/hadoop-2.7.2/sbin/start-dfs.sh
 ssh hadoop103 "source /etc/profile ; /opt/module/hadoop-2.7.2/sbin/
start-yarn.sh"

 #啟動 Zookeeper叢集
 zk.sh start

    #Zookeeper的啟動需要一定時間，此時間根據使用者電腦的效能而定，可適當調整
    sleep 4s;

 #啟動 Flume擷取叢集
 f1.sh start

 #啟動 Kafka擷取叢集
    #Kafka的啟動需要一定時間，此時間根據使用者電腦的效能而定，可適當調整
    kf.sh start

    sleep 6s;

 #啟動 Flume消費叢集
 f2.sh start
};;
"stop"){
 echo " -------- 停止 叢集 -------"


 #停止 Flume消費叢集
 f2.sh stop

 #停止 Kafka擷取叢集
```

```
kf.sh stop

sleep 6s;

#停止 Flume擷取叢集
f1.sh stop

#停止 Zookeeper叢集
zk.sh stop

echo " -------- 停止 hadoop叢集 -------"
ssh hadoop103 "/opt/module/hadoop-2.7.2/sbin/stop-yarn.sh"
/opt/module/hadoop-2.7.2/sbin/stop-dfs.sh
};;
esac
```

（2）增加指令稿執行許可權。

```
[atguigu@hadoop102 bin]$ chmod 777 cluster.sh
```

（3）cluster 叢集啟動指令稿。

```
[atguigu@hadoop102 module]$ cluster.sh start
```

（4）cluster 叢集停止指令稿。

```
[atguigu@hadoop102 module]$ cluster.sh stop
```

4.6 本章歸納

本章主要對使用者行為資料獲取模組的架設進行了說明，包含擷取架構 Flume 的安裝設定、Kafka 的安裝部署和 Zookeeper 的安裝部署，並對整個擷取系統的整體架構進行了詳細說明。在本章中，讀者除了需要學會架設完整的巨量資料擷取系統，還需要掌握資料獲取架構 Flume 的基本用法。舉例來說，如何編輯 Flume 的 Agent 設定檔，以及如何設定 Flume 的各項屬性，此外，還應具備一定的 Shell 指令稿撰寫能力，學會撰寫基本的程式啟動、停止指令稿。

業務資料獲取模組

在資料倉儲技術出現之前，人們對業務資料的分析處於直接存取查詢階段，直接存取查詢業務資料雖然速度快，但也存在很多問題。舉例來說，業務資料的表結構為交易處理的效能而最佳化，有時並不適合查詢與分析，交易處理的優先順序通常高於分析系統，若二者執行於同一硬體上，分析系統的效能常常很差，而且很有可能影響業務系統的效能。所以將業務資料獲取進資料倉儲系統是非常有必要的。

本章主要說明如何將業務資料獲取進資料倉儲系統，以及在業務資料的擷取過程中需要注意的問題。

5.1 電子商務業務概述

在進行需求的實現之前，本節先對業務資料倉儲的基礎理論說明，包含本專案主要有關的電子商務業務流程、電子商務常識及電子商務表結構等。

5.1.1 電子商務業務流程

如圖 5-1 所示，下面以一個普通使用者的瀏覽足跡為例對電子商務的業務流程說明。使用者開啟電子商務網站首頁開始瀏覽，可能透過分類查詢或全文檢索搜尋自己喜歡的商品，這些商品都儲存在後台的管理系統中。

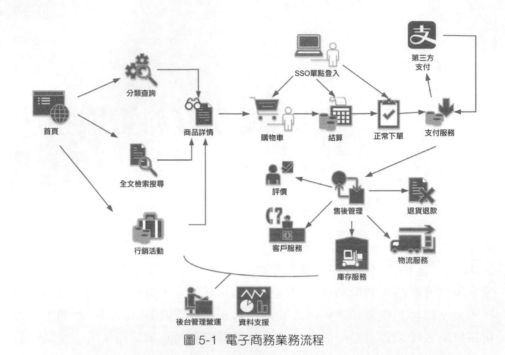

圖 5-1 電子商務業務流程

當使用者找到自己喜歡的商品並想要購買時，可能會將商品加到購物車，此時發現需要登入。登入後對商品進行結算，這時候購物車的管理和商品訂單資訊的產生都會對業務資料倉儲產生影響，會產生對應的訂單資料和支付資料。

訂單正式產生之後，系統還會對訂單進行追蹤處理，直到訂單全部完成。

電子商務的業務流程主要包含使用者在前台瀏覽商品時的商品詳情管理、使用者將商品加入購物車進行支付時的使用者個人中心和支付服務管理，以及使用者支付完成後的訂單後台服務管理，這些流程涉及十幾張或幾十張業務資料表，甚至更多。

資料倉儲用於輔助管理者決策，與業務流程息息相關，建設資料模型的首要前提是了解業務流程，只有了解了業務流程，才能為資料倉儲的建立提供指導方向，進一步反過來為業務提供更好的決策資料支撐，讓資料倉儲的價值最大化。

5.1.2 電子商務常識

SKU 是 Stock Keeping Unit（庫存量基本單位）的縮寫，現在已經被引申為產品統一編號的簡稱，每種產品均對應唯一的 SKU。SPU 是 Standard Product Unit（標準產品單位）的縮寫，是商品資訊聚合的最小單位，是一組可重複使用、易檢索的標準化資訊集合。透過 SPU 表示一種商品的好處是可以共用商品的圖片、海報、銷售屬性等。

舉例來説，iPhone11 手機就是 SPU。一部白色、128GB 記憶體的 iPhone11，就是 SKU。在電子商務網站的商品詳情頁面，所有不同類型的 iPhone11 手機可以共用商品海報和商品圖片等資訊，避免了資料的容錯。

5.1.3 電子商務表結構

如圖 5-2 所示為本電子商務資料倉儲系統有關的業務資料表結構，以訂單表、使用者表、SKU 商品表、活動表和優惠券表為中心，延伸出優惠券領用表、支付流水錶、活動訂單連結表、訂單詳情表、訂單狀態表、商品評價表、編碼字典表、退款表、SPU 商品表等。

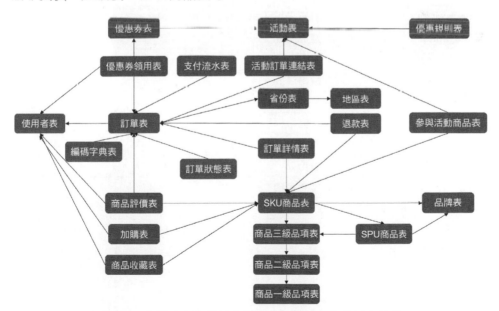

圖 5-2 本電子商務資料倉儲系統有關的業務資料表結構

使用者表提供使用者的詳細資訊,支付流水錶提供訂單的支付詳情,訂單詳情表提供訂單的商品數量等資訊,SKU 商品表為訂單詳情表提供商品的詳細資訊。本章只以圖 5-2 中的 24 張表為例説明,在實際專案中,業務資料庫中的表格遠遠不止這些。

各張表的表結構如表 5-1 ～表 5-24 所示。

表 5-1 訂單表

標籤	含義
id	編號
consignee	收件人
consignee_tel	收件人電話
final_total_amount	總金額
order_status	訂單狀態
user_id	使用者 id
delivery_address	送貨地址
order_comment	訂單備註
out_trade_no	訂單交易編號(第三方支付用)
trade_body	訂單描述(第三方支付用)
create_time	建立時間
operate_time	操作時間
expire_time	故障時間
tracking_no	物流訂單編號
parent_order_id	父訂單編號
img_url	圖片路徑
province_id	省份 id
benefit_reduce_amount	優惠金額
original_total_amount	原價金額
feight_fee	運費

表 5-2 訂單詳情表

標籤	含義
id	編號
order_id	訂單編號
sku_id	商品 id
sku_name	商品名稱（容錯）
img_url	圖片名稱（容錯）
order_price	商品價格（下單時商品的價格）
sku_num	商品數量
create_time	建立時間
source_type	來源類型
source_id	來源編號

表 5-3 SKU 商品表

標籤	含 義
id	商品 id
spu_id	標準產品單位 id
price	價格
sku_name	商品名稱
sku_desc	商品描述
weight	重量
tm_id	品牌 id
category3_id	三級品項 id
sku_default_img	預設顯示圖片（容錯）
create_time	建立時間

表 5-4 使用者表

標籤	含 義
id	使用者 id
login_name	使用者名稱
nick_name	使用者暱稱

標籤	含　義
passwd	使用者密碼
name	真實姓名
phone_num	手機號碼
email	電子郵件
head_img	圖示
user_level	使用者等級
birthday	生日
gender	性別：男 =M，女 =F
create_time	建立時間
operate_time	操作時間

表 5-5　商品一級品項表

標籤	含　義
id	id
name	名稱

表 5-6　商品二級品項表

標籤	含　義
id	id
name	名稱
category1_id	一級品項 id

表 5-7　商品三級品項表

標籤	含　義
id	id
name	名稱
category2_id	二級品項 id

表 5-8 支付流水錶

標籤	含　義
id	編號
out_trade_no	對外業務編號
order_id	訂單編號
user_id	使用者 id
alipay_trade_no	支付寶交易流水編號
total_amount	支付金額
subject	交易內容
payment_type	支付類型
payment_time	支付時間

表 5-9 省份表

標籤	含　義
id	編號
name	省份名稱
region_id	地區 id
area_code	地區編碼
iso_code	國際編碼

表 5-10 地區表

標籤	含　義
id	編號
name	地區名稱

表 5-11 品牌表

標籤	含　義
id	id
tm_name	品牌名稱

表 5-12 訂單狀態表

標籤	含　義
id	編號
order_id	訂單編號
order_status	訂單狀態
operate_time	操作時間

表 5-13 SPU 商品表

標籤	含　義
id	商品 id
spu_name	標準產品單位名稱
description	商品描述（後台簡述）
category3_id	三級品項 id
tm_id	品牌 id

表 5-14 商品評價表

標籤	含　義
id	編號
user_id	使用者 id
sku_id	商品 id
spu_id	標準產品單位 id
order_id	訂單編號
appraise	評價：好評 =1，中評 =2，差評 =3
comment_txt	評價內容
create_time	建立時間

表 5-15 退款表

標籤	含　義
id	編號
user_id	使用者 id
order_id	訂單編號
sku_id	商品 id

標籤	含　義
refund_type	退款類型
refund_amount	退款金額
refund_reason_type	退款原因類型
refund_reason_txt	退款原因內容
create_time	建立時間

表 5-16　加購表

標籤	含　義
id	編號
user_id	使用者 id
sku_id	商品 id
cart_price	放入購物車時的價格
sku_num	數量
img_url	圖片檔案
sku_name	商品名稱（容錯）
create_time	建立時間
operate_time	操作時間
is_ordered	是否已經下單
order_time	下單時間
source_type	來源類型
source_id	來源編號

表 5-17　商品收藏表

標籤	含　義
id	編號
user_id	使用者 id
sku_id	商品 id
spu_id	標準產品單位 id
is_cancel	是否取消：正常 =0，已取消 =1
create_time	收藏時間
cancel_time	取消時間

表 5-18 優惠券領用表

標籤	含 義
id	編號
coupon_id	優惠券 id
user_id	使用者 id
order_id	訂單編號
coupon_status	優惠券狀態
get_time	領券時間
using_time	使用（下單）時間
used_time	使用（支付）時間
expire_time	過期時間

表 5-19 優惠券表

標籤	含 義
id	優惠券編號
coupon_name	優惠券名稱
coupon_type	優惠券類型：現金券 =1，折扣券 =2，滿減券 =3，滿件打折券 =4
condition_amount	滿減金額
condition_num	滿減件數
activity_id	活動編號
benefit_amount	優惠金額
benefit_discount	優惠折扣
create_time	建立時間
range_type	範圍類型：商品 =1，品項 =2，品牌 =3
spu_id	標準產品單位 id
tm_id	品牌 id
category3_id	三級品項 id
limit_num	最多領用次數
operate_time	操作時間
expire_time	過期時間

表 5-20　活動表

標籤	含　義
id	活動 id
activity_name	活動名稱
activity_type	活動類型
activity_desc	活動描述
start_time	開始時間
end_time	結束時間
create_time	建立時間

表 5-21　活動訂單連結表

標籤	含　義
id	編號
activity_id	活動 id
order_id	訂單編號
create_time	建立時間

表 5-22　優惠規則表

標籤	含　義
id	編號
activity_id	活動 id
condition_amount	滿減金額
condition_num	滿減件數
benefit_amount	優惠金額
benefit_discount	優惠折扣
benefit_level	優惠等級

表 5-23　編碼字典表

標籤	含　義
dic_code	編號
dic_name	編碼名稱

標籤	含　義
parent_code	父編號
create_time	建立時間
operate_time	操作時間

表 5-24　參與活動商品表

標籤	含　義
id	編號
activity_id	活動 id
sku_id	商品 id
create_time	建立時間

5.1.4　資料同步策略

資料同步是指將資料從關聯式資料庫同步到巨量資料的儲存系統中，針對不同類型的表應該有不同的同步策略。

表的類型包含每日全量表、每日增量表、每日新增及變化表和拉鏈表 (Zipper Table)。

- 每日全量表：儲存完整的資料。
- 每日增量表：儲存新增加的資料。
- 每日新增及變化表：儲存新增加的資料和變化的資料。
- 拉鏈表：對新增及變化表進行定期合併。

資料同步策略如下。

1. 每日全量同步策略

每日全量同步策略是指每天儲存一份完整資料作為一個分區，適用於表資料量不大，且每天既會有新資料插入，又會有舊資料修改的場景。

維度資料表資料量通常比較小，可以進行每日全量同步，即每天儲存一份完整資料。

2. 每日增量同步策略

每日增量同步策略是指每天儲存一份增量資料作為一個分區，適用於表資料量大，且每天只會有新資料插入的場景。

3. 新增及變化策略

新增及變化策略只同步每日的新增及變化資料，適用於表的資料量大，有新增也有修改，但修改頻率不高（緩慢變化維度）的場景。這種表從資料量的角度考慮，若採用每日全量同步策略，則資料量太大，容錯也太大；若採用每日增量同步策略，則無法反映資料的變化情況。利用每日新增及變化表，製作一張拉鏈表，可以方便地取得某個時間切片的快照資料。

拉鏈表是指在來源表欄位的基礎上，增加一個開始時間和一個結束時間，該時間段用來表示一個狀態的生命週期。製作拉鏈表，每天只需要同步新增和修改的資料即可，如表 5-25 所示，可以透過類似 "select * from user where start =<'2019-01-02' and end>='2019-01-02'" 的查詢敘述來取得 2019-01-02 當天的所有訂單狀態資訊。

表 5-25　拉鏈表

name	start	end
張三	1990-01-01	2018-12-31
張小三	2019-01-01	2019-04-30
張大三	2019-05-01	9999-99-99
……	……	……

4. 特殊維度同步策略

某些特殊的維度資料表可以不遵循上述資料同步策略。

1）客觀世界維度
對沒有變化的客觀世界的維度（舉例來說，性別、地區、民族等），可以只儲存一份固定值。

2）日期維度
對於日期維度，可以一次性匯入一年或許多年的資料。

5.2 業務資料獲取

業務資料通常儲存在關聯式資料庫中，為了進行資料獲取，我們首先要產生業務資料，然後選用合適的資料獲取工具。在本專案中，我們選用 MySQL 作為業務資料的產生和儲存資料庫，選用 Sqoop 作為資料獲取工具。本節主要說明 MySQL 和 Sqoop 的安裝部署、業務資料的產生和建模，以及業務資料匯入資料倉儲的相關內容。

5.2.1 MySQL 安裝

1. 安裝套件準備

（1）使用 rpm 指令配合管線符號檢視 MySQL 是否已經安裝，其中，-q 選項為 query，-a 選項為 all，意思為查詢全部安裝，如果已經安裝 MySQL，則將其移除。

① 檢視 MySQL 是否已經安裝。

```
[root@hadoop102 桌面]# rpm -qa|grep MySQL
mysql-libs-5.1.73-7.el6.x86_64
```

② 移除，-e 選項表示移除，--nodeps 選項表示忽略所有依賴強制移除。

```
[root@hadoop102 桌面]# rpm -e --nodeps mysql-libs-5.1.73-7.el6.x86_64
```

（2）將 MySQL 安裝套件 mysql-libs.zip 上傳至 /opt/software 目錄下並解壓。

```
[root@hadoop102 software]# unzip mysql-libs.zip
[root@hadoop102 software]# ls
mysql-libs.zip
mysql-libs
```

2. 安裝 MySQL 伺服器端

（1）安裝 MySQL 伺服器端，使用 rpm 指令安裝 MySQL，-i 選項為 install，-v 選項為 vision，-h 選項用於展示安裝過程。

```
[root@hadoop102 mysql-libs]# rpm -ivh MySQL-server-5.6.24-1.el6.x86_64.rpm
```

（2）伺服器端安裝完成後會產生一個預設隨機密碼，儲存在 /root/.mysql_secret 檔案中，root 使用者可以直接使用 cat 指令或 sudo cat 指令檢視產生的隨機密碼，登入 MySQL 後需要立即更改密碼。

```
[root@hadoop102 mysql-libs]# cat /root/.mysql_secret
 OEXaQuS8IWkG19Xs
```

（3）以 root 使用者身份登入，或使用 sudo 指令檢視 MySQL 服務的執行狀態，可以看到現在是 "MySQL is not running…" 狀態。

```
[root@hadoop102 mysql-libs]# service mysql status
MySQL is not running…
```

（4）以 root 使用者身份登入，或使用 sudo 指令啟動 MySQL。

```
[root@hadoop102 mysql-libs]# service mysql start
```

3. 安裝 MySQL 用戶端

（1）使用 rpm 指令安裝 MySQL 用戶端。

```
[root@hadoop102 mysql-libs]# rpm -ivh MySQL-client-5.6.24-1.el6.x86_64.rpm
```

（2）登入 MySQL，以 root 使用者身份登入，密碼為安裝伺服器端時自動產生的隨機密碼。

```
[root@hadoop102 mysql-libs]# mysql -uroot -pOEXaQuS8IWkG19Xs
```

（3）登入 MySQL 後，立即修改密碼，修改密碼後記住該密碼。

```
mysql>SET PASSWORD=PASSWORD('000000');
```

（4）退出 MySQL。

```
mysql>exit
```

4. MySQL 中 user 表的主機設定

由於 MySQL 需要被 Hive 連接存取，為避免出現許可權問題，我們可以設定在任何主機上只要使用 root 使用者身份就可登入 MySQL。為此，我們需要修改 MySQL 的 user 表，實際步驟如下。

（1）以 root 使用者身份輸入設定的密碼，登入 MySQL。

```
[root@hadoop102 mysql-libs]# mysql -uroot -p000000
```

（2）顯示目前 MySQL 的所有資料庫。

```
mysql>show databases;
```

（3）使用 MySQL 資料庫。

```
mysql>use mysql;
```

（4）顯示 MySQL 資料庫中的所有表。

```
mysql>show tables;
```

（5）user 表中儲存了允許登入 MySQL 的使用者、密碼等資訊，展開 user 表，可以發現 user 表的欄位非常多。

```
mysql>desc user;
```

（6）查詢 user 表中的部分欄位 host、user、password，顯示關鍵資訊，可以看到 4 筆資訊，host 分別是 localhost、hadoop102、127.0.0.1 和 ::1，這 4 個 host 都表明只有本機可以連接 MySQL。

```
mysql>select user, host, password from user;
```

（7）修改 user 表，把 host 欄位值修改為 %，使任何 host 都可以透過 root 使用者＋密碼連接 MySQL。

```
mysql>update user set host='%' where host='localhost';
```

（8）只有刪除 root 使用者的其他 host，才能使 "%" 生效。

```
mysql>delete from user where Host='hadoop102 ';
mysql>delete from user where Host='127.0.0.1';
mysql>delete from user where Host='::1';
```

（9）更新許可權，使修改生效。

```
mysql>flush privileges;
```

（10）退出。

```
mysql> quit;
```

5.2.2 業務資料產生

業務資料的建資料庫、建表和資料產生透過匯入指令稿完成，建議讀者安裝一個資料庫視覺化工具。本節以 SQLyog 為例説明，資料產生步驟如下。

1. 建表敘述的匯入

（1）透過 SQLyog 建立資料庫 gmall。
（2）設定資料庫編碼，如圖 5-3 所示。

圖 5-3 設定資料庫編碼

2. 資料產生

（1）在 hadoop102 的 /opt/module 目錄下建立 db_log 資料夾。

```
[atguigu@hadoop102 module]$ mkdir db_log
```

（2）在本書資料中找到 gmall-mock-db-SNAPSHOT.jar 和 application.properties 檔案，把 gmall-mock-db-SNAPSHOT.jar 和 application.properties 上傳到 hadoop102 的 /opt/module/db_log 目錄下。

（3）根據需求修改 application.properties 的相關設定，透過修改業務日期的設定可以產生不同日期的業務資料。

```
logging.level.root=info

spring.datasource.driver-class-name=com.mysql.jdbc.Driver
spring.datasource.url=jdbc:mysql://hadoop102:3306/gmall?characterEncoding
=utf-8&useSSL=false&serverTimezone=GMT%2B8
spring.datasource.username=root
spring.datasource.password=000000

logging.pattern.console=%m%n
mybatis-plus.global-config.db-config.field-strategy=not_null

#業務日期
mock.date=2020-03-10
#是否重置
mock.clear=1

#產生新使用者個數
mock.user.count=50
#產生新使用者中男性的比例
mock.user.male-rate=20

#收藏和取消收藏的操作比例
mock.favor.cancel-rate=10
#收藏操作次數
mock.favor.count=100

#加入購物車操作次數
mock.cart.count=10
#每件商品最多加入購物車的數量
mock.cart.sku-maxcount-per-cart=3
```

```
#使用者下單比例
mock.order.user-rate=80
#使用者從購物車中結算商品的比例
mock.order.sku-rate=70
#是否參加活動
mock.order.join-activity=1
#是否使用優惠券
mock.order.use-coupon=1
#優惠券領取人數
mock.coupon.user-count=10


#使用者提交訂單後的支付比例
mock.payment.rate=70
#不同支付方式比例，支付寶 微信 銀聯
mock.payment.payment-type=30:60:10


#不同評價比例，好 中 差 自動
mock.comment.appraise-rate=30:10:10:50


#不同退款原因比例，品質問題 商品描述與實際描述不一致 缺貨 號碼不合適 拍錯 不想
買了 其他
mock.refund.reason-rate=30:10:20:5:15:5:5
```

（4）在 /opt/module/db_log 目錄下執行以下指令，產生 2020-03-10 的資料。

```
[atguigu@hadoop102 db_log]$ java -jar gmall-mock-db-SNAPSHOT.jar
```

（5）在設定檔 application.properties 中修改以下設定。

```
mock.date=2020-03-11
mock.clear=05
```

（6）再次執行指令，產生 2020-03-11 的資料。

```
[atguigu@hadoop102 db_log]$ java -jar gmall-mock-db-SNAPSHOT.jar
```

5.2.3 業務資料建模

有時候，我們需要處理的業務資料並沒有表與表之間的關係圖，所以需要自己在 SQLyog 上建模。其他資料庫視覺化工具也有類似功能，此處不再贅述。建模具體過程如下。

（1）在「檔案」下拉式功能表中選擇「新架構設計器」指令，如圖 5-4 所示。

圖 5-4 選擇「新架構設計器」指令

（2）打開架構設計器頁面，如圖 5-5 所示。

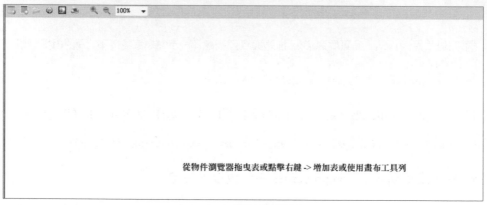

圖 5-5 架構設計器頁面

（3）依次將左側 gmall 資料庫中的表拖曳到右側的架構設計器頁面，如圖 5-6 所示。

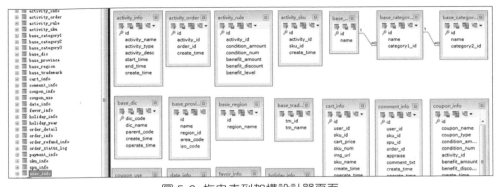
圖 5-6 拖曳表到架構設計器頁面

（4）根據外鍵關係，將所有表連接在一起。

舉例來說，商品三級品項表連結商品二級品項表，如圖 5-7 所示。

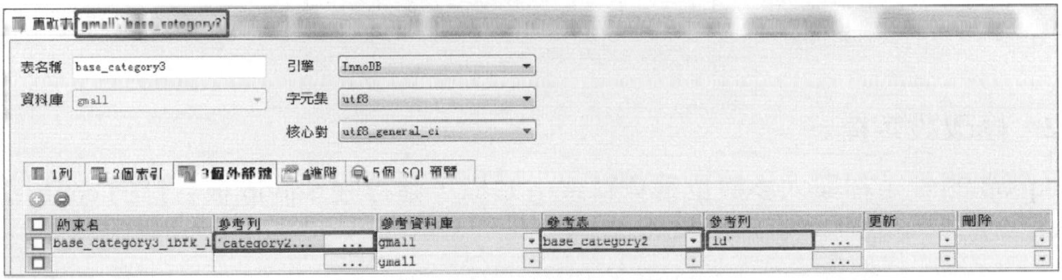

圖 5-7 設定商品三級品項表外鍵

又如，商品二級品項表連結商品一級品項表，如圖 5-8 所示。

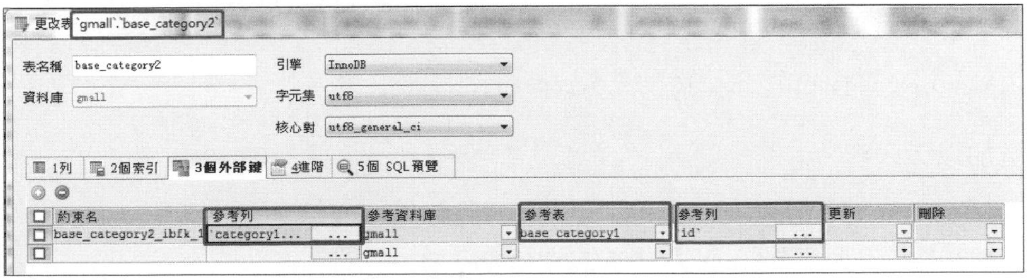

圖 5-8 設定商品二級品項表外鍵

採用以上方法，將所有表連結起來，就可以獲得與圖 5-2 相似的電子商務業務
資料表結構，方便使用者清晰明瞭地看出所需處理的業務資料表之間的建模
關係。

5.2.4 Sqoop 安裝

Sqoop 是一個用於將關聯式資料庫和 Hadoop 中的資料進行相互傳輸的工具，可以將一個關聯式資料庫（如 MySQL、Oracle）中的資料匯入 Hadoop（如 HDFS、Hive、HBase）中，也可以將 Hadoop（如 HDFS、Hive、HBase）中的資料匯入關聯式資料庫（如 MySQL、Oracle）中。Sqoop 的安裝步驟如下。

1. 下載並解壓

（1）下載安裝套件。

（2）上傳安裝套件 sqoop-1.4.6.bin_hadoop-2.0.4-alpha.tar.gz 到虛擬機器中。

（3）解壓 Sqoop 安裝套件到指定目錄。

```
[atguigu@hadoop102 software]$ tar -zxf sqoop-1.4.6.bin_hadoop-2.0.4-alpha.tar.
gz -C /opt/module/
```

2. 修改設定檔

Sqoop 的設定檔與大多數巨量資料架構類似，儲存在 Sqoop 根目錄的 conf 目錄下。

（1）重新命名設定檔。

```
[atguigu@hadoop102 conf]$ mv sqoop-env-template.sh sqoop-env.sh
```

（2）修改設定檔 sqoop-env.sh。

```
[atguigu@hadoop102 conf]$ vim sqoop-env.sh
```

增加以下內容。

```
export HADOOP_COMMON_HOME=/opt/module/hadoop-2.7.2
export HADOOP_MAPRED_HOME=/opt/module/hadoop-2.7.2
export ZOOKEEPER_HOME=/opt/module/zookeeper-3.4.10
export ZOOCFGDIR=/opt/module/zookeeper-3.4.10
```

3. 複製 JDBC 驅動

複製 JDBC 驅動到 Sqoop 的 lib 目錄下。

```
[atguigu@hadoop102 mysql]$ cp mysql-connector-java-5.1.27-bin.jar /opt/
module/sqoop-1.4.6.bin__hadoop-2.0.4-alpha/lib/
```

4. 驗證 Sqoop

我們可以透過執行以下指令來驗證 Sqoop 的設定是否正確。

```
[atguigu@hadoop102 sqoop]$ bin/sqoop help
```

然後會出現 Warning 警告（警告資訊已省略），並伴隨著幫助指令的輸出。

```
Available commands:
  codegen            Generate code to interact with database records
  create-hive-table     Import a table definition into Hive
  eval               Evaluate a SQL statement and display the results
  export             Export an HDFS directory to a database table
  help               List available commands
  import             Import a table from a database to HDFS
  import-all-tables     Import tables from a database to HDFS
  import-mainframe     Import datasets from a mainframe server to HDFS
  job                Work with saved jobs
  list-databases        List available databases on a server
  list-tables           List available tables in a database
  merge              Merge results of incremental imports
  metastore           Run a standalone Sqoop metastore
  version            Display version information
```

5. 測試 Sqoop 是否能夠成功連接資料庫

```
[atguigu@hadoop102 sqoop]$ bin/sqoop list-databases --connect
jdbc:mysql://hadoop102:3306/ --username root --password 000000
```

若出現以下輸出，則表示連接成功。

```
information_schema
metastore
mysql
```

```
oozie
performance_schema
```

5.2.5 業務資料匯入資料倉儲

業務資料表同步策略分析如圖 5-9 所示,根據表格性質的不同制定不同的資料同步策略。在後續的資料匯入過程中,將根據不同的資料同步策略執行不同的資料匯入指令。

圖 5-9 業務資料表同步策略分析

業務資料產生完畢之後,即可透過 Sqoop 資料匯入指令將資料匯入 HDFS 中,供後續資料倉儲的架設使用。在 Sqoop 中,「匯入」的概念是指使用 import 關鍵字從非巨量資料叢集(RDBMS)向巨量資料叢集(HDFS、Hive、HBase)中傳輸資料,資料的匯入指令範例如下。

```
/opt/module/sqoop/bin/sqoop import \
--connect \
--username \
--password \
--target-dir \
```

```
--delete-target-dir \
--num-mappers \
--fields-terminated-by \
--query "$2" "and $CONDITIONS;"
```

指令參數說明如下。

- --connect：指定 JDBC 的 url，需精確到資料庫；
- --username：指定連接資料庫使用的使用者名稱；
- --password：指定連接資料庫使用的密碼；
- --target-dir：指定 HDFS 匯入表的儲存目錄；
- --delete-target-dir：是否刪除存在的 import 目標目錄；
- --num-mappers：指定平行處理執行的資料匯入作業個數；
- --fields-terminated-by：指定每個欄位以什麼符號結束；
- --query：指定查詢敘述，其中，"and $CONDITIONS;" 必須存在。

1. Sqoop 定時匯入指令稿

（1）在 /home/atguigu/bin 目錄下建立指令稿 mysql_to_hdfs.sh。

```
[atguigu@hadoop102 bin]$ vim mysql_to_hdfs.sh
```

在指令稿中撰寫以下內容。

```
#!/bin/bash

sqoop=/opt/module/sqoop/bin/sqoop

if [ -n "$2" ] ;then
    do_date=$2
else
    do_date=`date -d '-1 day' +%F`
fi

#撰寫通用資料匯入指令，透過第一個參數傳入表名，透過第二個參數傳入查詢敘述，
#對匯入資料使用LZO壓縮格式，並對LZO壓縮檔建立索引
import_data(){
$sqoop import \
```

```
--connect jdbc:mysql://hadoop102:3306/gmall \
--username root \
--password 000000 \
--target-dir /origin_data/gmall/db/$1/$do_date \
--delete-target-dir \
--query "$2 and  \$CONDITIONS" \
--num-mappers 1 \
--fields-terminated-by '\t' \
--compress \
--compression-codec lzop \
--null-string '\\N' \
--null-non-string '\\N'

hadoop jar /opt/module/hadoop-2.7.2/share/hadoop/common/hadoop-lzo-
0.4.20.jar com.hadoop.compression.lzo.DistributedLzoIndexer/origin_
data/gmall/db/$1/$do_date
}

#針對不同表格分別呼叫不同的通用資料匯入指令，全量匯入資料的表格的查詢準則為
#where 1=1 ，增量匯入資料的表格的查詢準則為當天日期
import_order_info(){
  import_data order_info "select
                            id,
                            final_total_amount,
                            order_status,
                            user_id,
                            out_trade_no,
                            create_time,
                            operate_time,
                            province_id,
                            benefit_reduce_amount,
                            original_total_amount,
                            feight_fee
                          from order_info
                          where (date_format(create_time,'%Y-%m-
%d')='$do_date'
                          or date_format(operate_time,'%Y-%m-%d')=
'$do_date')"
}
```

```
import_coupon_use(){
  import_data coupon_use "select
                          id,
                          coupon_id,
                          user_id,
                          order_id,
                          coupon_status,
                          get_time,
                          using_time,
                          used_time
                        from coupon_use
                        where (date_format(get_time,'%Y-%m-%d')='$do_date'
                        or date_format(using_time,'%Y-%m-%d')='$do_date'
                        or date_format(used_time,'%Y-%m-%d')='$do_date')"
}

import_order_status_log(){
  import_data order_status_log "select
                                id,
                                order_id,
                                order_status,
                                operate_time
                              from order_status_log
                              where
date_format(operate_time,'%Y-%m-%d')='$do_date'"
}

import_activity_order(){
  import_data activity_order "select
                              id,
                              activity_id,
                              order_id,
                              create_time
                            from activity_order
                            where
date_format(create_time,'%Y-%m-%d')='$do_date'"
}
```

```
import_user_info(){
  import_data "user_info" "select
                           id,
                           name,
                           birthday,
                           gender,
                           email,
                           user_level,
                           create_time,
                           operate_time
                        from user_info
                        where (DATE_FORMAT(create_time,'%Y-%m-%d')=
'$do_date'
                        or DATE_FORMAT(operate_time,'%Y-%m-%d')=
'$do_date')"
}

import_order_detail(){
  import_data order_detail "select
                              od.id,
                              order_id,
                              user_id,
                              sku_id,
                              sku_name,
                              order_price,
                              sku_num,
                              od.create_time
                           from order_detail od
                           join order_info oi
                           on od.order_id=oi.id
                           where
DATE_FORMAT(od.create_time,'%Y-%m-%d')='$do_date'"
}

import_payment_info(){
  import_data "payment_info"  "select
                                 id,
                                 out_trade_no,
                                 order_id,
```

```
                                    user_id,
                                    alipay_trade_no,
                                    total_amount,
                                    subject,
                                    payment_type,
                                    payment_time
                            from payment_info
                            where
DATE_FORMAT(payment_time,'%Y-%m-%d')='$do_date'"
}


import_comment_info(){
  import_data comment_info "select
                                    id,
                                    user_id,
                                    sku_id,
                                    spu_id,
                                    order_id,
                                    appraise,
                                    comment_txt,
                                    create_time
                            from comment_info
                            where
date_format(create_time,'%Y-%m-%d')='$do_date'"
}


import_order_refund_info(){
  import_data order_refund_info "select
                                    id,
                                    user_id,
                                    order_id,
                                    sku_id,
                                    refund_type,
                                    refund_num,
                                    refund_amount,
                                    refund_reason_type,
                                    create_time
                            from order_refund_info
                            where
```

```
date_format(create_time,'%Y-%m-%d')='$do_date'"
}

import_sku_info(){
  import_data sku_info "select
                        id,
                        spu_id,
                        price,
                        sku_name,
                        sku_desc,
                        weight,
                        tm_id,
                        category3_id,
                        create_time
                     from sku_info where 1=1"
}

import_base_category1(){
  import_data "base_category1" "select
                                 id,
                                 name
                               from base_category1 where 1=1"
}

import_base_category2(){
  import_data "base_category2" "select
                                 id,
                                 name,
                                 category1_id
                               from base_category2 where 1=1"
}

import_base_category3(){
  import_data "base_category3" "select
                                 id,
                                 name,
                                 category2_id
                               from base_category3 where 1=1"
}
```

```
import_base_province(){
  import_data base_province "select
                                id,
                                name,
                                region_id,
                                area_code,
                                iso_code
                             from base_province
                             where 1=1"
}

import_base_region(){
  import_data base_region "select
                                id,
                                region_name
                             from base_region
                             where 1=1"
}

import_base_trademark(){
  import_data base_trademark "select
                                tm_id,
                                tm_name
                             from base_trademark
                             where 1=1"
}

import_spu_info(){
  import_data spu_info "select
                                id,
                                spu_name,
                                category3_id,
                                tm_id
                             from spu_info
                             where 1=1"
}

import_favor_info(){
```

```
   import_data favor_info "select
                          id,
                          user_id,
                          sku_id,
                          spu_id,
                          is_cancel,
                          create_time,
                          cancel_time
                      from favor_info
                      where 1=1"
}

import_cart_info(){
   import_data cart_info "select
                          id,
                          user_id,
                          sku_id,
                          cart_price,
                          sku_num,
                          sku_name,
                          create_time,
                          operate_time,
                          is_ordered,
                          order_time
                      from cart_info
                      where 1=1"
}

import_coupon_info(){
   import_data coupon_info "select
                          id,
                          coupon_name,
                          coupon_type,
                          condition_amount,
                          condition_num,
                          activity_id,
                          benefit_amount,
                          benefit_discount,
                          create_time,
```

```
                            range_type,
                            spu_id,
                            tm_id,
                            category3_id,
                            limit_num,
                            operate_time,
                            expire_time
                        from coupon_info
                        where 1=1"
}

import_activity_info(){
  import_data activity_info "select
                            id,
                            activity_name,
                            activity_type,
                            start_time,
                            end_time,
                            create_time
                        from activity_info
                        where 1=1"
}

import_activity_rule(){
    import_data activity_rule "select
                                id,
                                activity_id,
                                condition_amount,
                                condition_num,
                                benefit_amount,
                                benefit_discount,
                                benefit_level
                            from activity_rule
                            where 1=1"
}

import_base_dic(){
    import_data base_dic "select
                            dic_code,
```

```
                                dic_name,
                                parent_code,
                                create_time,
                                operate_time
                        from base_dic
                        where 1=1"
}
import_activity_sku(){
    import_data activity_sku "select
                                id,
                                activity_id,
                                sku_id,
                                create_time
                        from activity_sku
                        where 1=1"
}
#對傳入的第一個參數進行判斷，根據傳入參數的不同決定匯入哪種表資料，若傳入
#first，則表示初次執行指令稿，匯入所有表資料；若傳入all，則匯入除地區外的所
#有表資料
case $1 in
  "order_info")
     import_order_info
;;
  "base_category1")
     import_base_category1
;;
  "base_category2")
     import_base_category2
;;
  "base_category3")
     import_base_category3
;;
  "order_detail")
     import_order_detail
;;
  "sku_info")
     import_sku_info
;;
  "user_info")
```

```
        import_user_info
;;
  "payment_info")
      import_payment_info
;;
  "base_province")
      import_base_province
;;
  "base_region")
      import_base_region
;;
  "base_trademark")
      import_base_trademark
;;
  "activity_info")
      import_activity_info
;;
  "activity_order")
      import_activity_order
;;
  "cart_info")
      import_cart_info
;;
  "comment_info")
      import_comment_info
;;
  "coupon_info")
      import_coupon_info
;;
  "coupon_use")
      import_coupon_use
;;
  "favor_info")
      import_favor_info
;;
  "order_refund_info")
      import_order_refund_info
;;
  "order_status_log")
```

```
        import_order_status_log
;;
  "spu_info")
      import_spu_info
;;
  "activity_rule")
      import_activity_rule
;;
  "base_dic")
      import_base_dic
;;
"activity_sku")
      import_activity_sku
;;
"first")
    import_base_category1
    import_base_category2
    import_base_category3
    import_order_info
    import_order_detail
    import_sku_info
    import_user_info
    import_payment_info
    import_base_province
    import_base_region
    import_base_trademark
    import_activity_info
    import_activity_order
    import_cart_info
    import_comment_info
    import_coupon_use
    import_coupon_info
    import_favor_info
    import_order_refund_info
    import_order_status_log
    import_spu_info
    import_activity_rule
    import_base_dic
    import_activity_sku
```

```
;;
"all")
    import_base_category1
    import_base_category2
    import_base_category3
    import_order_info
    import_order_detail
    import_sku_info
    import_user_info
    import_payment_info
    import_base_trademark
    import_activity_info
    import_activity_order
    import_cart_info
    import_comment_info
    import_coupon_use
    import_coupon_info
    import_favor_info
    import_order_refund_info
    import_order_status_log
    import_spu_info
    import_activity_rule
    import_base_dic
    import_activity_sku
;;
esac
```

（2）增加指令稿執行許可權。

```
[atguigu@hadoop102 bin]$ chmod 777 mysql_to_hdfs.sh
```

（3）初次執行指令稿，傳入 first，匯入 2020-03-10 的資料。

```
[atguigu@hadoop102 bin]$ mysql_to_hdfs.sh first 2020-03-10
```

（4）再次執行指令稿，傳入 all，匯入 2020-03-11 的資料。

```
[atguigu@hadoop102 bin]$ mysql_to_hdfs.sh all 2020-03-11
```

2. Sqoop 匯入資料異常處理

（1）問題描述：執行 Sqoop 匯入資料指令稿時，發生以下異常。

```
java.sql.SQLException: Streaming result set com.mysql.jdbc.RowDataDynamic
@65d6b83b is still active. No statements may be issued when any streaming
result sets are open and in use on a given connection. Ensure that you
have called .close() on any active streaming result sets before
attempting more queries.
 at com.mysql.jdbc.SQLError.createSQLException(SQLError.java:930)
 at com.mysql.jdbc.MysqlIO.checkForOutstandingStreamingData(MysqlIO.
java:2646)
 at com.mysql.jdbc.MysqlIO.sendCommand(MysqlIO.java:1861)
 at com.mysql.jdbc.MysqlIO.sqlQueryDirect(MysqlIO.java:2101)
 at com.mysql.jdbc.ConnectionImpl.execSQL(ConnectionImpl.java:2548)
 at com.mysql.jdbc.ConnectionImpl.execSQL(ConnectionImpl.java:2477)
 at com.mysql.jdbc.StatementImpl.executeQuery(StatementImpl.java:1422)
 at com.mysql.jdbc.ConnectionImpl.getMaxBytesPerChar(ConnectionImpl.
java:2945)
 at com.mysql.jdbc.Field.getMaxBytesPerCharacter(Field.java:582)
```

（2）問題解決方案：增加以下匯入參數。

```
--driver com.mysql.jdbc.Driver \
```

5.3 本章歸納

本章主要對業務資料獲取模組進行了架設，在架設過程中，讀者可以發現，業務資料具有很多數量與種類的資料表，所以需要針對不同類型的資料表制定不同的資料同步策略，然後在制定好策略的前提下選用合適的資料獲取工具。閱讀本章後，希望讀者對電子商務業務資料的擷取工作有更多的了解。

資料倉儲架設模組

經過對資料獲取模組的架設，現在已經有了可以分析的資料來源，但這些資料僅停留在巨量資料儲存系統中是發揮不了任何作用的，我們只有分析重構才能發現其中的價值，也就是要架設資料倉儲。

資料倉儲（Data Warehouse）是一個針對主題（Subject Oriented）的、整合（Integrate）的、相對穩定（Non-Volatile）的、反映歷史變化（Time Variant）的資料集合，用於支援管理決策。我們可以從兩個層次了解資料倉儲的概念：首先，資料倉儲用於支援管理決策，針對分析類型資料處理，它不同於企業現有的操作類型資料庫；其次，資料倉儲是對多個異質的資料來源的有效整合，整合後按照主題進行重組，並包含歷史資料，而且儲存在資料倉儲中的資料一般不再修改。

只有對資料倉儲和資料倉儲架設工具有一定的了解，才能實現最後需求，本章從資料倉儲的基礎知識、環境架設、需求實現幾個方面多作說明。

6.1 資料倉儲理論準備

在架設資料倉儲之前，先對資料倉儲的基礎理論知識介紹。資料倉儲包含的內容很多，例如架構、建模、方法論等，對應到實際工作中，可以概括為以下內容。

- 以 Hadoop、Spark、Hive 等元件為中心的資料架構系統。
- 各種資料建模方法,如維度建模。
- 資料同步策略。
- 資料倉儲分層的相關理論。

無論資料倉儲的規模有多大,在資料倉儲架設之初,只有對基礎理論知識有一定的掌握,對資料倉儲的整體架構有所規劃,才能架設出合理高效的資料倉儲系統。

本節將圍繞資料建模、資料同步策略、資料倉儲分層理論幾個方面,介紹資料倉儲的深層核心知識。

6.1.1 範式理論

關聯式資料庫在設計時,要遵守一定的標準要求,目前業界的範式包含第一範式(1NF)、第二範式(2NF)、第三範式(3NF)、巴斯 - 科德範式(BCNF)、第四範式(4NF)和第五範式(5NF)。範式可以視為一張資料表的表結構符合的設計標準的等級。使用範式的根本目的包含以下兩點。

(1)減少資料容錯,儘量讓每個資料只出現一次。
(2)保障資料的一致性。

其缺點是在取得資料時,需要透過 join 連接出最後的資料。

1. 什麼是函數依賴

函數依賴範例如表 6-1 所示。

表 6-1 函數依賴範例:學生成績表

學號	姓名	系名	系主任	課名	分數
1	李小明	經濟系	王強	高等數學	95
1	李小明	經濟系	王強	大學英文	87
1	李小明	經濟系	王強	普通化學	76
2	張莉莉	經濟系	王強	高等數學	72

學號	姓名	系名	系主任	課名	分數
2	張莉莉	經濟系	王強	大學英文	98
2	張莉莉	經濟系	王強	電腦基礎	82
3	高芳芳	法律系	劉玲	高等數學	88
3	高芳芳	法律系	劉玲	法學基礎	84

函數依賴分為完全函數依賴、部分函數依賴和傳遞函數依賴。

1）完全函數依賴

設（X, Y）是關係 R 的兩個屬性集合，X' 是 X 的真子集，存在 $X \rightarrow Y$，但對每一個 X' 都有 $X'! \rightarrow Y$，則稱 Y 完全依賴於 X。

舉例來說，透過（學號，課名）可推出分數，但單獨用學號推不出分數，那麼就可以說，分數完全依賴於（學號，課名）。

透過（A, B）能得出 C，但單獨透過 A 或 B 得不出 C，那麼就可以說，C 完全依賴於（A, B）。

2）部分函數依賴

假如 Y 依賴於 X，但同時 Y 並不完全依賴於 X，那麼就可以說，Y 部分依賴於 X。

舉例來說，透過（學號，課名）可推出姓名，但直接透過學號也可以推出姓名，所以姓名部分依賴於（學號，課名）。

透過（A, B）能得出 C，透過 A 也能得出 C，或透過 B 也能得出 C，那麼就可以說，C 部分依賴於（A, B）。

3）傳遞函數依賴

設（X, Y, Z）是關係 R 中互不相同的屬性集合，存在 $X \rightarrow Y(Y! \rightarrow X), Y \rightarrow Z$，則稱 Z 傳遞依賴於 X。

舉例來說，透過學號可推出系名，透過系名可推出系主任，但透過系主任推不出學號，系主任主要依賴於系名。這種情況可以說，系主任傳遞依賴於學號。

透過 A 可獲得 B，透過 B 可獲得 C，但透過 C 得不到 A，那麼就可以說，C 傳遞依賴於 A。

2. 第一範式

第一範式（1NF）的核心原則是屬性不可分割。如表 6-2 所示，「商品」列中的資料不是原子資料項目，是可以分割的，明顯不符合第一範式。

表 6-2　不符合第一範式的表格設計

id	商品	商家 id	使用者 id
001	5 台電腦	××× 旗艦店	00001

對表 6-2 進行修改，使表格符合第一範式的要求，如表 6-3 所示。

表 6-3　符合第一範式的表格設計

id	商品	數量（台）	商家 id	使用者 id
001	電腦	5	××× 旗艦店	00001

實際上，第一範式是所有關聯式資料庫最基本的要求，在關聯式資料庫（RDBMS），如 SQL Server、Oracle、MySQL 中建立資料表時，如果資料表的設計不符合這個最基本的要求，那麼操作一定是不能成功的。也就是說，只要在 RDBMS 中已經存在的資料表，一定是符合第一範式的。

3. 第二範式

第二範式（2NF）的核心原則是不能存在部分函數依賴。

如表 6-1 所示，該表格明顯存在部分函數依賴。這張表的主鍵是（學號，課名），分數確實完全依賴於（學號，課名），但姓名並不完全依賴於（學號，課名）。

將表 6-1 進行調整，結果如表 6-4 和表 6-5 所示，即去掉部分函數依賴，使其符合第二範式。

表 6-4　學號 - 課名 - 分數表

學號	課名	分數
1	高等數學	95
1	大學英文	87
1	普通化學	76
2	高等數學	72
2	大學英文	98
2	電腦基礎	82
3	高等數學	88
3	法學基礎	84

表 6-5　學號 - 姓名 - 系明細表

學號	姓名	系名	系主任
1	李小明	經濟系	王強
2	張莉莉	經濟系	王強
3	高芳芳	法律系	劉玲

4. 第三範式

第三範式（3NF）的核心原則是不能存在傳遞函數依賴。

表 6-5 中存在傳遞函數依賴，透過系主任不能推出學號，將表格進一步拆分，使其符合第三範式，如表 6-6 和表 6-7 所示。

表 6-6　學號 - 姓名 - 系名表

學號	姓名	系名
1	李小明	經濟系
2	張莉莉	經濟系
3	高芳芳	法律系

表 6-7 系名 - 系主任表

系名	系主任
經濟系	王強
法律系	劉玲

6.1.2 關係模型與維度模型

當今的資料處理大致可以分成兩大類：連線交易處理（On-Line Transaction Processing，OLTP）、連線分析處理（On-Line Analytical Processing，OLAP）。OLTP 是傳統關聯式資料庫的主要應用，主要是基本的、日常的交易處理，如銀行交易。OLAP 是資料倉儲系統的主要應用，支援複雜的分析操作，偏重決策支援，並且可提供直觀、容易的查詢結果。二者的主要區別如表 6-8 所示。

表 6-8 OLTP 與 OLAP 的主要區別

對比屬性	OLTP	OLAP
讀特性	每次查詢只傳回少量記錄	對大量記錄進行整理
寫特性	隨機、低延遲時間寫入使用者的輸入	批次匯入
使用場景	使用者，Java EE 專案	內部分析師，為決策提供支援
資料表徵	最新資料狀態	隨時間變化的歷史狀態
資料規模	GB	TP 到 PB

關係模型示意如圖 6-1 所示，嚴格遵循第三範式（3NF）。從圖 6-1 中可以看出，模型較為鬆散、零碎，物理表數量多，但資料容錯程度低。由於資料分佈於許多的表中，因此這些資料可以更為靈活地被應用，功能性較強。關係模型主要應用於 OLTP 中，為了確保資料的一致性及避免容錯，大部分業務系統的表都是遵循第三範式的。

維度模型示意如圖 6-2 所示，其主要應用於 OLAP 中，通常以某一張事實資料表為中心進行表的組織，主要針對業務，其特徵是可能存在資料的容錯，但使用者能方便地獲得資料。

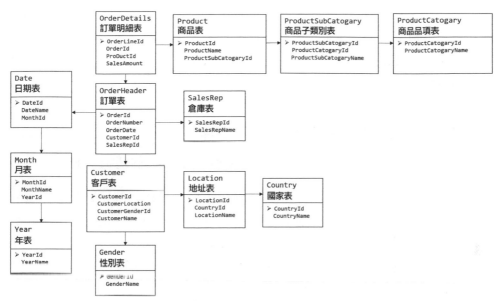

圖 6-1 關係模型示意

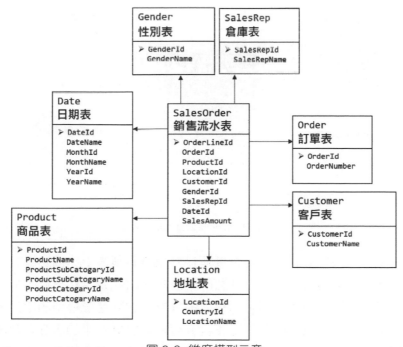

圖 6-2 維度模型示意

關係模型雖然資料容錯程度低，但在大規模資料中進行跨表分析、統計、查詢時，會造成多表連結，這會大幅降低執行效率。所以通常我們採用維度模型建模，把各種相關表整理成事實資料表和維度資料表兩種。所有的維度資料表圍繞事實資料表進行解釋。

6.1.3 星形模型、雪花模型與星座模型

在維度建模的基礎上，有三種模型：星形模型、雪花模型與星座模型。

當所有維度資料表都直接連接到事實資料表上時，整個圖解就像星星一樣，故該模型稱為星形模型，如圖 6-3 所示。星形模型是一種非正規化的結構，多維資料集的每個維度都直接與事實資料表相連接，不存在漸層維度，所以資料有一定的容錯。舉例來說，在地域維度資料表中，存在國家 A 省 B 的城市 C 及國家 A 省 B 的城市 D 兩筆記錄，那麼國家 A 和省 B 的資訊分別儲存了兩次，即存在容錯。

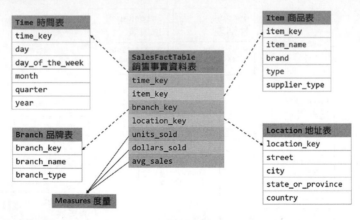

圖 6-3　星形模型建模示意

當有一張或多張維度資料表沒有直接連接到事實資料表上，而透過其他維度資料表連接到事實資料表上時，其圖解就像多個雪花連接在一起，故該模型稱為雪花模型。雪花模型是對星形模型的擴充。它對星形模型的維度資料表進行進一步層次化，原有的各維度資料表可能被擴充為小的事實資料表，形成一些局部的「層次」區域，這些被分解的表都連接到主維度資料表而非事

實資料表上,如圖 6-4 所示。雪花模型的優點是可透過大幅地減少資料儲存量及聯合較小的維度資料表來改善查詢效能。雪花模型去除了資料容錯,比較接近第三範式,但無法完全遵守,因為遵守第三範式的成本太高。

圖 6-4 雪花模型建模示意

星座模型與前兩種模型的區別是事實資料表的數量,星座模型是以多張事實資料表為基礎的,且事實資料表之間共用一些維度資料表。星座模型與前兩種模型並不衝突,如圖 6-5 所示為星座模型建模示意。星座模型基本上是很多資料倉儲的常態,因為很多資料倉儲都有多張事實資料表。

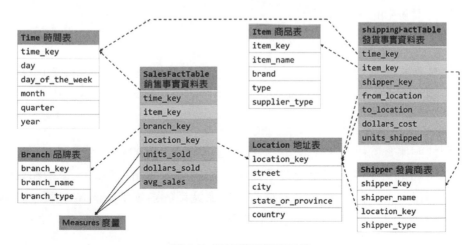

圖 6-5 星座模型建模示意

目前在企業實際開發中，其不會只選擇一種模型，而會根據情況靈活組合，甚至使各模型並存（一層維度和多層維度都儲存）。但是，從整體來看，企業更偏好選擇維度更少的星形模型。尤其是對於 Hadoop 系統，減少 join 就是減少中間資料的傳輸和計算，可明顯改善效能。

6.1.4 表的分類

在資料倉儲建模理論中，通常將表分為事實資料表和維度資料表兩種。事實資料表加維度資料表，能夠描述一個完整的業務事件。舉例來說，昨天早上張三在某電子商務平台花費 200 元購買了一個皮包，描述該業務事件需要三個維度，分別是時間維度（昨天早上）、商家維度（電子商務平台）和商品維度（皮包）。實際分類原則如下。

1. 事實資料表

事實資料表中的每行資料代表一個業務事件。「事實」這個術語表示的是業務事件的度量值。舉例來說，訂單事件中的下單金額。並且事實資料表會不斷擴大，資料量一般較大。

1）交易性事實資料表
交易性事實資料表以每個交易或事件為單位，舉例來說，以一筆支付記錄作為事實資料表中的一行資料。

2）週期型快照事實資料表
週期型快照事實資料表中不會保留所有資料，只保留固定時間間隔的資料，舉例來說，每天或每月的銷售額，以及每月的帳戶餘額等。

3）累積型快照事實資料表
累積型快照事實資料表（見表 6-9）用於追蹤業務事實的變化。舉例來說，資料倉儲中可能需要累計或儲存從下訂單開始到訂單商品被包裝、運輸和簽收的各業務階段的時間點數據來追蹤訂單生命週期的進展情況。當這個業務過程進行時，事實資料表的記錄也要不斷更新。

表 6-9　累積型快照事實資料表

訂單 id	使用者 id	下單時間	打包時間	發貨時間	簽收時間
001	000001	2019-02-12 10:10	2019-02-12 11:10	2019-02-12 12:10	2019-02-12 13:10

2. 維度資料表

維度資料表一般是指對應業務狀態編號的解釋表，也可以稱為碼表。舉例來說，訂單狀態表、商品品項表等，如表 6-10 和表 6-11 所示。

表 6-10　訂單狀態表

訂單狀態編號	訂單狀態名稱
1	未支付
2	支付
3	發貨中
4	已發貨
5	已完成

表 6-11　商品品項表

商品品項編號	品類別名稱
1	服裝
2	保健
3	電器
4	圖書

6.1.5　為什麼要分層

資料倉儲中的資料要想真正發揮最大的作用，必須對資料倉儲進行分層，資料倉儲分層的優點如下。

■ 把複雜問題簡單化。可以將一個複雜的任務分解成多個步驟來完成，每層只處理單一的步驟。
■ 減少重複開發。標準資料分層，透過使用中間層資料，可以大幅減少重複計算量，增加計算結果的重複使用性。

- 隔離原始資料。使真實資料與最後統計資料解耦。
- 資料倉儲實際如何分層取決於設計者對資料倉儲的整體規劃，不過大部分的想法是相似的。本書將資料倉儲分為五層，如下所述。
- ODS 層：原始資料層，儲存原始資料，直接載入原始記錄檔、資料，資料保持原貌不做處理。
- DWD 層：明細資料層，對 ODS 層資料進行清洗（去除空值、無效資料、超過極限範圍的資料）、維度退化、脫敏等。
- DWS 層：服務資料層，以 DWD 層的資料為基礎，按天進行輕度整理。
- DWT 層：主題資料層，以 DWS 層的資料為基礎，按主題進行整理，獲得每個主題的全量資料表。
- ADS 層：資料應用層，針對實際的資料需求，為各種統計報表提供資料。

資料倉儲分層後要遵守一定的資料倉儲命名標準，本專案中的標準如下。

1. 表命名

ODS 層命名為 ods_ 表名。
DWD 層命名為 dwd_dim/fact_ 表名。
DWS 層命名為 dws_ 表名。
DWT 層命名為 dwt_ 表名。
ADS 層命名為 ads_ 表名。
臨時表命名為 tmp_×××。
使用者行為表以 .log 為副檔名。

2. 指令稿命名

指令稿命名格式為資料來源 _to_ 目標 _db/log.sh。
使用者行為需求相關指令稿以 .log 為副檔名；業務資料需求相關指令稿以 .db 為副檔名。

6.1.6　資料倉儲建模

在本專案中，資料倉儲分為五層，分別為 ODS 層（原始資料層）、DWD 層
（明細資料層）、DWS 層（服務資料層）、DWT 層（主題資料層）和 ADS 層
（資料應用層）。本節主要探討五層架構中的資料倉儲建模思想。

1.　ODS 層

ODS 層主要進行以下處理。

（1）資料保持原貌不進行任何修改，造成備份資料的作用。
（2）資料採用壓縮格式，以減少磁碟儲存空間（舉例來説，100GB 的原始資
　　　料，可以被壓縮到 10GB 左右）。
（3）建立分區表，防止後續進行全資料表掃描。

2.　DWD 層

DWD 層需要建置維度模型，一般採用星形模型，而呈現的狀態一般為星座模
型。

維度建模　般按照以下四個步驟進行。

1）選擇業務過程
在業務系統中，若業務表過多，則可挑選我們有興趣的業務線，以下單業
務、支付業務、退款業務及物流業務，一筆業務線對應一張事實資料表。小
型公司的業務表可能比較少，建議選擇所有業務線。

2）宣告粒度
資料粒度是指資料倉儲中儲存資料的細化程度或綜合程度的等級。

宣告粒度表示精確定義事實資料表中的一行資料所表示的內容，應該盡可能
選擇最細粒度，以此來滿足各種各樣的需求。

典型的粒度宣告如下。

- 將訂單中的每個商品項作為下單事實資料表中的一行資料，粒度為每次。
- 將每週的訂單次數作為一行資料，粒度為每週。
- 將每月的訂單次數作為一行資料，粒度為每月。

如果在 DWD 層的粒度就是每週或每月，那麼後續就沒有辦法統計更細粒度（如每天）的指標了。所以建議採用最細粒度。

3）確定維度

維度的主要作用是描述業務事實，主要表示的是「誰、何處、何時」等資訊，如時間維度、使用者維度、地區維度等常見維度。

4）確定事實

此處的「事實」一詞指的是業務中的度量值。舉例來說，訂單表的度量值就是訂單件數、訂單金額等。

在 DWD 層中，以業務過程為建模驅動，以每個實際業務過程的特點為基礎，建置最細粒度的明細資料層事實資料表。事實資料表可進行適當的寬表化處理。

透過以上步驟，並結合本資料倉儲專案的業務事實，得出業務匯流排矩陣表，如表 6-12 所示。業務匯流排矩陣的原則主要是判斷維度資料表和事實資料表之間的關係，若兩者有連結，則使用 √ 標記。

表 6-12 業務匯流排矩陣表

	時間	用戶	地區	商品	優惠券	活動	編碼	度量值
訂單	√	√	√			√		件數 / 金額
訂單詳情	√	√	√	√				件數 / 金額
支付	√	√	√					次數 / 金額
加購	√	√		√				件數 / 金額
收藏	√	√		√				個數
評價	√	√		√				筆數
退款	√	√		√				件數 / 金額
優惠券領用	√	√			√			個數

根據維度建模中的星形模型思想,將維度進行退化。如圖 6-6 所示,地區表和省份表退化為地區維度資料表,SKU 商品表、品牌表、SPU 商品表、商品三級品項表、商品二級品項表、商品一級品項表退化為商品維度資料表,活動訂單連結表和優惠規則表退化為活動維度資料表。

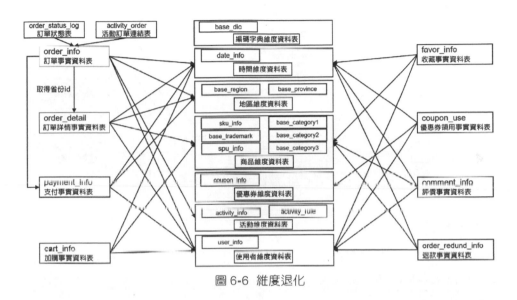

圖 6-6 維度退化

至此,資料倉儲的維度建模已經完畢,DWS 層、DWT 層、ADS 層和維度建模已經沒有關係了。

DWS 層和 DWT 層都是按照主題來建立寬表的,而主題相當於觀察問題的角度,不同的維度資料表表示不同的角度。

3. DWS 層

DWS 層用於統計各主題物件的當天行為,服務於 DWT 層的主題寬表。如圖 6-7 所示,DWS 層的寬表欄位是站在不同維度的角度去看事實資料表的,特別注意事實資料表的度量值,透過與之連結的事實資料表,獲得不同事實資料表的度量值。

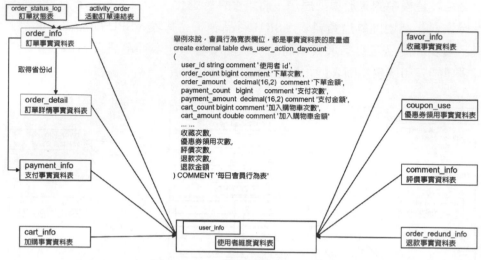

圖 6-7 DWS 層寬表欄位取得想法

4. DWT 層

DWT 層以分析的主題物件為建模驅動，以上層的應用和產品的指標需求為基礎，建置主題物件的全量寬表。

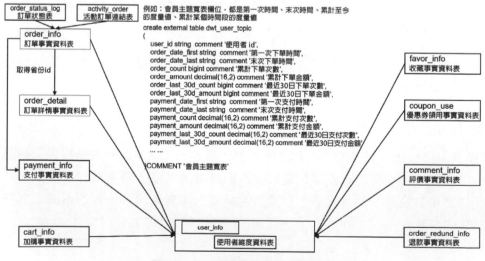

圖 6-8 DWT 層寬表欄位取得想法

DWT 層主題寬表都記錄什麼欄位？

DWT 層的寬表欄位站在維度資料表的角度去看事實資料表，特別注意事實資料表度量值的累計值、事實資料表行為的第一次時間和末次時間，如圖 6-8 所示。舉例來說，訂單事實資料表的度量值是下單次數、下單金額。訂單事實資料表的行為是下單。我們站在使用者維度資料表的角度去看訂單事實資料表，特別注意訂單事實資料表至今的累計下單次數、累計下單金額和某時間段內的累計下單次數、累計下單金額，以及關注下單行為的第一次時間和末次時間。

5. ADS 層

ADS 層分別對裝置主題、會員主題、商品主題和行銷主題進行指標分析，其中，行銷主題是會員主題和商品主題的跨主題分析案例。

6.1.7 業務術語

在進行需求的實現之前，我們先對相關的業務術語介紹，只有了解這些業務術語的含義，才能找到實現需求的想法。

1）使用者
使用者以裝置為判斷標準，在行動統計中，每台獨立裝置被認為是一個獨立使用者。Android 系統根據 IMEI 號，iOS 系統根據 OpenUDID 來標識一個獨立使用者，每部手機是一個使用者。

2）新增裝置
新增裝置是指第一次聯網使用應用的裝置。如果在一台裝置上第一次開啟某應用，那麼這台裝置被定義為新增裝置；移除再安裝裝置，不會算作一次新增。新增裝置包含日新增裝置、周新增裝置及月新增裝置。

3）活躍裝置
開啟應用的裝置即活躍裝置，不考慮裝置的使用情況。一台裝置一天內多次開啟某應用會被記為一台活躍裝置。

4）周（月）活躍裝置

周（月）活躍裝置是指某個自然周（月）內啟動過應用的裝置，該周（月）內多次啟動某裝置只記為一台活躍裝置。

5）月活躍率

月活躍率是指月活躍裝置佔截至該月累計的裝置總和的比例。

6）沉默裝置

沉默裝置是指裝置僅在安裝當天（次日）啟動過一次應用，後續沒有再次啟動的行為。該指標可以反映新增裝置品質和裝置與應用的比對程度。

7）版本分佈

版本分佈是指不同版本的應用周內各天的新增裝置數、活躍裝置數和啟動次數。其有利於判斷應用的各版本之間的優劣和使用者行為習慣。

8）本周回流裝置

本周回流裝置是指上周未啟動過應用而本周啟動了應用的裝置。

9）連續 *n* 周活躍裝置

連續 *n* 周活躍裝置是指連續 *n* 周，每週至少啟動一次應用的裝置。

10）忠誠使用者

忠誠使用者是指連續活躍 5 周以上的使用者。

11）連續活躍裝置

連續活躍裝置是指連續 2 周及 2 周以上活躍的裝置。

12）留存裝置

某段時間內的新增裝置，經過一段時間後，仍然使用應用的被認為是留存裝置；這部分裝置佔當時新增裝置的比例即留存率。

舉例來説，5 月份新增裝置 200 台，這 200 台裝置在 6 月份啟動過應用的有 100 台，7 月份啟動過應用的有 80 台，8 月份啟動過應用的有 50 台；則 5 月份新增裝置一個月後的留存率是 50%，兩個月後的留存率是 40%，三個月後的留存率是 25%。

13）使用者新鮮度

使用者新鮮度是指每天啟動應用的新舊使用者比例，即新增使用者數佔活躍使用者數的比例。

14）單次使用時長

單次使用時長是指裝置每次啟動使用的時間長度。

15）日使用時長

日使用時長是指裝置累計一天內的使用時間長度。

16）啟動次數計算標準

iOS 平台應用退到後台就算一次獨立的啟動；Android 平台規定，若兩次啟動之間的時間間隔小於 30 秒，則計算一次啟動。舉例來說，使用者在使用應用過程中，若因收發簡訊或接電話等退出應用且 30 秒內又傳回應用中，這兩次行為應該是延續而非獨立的，可以算作一次使用行為，即一次啟動。業內大多使用 30 秒這個標準，但使用者是可以自訂此時間間隔的。

6.2 資料倉儲架設環境準備

在第 2 章中，我們選取設定了 Tez 引擎的 Hive 作為資料倉儲架設工具，本節我們來進行 Hive 的安裝部署。

Hive 是一款用類別 SQL 敘述來協助讀 / 寫、管理那些儲存在分散式儲存系統上的巨量資料集的資料倉儲軟體。Hive 可以將類別 SQL 敘述解析成 MapReduce 程式，進一步避免撰寫繁雜的 MapReduce 程式，讓使用者分析資料變得容易。Hive 要分析的資料儲存在 HDFS 上，所以它本身不提供資料儲存功能。Hive 將資料對映成一張張的表，而將表的結構資訊儲存在關聯式資料庫（如 MySQL）中，所以在安裝 Hive 之前，我們需要先安裝 MySQL。

安裝 MySQL 的實際操作在第 5 章中已經介紹過，讀者需要在 hadoop103 和 hadoop104 兩台節點伺服器上分別操作一遍，以便為下一節內容做準備。

6.2.1 MySQL HA

MySQL 中儲存了 Hive 所有表格的中繼資料資訊，一旦 MySQL 中的資料遺失或損壞，會對整個資料倉儲系統造成不可挽回的損失，為避免這種情況的發生，我們可以選擇每天對中繼資料進行備份，進而實現 MySQL HA（High Availability，高可用）。

MySQL 的 HA 方案不止一種，本章介紹較為常用的一種──以 Keepalived 為基礎的 MySQL HA。

MySQL 的 HA 離不開其主從複製技術。主從複製是指一台伺服器充當主要資料庫伺服器（master），另一台或多台伺服器充當從資料庫伺服器（slave），從資料庫伺服器自動向主要資料庫伺服器同步資料。實現 MySQL 的 HA，需要使兩台伺服器互為主從關係。

Keepalived 是基於 VRRP（Virtual Router Redundancy Protocol，虛擬路由器容錯協定）的一款高可用軟體。Keepalived 有一台主要資料庫伺服器和多台從資料庫伺服器，在主要資料庫伺服器和從資料庫伺服器上面部署相同的服務設定，使用一個虛擬 IP 位址對外提供服務，當主要資料庫伺服器出現故障時，虛擬 IP 位址會自動漂移到從資料庫伺服器上。

實際操作步驟如下。

1. MySQL 主從設定

1）主從叢集規劃
主從叢集規劃如表 6-13 所示。

表 6-13 主從叢集規劃

hadoop102	hadoop103	hadoop104
	MySQL（master）	MySQL（slave）

★ 注意：MySQL 的安裝步驟參考 5.2.1 節。

2）設定 master

修改 hadoop103 中 MySQL 的 /usr/my.cnf 設定檔。

```
[mysqld]

#開啟binlog
log_bin = mysql-bin
#binlog記錄檔類型
binlog_format = row
#MySQL伺服器的唯一id
server_id = 1
```

重新啟動 hadoop103 的 MySQL 服務。

```
[atguigu@hadoop103 ~]$ sudo service mysql restart
```

進入 MySQL 用戶端，執行以下指令，檢視 master 狀態，如圖 6-9 所示。

```
mysql> show master status;
+------------------+----------+--------------+------------------+-------------------+
| File             | Position | Binlog_Do_DB | Binlog_Ignore_DB | Executed_Gtid_Set |
+------------------+----------+--------------+------------------+-------------------+
| mysql-bin.000001 |      120 |              |                  |                   |
+------------------+----------+--------------+------------------+-------------------+
1 row in set (0.00 sec)
```

圖 6-9 檢視 MySQL 用戶端的 master 狀態（1）

3）設定 slave

修改 hadoop104 中 MySQL 的 /usr/my.cnf 設定檔。

```
[mysqld]
#MySQL伺服器的唯一id
server_id = 2
#開啟slave中繼記錄檔
relay_log=mysql-relay
```

重新啟動 hadoop104 的 MySQL 服務。

```
[atguigu@hadoop104 ~]$ sudo service mysql restart
```

進入 hadoop104 的 MySQL 用戶端，執行以下指令。

```
mysql>
CHANGE MASTER TO
MASTER_HOST='hadoop103',
MASTER_USER='root',
MASTER_PASSWORD='000000',
MASTER_LOG_FILE='mysql-bin.000001',
MASTER_LOG_POS=120;
```

啟動 slave。

```
mysql> start slave;
```

檢視 slave 狀態，如圖 6-10 所示。

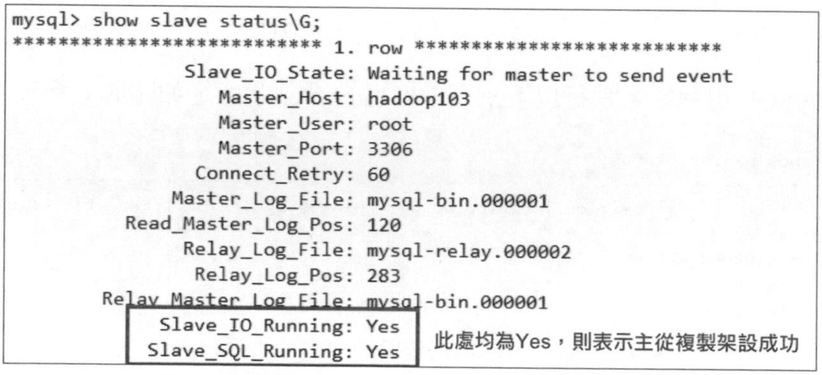

圖 6-10 檢視 MySQL 用戶端的 slave 狀態（1）

2. MySQL 雙主設定

1）雙主叢集規劃

雙主叢集規劃如表 6-14 所示。

表 6-14 雙主叢集規劃

hadoop102	hadoop103	hadoop104
	MySQL（master，slave）	MySQL（slave，master）

2）設定 master

修改 hadoop104 中 MySQL 的 /usr/my.cnf 設定檔。

```
[mysqld]

#開啟binlog
log_bin = mysql-bin
#binlog記錄檔類型
binlog_format = row
#MySQL伺服器的唯一id
server_id = 2
#開啟slave中繼記錄檔
relay_log=mysql-relay
```

重新啟動 hadoop104 的 MySQL 服務。

```
[atguigu@hadoop104 ~]$ sudo service mysql restart
```

進入 hadoop104 的 MySQL 用戶端，執行以下指令，檢視 master 狀態，如圖 6-11 所示。

```
mysql> show master status;
+------------------+----------+--------------+------------------+-------------------+
| File             | Position | Binlog_Do_DB | Binlog_Ignore_DB | Executed_Gtid_Set |
+------------------+----------+--------------+------------------+-------------------+
| mysql-bin.000001 |      120 |              |                  |                   |
+------------------+----------+--------------+------------------+-------------------+
1 row in set (0.00 sec)
```

圖 6-11 檢視 MySQL 用戶端的 master 狀態（2）

3）設定 slave

修改 hadoop103 中 MySQL 的 /usr/my.cnf 設定檔。

```
[mysqld]
#MySQL伺服器的唯一id
server_id = 1

#開啟binlog
log_bin = mysql-bin
#binlog記錄檔類型
binlog_format = row
#開啟slave中繼記錄檔
relay_log=mysql-relay
```

重新啟動 hadoop103 的 MySQL 服務。

```
[atguigu@hadoop103 ~]$ sudo service mysql restart
```

進入 hadoop103 的 MySQL 用戶端，執行以下指令。

```
mysql>
CHANGE MASTER TO
MASTER_HOST='hadoop104',
MASTER_USER='root',
MASTER_PASSWORD='000000',
MASTER_LOG_FILE='mysql-bin.000001',
MASTER_LOG_POS=120;
```

啟動 slave。

```
mysql> start slave;
```

檢視 slave 狀態，如圖 6-12 所示。

```
mysql> show slave status\G;
*************************** 1. row ***************************
               Slave_IO_State: Waiting for master to send event
                  Master_Host: hadoop104
                  Master_User: root
                  Master_Port: 3306
                Connect_Retry: 60
              Master_Log_File: mysql-bin.000001
          Read_Master_Log_Pos: 120
               Relay_Log_File: mysql-relay.000002
                Relay_Log_Pos: 283
        Relay_Master_Log_File: mysql-bin.000001
             Slave_IO_Running: Yes
            Slave_SQL_Running: Yes      此處均為Yes，則表示主從複製架設成功
```

圖 6-12 檢視 MySQL 用戶端的 slave 狀態（2）

3. Keepalived 安裝部署

Keepalived 需要分別安裝部署在 hadoop103 和 hadoop104 兩台節點伺服器上，
實際操作步驟如下。

1）在 hadoop103 上安裝部署

（1）執行 yum 指令，安裝 Keepalived。

```
[atguigu@hadoop103 ~]$ sudo yum install -y keepalived
```

（2）修改 Keepalived 的設定檔 /etc/keepalived/keepalived.conf。

```
! Configuration File for keepalived
global_defs {
    router_id MySQL-ha
}
vrrp_instance VI_1 {
    state master          #初始狀態
    interface eth0        #網路卡
    virtual_router_id 51  #虛擬路由id
    priority 100          #優先順序
    advert_int 1          #Keepalived心跳間隔
    nopreempt             #只在高優先順序設定，原master恢復之後不重新上位
    authentication {
        auth_type PASS    #認證方式
        auth_pass 1111
    }
    virtual_ipaddress {
        192.168.1.100  #虛擬IP位址
    }
}

#宣告虛擬伺服器
virtual_server 192.168.1.100 3306 {
    delay_loop 6
    persistence_timeout 30
    protocol TCP
    #宣告真實伺服器
    real_server 192.168.1.103 3306 {
        notify_down /var/lib/mysql/killkeepalived.sh
        #真實伺服器出現故障後呼叫指令稿
        TCP_CHECK {
            connect_timeout 3       #逾時
            nb_get_retry 3          #重試次數
```

```
            delay_before_retry 2         #重試時間間隔
        }
    }
}
```

（3）編輯指令檔 /var/lib/mysql/killkeepalived.sh。

```
#! /bin/bash
sudo service keepalived stop
```

（4）增加指令稿執行許可權。

```
[atguigu@hadoop103 ~]$ sudo chmod +x /var/lib/mysql/killkeepalived.sh
```

（5）啟動 Keepalived 服務。

```
[atguigu@hadoop103 ~]$ sudo service keepalived start
```

（6）設定 Keepalived 服務開機自啟。

```
[atguigu@hadoop103 ~]$ sudo chkconfig keepalived on
```

2）在 hadoop104 上安裝部署

（1）執行 yum 指令，安裝 Keepalived。

```
[atguigu@hadoop104 ~]$ sudo yum install -y keepalived
```

（2）修改 Keepalived 的設定檔 /etc/keepalived/keepalived.conf。

```
! Configuration File for keepalived
global_defs {
    router_id MySQL-ha
}
vrrp_instance VI_1 {
    state master              #初始狀態
    interface eth0            #網路卡
    virtual_router_id 51      #虛擬路由id
    priority 100              #優先順序
    advert_int 1              #Keepalived心跳間隔
    authentication {
        auth_type PASS        #認證方式
```

```
        auth_pass 1111
    }
    virtual_ipaddress {
        192.168.1.100                #虛擬IP位址
    }
}

#宣告虛擬伺服器
virtual_server 192.168.1.100 3306 {
    delay_loop 6
    persistence_timeout 30
    protocol TCP
    #宣告真實伺服器
    real_server 192.168.1.104 3306 {
        notify_down /var/lib/mysql/killkeepalived.sh
        #真實伺服器出現故障後呼叫指令稿
        TCP_CHECK {
            connect_timeout 3        #逾時
            nb_get_retry 3           #重試次數
            delay_before_retry 3     #重試時間間隔
        }
    }
}
```

（3）編輯指令檔 /var/lib/mysql/killkeepalived.sh。

```
#! /bin/bash
sudo service keepalived stop
```

（4）增加指令稿執行許可權。

```
[atguigu@hadoop104 ~]$ sudo chmod +x /var/lib/mysql/killkeepalived.sh
```

（5）啟動 Keepalived 服務。

```
[atguigu@hadoop104 ~]$ sudo service keepalived start
```

（6）設定 Keepalived 服務開機自啟。

```
[atguigu@hadoop104 ~]$ sudo chkconfig keepalived on
```

4. 確保開機時 MySQL 先於 Keepalived 啟動

需要分別在 hadoop103、hadoop104 節點伺服器上進行以下操作。

第一步:檢視開機時 MySQL 的啟動次序,結果如圖 6-13 所示。

```
[atguigu@hadoop104 ~]$ sudo vim /etc/init.d/mysql
```

```
#!/bin/sh
# Copyright Abandoned 1996 TCX DataKonsult AB & Monty Program KB & Detron HB
# This file is public domain and comes with NO WARRANTY of any kind

# MySQL daemon start/stop script.

# Usually this is put in /etc/init.d (at least on machines SYSV R4 based
# systems) and linked to /etc/rc3.d/S99mysql and /etc/rc0.d/K01mysql.
# When this is done the mysql server will be started when the machine is
# started and shut down when the systems goes down.
開機時的啟動次序,數值越小,越先啟動
# Comments to support chkconfig on RedHat Linux
# chkconfig: 2345 86 36
# description: A very fast and reliable SQL database engine.
```

圖 6-13 檢視開機時 MySQL 的啟動次序

第二步:檢視開機時 Keepalived 的啟動次序,結果如圖 6-14 所示。

```
[atguigu@hadoop104 ~]$ sudo vim /etc/init.d/keepalived
```

```
#!/bin/sh
#
# keepalived    High Availability monitor built upon LVS and VRRP
#
# chkconfig:    - 64 14
# description: Robust keepalive facility to the Linux Virtual Server project \
#              with multilayer TCP/IP stack checks.
```

圖 6-14 檢視開機時 Keepalived 的啟動次序

第三步:若 Keepalived 先於 MySQL 啟動,則需要按照以下步驟設定二者的啟動次序。

(1)修改開機時 MySQL 的啟動次序,如圖 6-15 所示。

```
[atguigu@hadoop104 ~]$ sudo vim /etc/init.d/mysql
```

```
#!/bin/sh
# Copyright Abandoned 1996 TCX DataKonsult AB & Monty Program KB & Detron HB
# This file is public domain and comes with NO WARRANTY of any kind

# MySQL daemon start/stop script.

# Usually this is put in /etc/init.d (at least on machines SYSV R4 based
# systems) and linked to /etc/rc3.d/S99mysql and /etc/rc0.d/K01mysql.
# When this is done the mysql server will be started when the machine is
# started and shut down when the systems goes down.

# Comments to support chkconfig on RedHat Linux
# chkconfig: 2345 64 36
# description: A very fast and reliable SQL database engine.
```

圖 6-15 修改開機時 MySQL 的啟動次序

（2）重新設定 MySQL 開機自啟。

```
[atguigu@hadoop104 ~]$ sudo chkconfig --del mysql
[atguigu@hadoop104 ~]$ sudo chkconfig --add mysql
[atguigu@hadoop104 ~]$ sudo chkconfig mysql on
```

（3）修改開機時 Keepalived 的啟動次序，如圖 6-16 所示。

```
[atguigu@hadoop104 ~]$ sudo vim /etc/init.d/keepalived
```

```
#!/bin/sh
#
# keepalived    High Availability monitor built upon LVS and VRRP
#
# chkconfig:    - 86 14
# description: Robust keepalive facility to the Linux Virtual Server project \
#              with multilayer TCP/IP stack checks.
```

圖 6-16 修改開機時 Keepalived 的啟動次序

（4）重新設定 Keepalived 開機自啟。

```
[atguigu@hadoop104 ~]$ sudo chkconfig --del keepalived
[atguigu@hadoop104 ~]$ sudo chkconfig --add keepalived
[atguigu@hadoop104 ~]$ sudo chkconfig keepalived on
```

6.2.2 Hive 安裝

在安裝了 MySQL 後，接下來可以著手對 Hive 進行正式的安裝部署。

1. 安裝及設定 Hive

（1）把 Hive 的安裝套件 apache-hive-3.1.2-bin.tar.gz 上傳到 Linux 的 /opt/software 目錄下，然後解壓 apache-hive-3.1.2-bin.tar.gz 到 /opt/module 目錄下。

```
[atguigu@hadoop102 software]$ tar -zxvf apache-hive-3.1.2-bin.tar.gz -C
/opt/module
```

（2）修改 apache-hive-3.1.2-bin 的名稱為 hive。

```
[atguigu@hadoop102 module]$ mv apache-hive-3.1.2-bin/ hive
```

（3）修改 /etc/profile 檔案，增加環境變數。

```
[atguigu@hadoop102 software]$ sudo vim /etc/profile
```

增加以下內容。

```
#HIVE_HOME
export HIVE_HOME=/opt/module/hive
export PATH=$PATH:$HIVE_HOME/bin
```

執行以下指令，使環境變數生效。

```
[atguigu@hadoop102 software]$ source /etc/profile
```

（4）進入 /opt/module/hive/lib 目錄下執行以下指令，解決記錄檔 jar 套件衝突。

```
[atguigu@hadoop102 lib]$ mv log4j-slf4j-impl-2.10.0.jar log4j-slf4j-impl-
2.10.0.jar.bak
```

2. 複製驅動

（1）在 /opt/software/mysql-libs 目錄下解壓 mysql-connector-java-5.1.27.tar.gz 驅動套件。

```
[root@hadoop102 mysql-libs]# tar -zxvf mysql-connector-java-5.1.27.tar.gz
```

（2）將 /opt/software/mysql-libs/mysql-connector-java-5.1.27 目錄下的 mysql-connector-java-5.1.27-bin.jar 複製到 /opt/module/hive/lib 目錄下，用於稍後啟動 Hive 時連接 MySQL。

```
[root@hadoop102 mysql-connector-java-5.1.27]# cp mysql-connector-java-
5.1.27-bin.jar
 /opt/module/hive/lib
```

3. 設定 Metastore 到 MySQL

（1）在 /opt/module/hive/conf 目錄下建立一個 hive-site.xml 檔案。

```
[atguigu@hadoop102 conf]$ touch hive-site.xml
[atguigu@hadoop102 conf]$ vim hive-site.xml
```

（2）根據官方文件設定參數，並複製資料到 hive-site.xml 檔案中。

```xml
<?xml version="1.0"?>
<?xml-stylesheet type="text/xsl" href="configuration.xsl"?>
<configuration>
<!--設定Hive儲存中繼資料資訊所需的MySQL URL位址，此處使用Keepalived服務對
外提供的虛擬IP位址-->
<property>
    <name>javax.jdo.option.ConnectionURL</name>
    <value>jdbc:mysql://hadoop100:3306/metastore?createDatabaseIfNot
Exist=true
</value>
    <description>JDBC connect string for a JDDC metastore</description>
 </property>
<!--設定Hive連接MySQL的驅動全類別名稱-->
 <property>
    <name>javax.jdo.option.ConnectionDriverName</name>
    <value>com.mysql.jdbc.Driver</value>
    <description>Driver class name for a JDBC metastore</description>
 </property>
<!--設定Hive連接MySQL的使用者名稱 -->
 <property>
    <name>javax.jdo.option.ConnectionUserName</name>
    <value>root</value>
    <description>username to use against metastore database</description>
 </property>
<!--設定Hive連接MySQL的密碼 -->
 <property>
```

```
    <name>javax.jdo.option.ConnectionPassword</name>
    <value>000000</value>
    <description>password to use against metastore database</description>
 </property>
<property>
        <name>hive.metastore.warehouse.dir</name>
        <value>/user/hive/warehouse</value>
    </property>

    <property>
        <name>hive.metastore.schema.verification</name>
        <value>false</value>
    </property>

    <property>
        <name>hive.metastore.uris</name>
        <value>thrift://hadoop102:9083</value>
    </property>

    <property>
    <name>hive.server2.thrift.port</name>
    <value>10000</value>
    </property>

    <property>
        <name>hive.server2.thrift.bind.host</name>
        <value>hadoop102</value>
    </property>

    <property>
        <name>hive.metastore.event.db.notification.api.auth</name>
        <value>false</value>
    </property>

    <property>
        <name>hive.cli.print.header</name>
        <value>true</value>
    </property>
```

```
    <property>
        <name>hive.cli.print.current.db</name>
        <value>true</value>
    </property>
</configuration>
```

4. 初始化中繼資料庫

（1）啟動 MySQL。

```
[atguigu@hadoop103 mysql-libs]$ mysql -uroot -p000000
```

（2）新增 Hive 中繼資料庫。

```
mysql> create database metastore;
mysql> quit;
```

（3）初始化 Hive 中繼資料庫。

```
[atguigu@hadoop102 conf]$ schematool -initSchema -dbType mysql -verbose
```

5. 啟動 Hive

（1）Hive 2.x 以上版本，需要先啟動 Metastore 和 HiveServer2 服務，否則會顯示出錯。

```
FAILED: HiveException java.lang.RuntimeException: Unable to instantiate org.
apache.hadoop.hive.ql.metadata.SessionHiveMetaStoreClient
```

（2）在 /opt/module/hive/bin 目錄下撰寫 Hive 服務啟動指令稿，在指令稿中啟動 Metastore 和 HiveServer2 服務。

```
[atguigu@hadoop102 bin]$ vim hiveservices.sh
```

指令稿內容如下。

```
#!/bin/bash
HIVE_LOG_DIR=$HIVE_HOME/logs

mkdir -p $HIVE_LOG_DIR
```

```
#檢查處理程序是否執行正常,參數1為處理程序名稱,參數2為處理程序通訊埠
function check_process()
{
    pid=$(ps -ef 2>/dev/null | grep -v grep | grep -i $1 | awk '{print $2}')
    ppid=$(netstat -nltp 2>/dev/null | grep $2 | awk '{print $7}' | cut
-d '/' -f 1)
    echo $pid
    [[ "$pid" =~ "$ppid" ]] && [ "$ppid" ] && return 0 || return 1
}

function hive_start()
{
    metapid=$(check_process HiveMetastore 9083)
    cmd="nohup hive --service metastore >$HIVE_LOG_DIR/metastore.log 2>&1 &"
    cmd=$cmd" sleep 4; hdfs dfsadmin -safemode wait >/dev/null 2>&1"
    [ -z "$metapid" ] && eval $cmd || echo "Metastore服務已啟動"
    server2pid=$(check_process HiveServer2 10000)
    cmd="nohup hive --service hiveserver2 >$HIVE_LOG_DIR/hiveServer2.
log 2>&1 &"
    [ -z "$server2pid" ] && eval $cmd || echo "HiveServer2服務已啟動"
}

function hive_stop()
{
    metapid=$(check_process HiveMetastore 9083)
    [ "$metapid" ] && kill $metapid || echo "Metastore服務未啟動"
    server2pid=$(check_process HiveServer2 10000)
    [ "$server2pid" ] && kill $server2pid || echo "HiveServer2服務未啟動"
}

case $1 in
"start")
    hive_start
    ;;
"stop")
    hive_stop
    ;;
"restart")
```

```
    hive_stop
    sleep 2
    hive_start
    ;;
"status")
    check_process HiveMetastore 9083 >/dev/null && echo "Metastore服務
執行正常" || echo "Metastore服務執行異常"
    check_process HiveServer2 10000 >/dev/null && echo "HiveServer2服務
執行正常" || echo "HiveServer2服務執行異常"
    ;;
*)
    echo Invalid Args!
    echo 'Usage: '$(basename $0)' start|stop|restart|status'
    ;;
esac
```

（3）增加指令稿執行許可權。

```
[atguigu@hadoop102 bin]$ chmod +x hiveservices.sh
```

（4）啟動 Hive 後台服務。

```
[atguigu@hadoop102 bin]$ hiveservices.sh start
```

（5）檢視 Hive 後台服務執行情況。

```
[atguigu@hadoop102 bin]$ hiveservices.sh status
Metastore服務執行正常
HiveServer2服務執行正常
```

（6）啟動 Hive 用戶端。

```
[atguigu@hadoop102 hive]$ bin/hive
```

6.2.3 Tez 引擎安裝

Tez 是 Hive 的執行引擎，效能優於 MR。為什麼 Tez 的效能優於 MR 呢？從如圖 6-17 所示的 Tez 引擎原理中可得出結論。

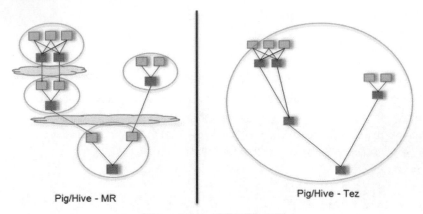

Pig/Hive - MR　　　　Pig/Hive - Tez

圖 6-17　Tez 引擎原理示意

圖 6-17 左圖表示用 Hive 直接撰寫 MR 程式，假設形成了四個有相依關係的 MR 作業，其中每一個橢圓形代表一個 MR 作業，淺色方塊代表 Map Task，深色方塊代表 Reduce Task，雲狀表示需要將上一步產生的結果資料持久化到 HDFS 中才能供下游作業使用，這會產生大量的磁碟 IO。

Tez 可以將多個有相依關係的作業轉為一個作業，這樣只需寫一次 HDFS 即可，且中間節點較少，進一步大幅提升了作業的計算效能。

1. 安裝套件準備

（1）下載 Tez 的相依套件。

（2）將 apache-tez-0.9.1-bin.tar.gz 複製到 hadoop102 的 /opt/software 目錄下。

```
[atguigu@hadoop102 software]$ ls
apache-tez-0.9.1-bin.tar.gz
```

（3）將 apache-tez-0.9.1-bin.tar.gz 解壓至 /opt/module 目錄下。

```
[atguigu@hadoop102 module]$ tar -zxvf apache-tez-0.9.1-bin.tar.gz -C
/opt/module
```

（4）修改名稱。

```
[atguigu@hadoop102 module]$ mv apache-tez-0.9.1-bin/ tez-0.9.1
```

2. 在 Hive 中設定 Tez

（1）進入 Hive 的設定目錄 /opt/module/hive/conf。

```
[atguigu@hadoop102 conf]$ pwd
/opt/module/hive/conf
```

（2）在 hive-env.sh 檔案中增加 Tez 環境變數設定和相依套件環境變數設定。

```
[atguigu@hadoop102 conf]$ vim hive-env.sh
```

增加的設定如下。

```
# 設定Hadoop叢集的安裝目錄
export HADOOP_HOME=/opt/module/hadoop-2.7.2

# 設定Hive的設定檔目錄
export HIVE_CONF_DIR=/opt/module/hive/conf

# 設定執行Tez環境所需的jar套件路徑
export TEZ_HOME=/opt/module/tez-0.9.1          #此處是讀者自己的Tez的解壓目錄
export TEZ_JARS=""
for jar in `ls $TEZ_HOME |grep jar`; do
    export TEZ_JARS=$TEZ_JARS:$TEZ_HOME/$jar
done
for jar in `ls $TEZ_HOME/lib`; do
    export TEZ_JARS=$TEZ_JARS:$TEZ_HOME/lib/$jar
done

export HIVE_AUX_JARS_PATH=/opt/module/hadoop-2.7.2/share/hadoop/common/
hadoop-lzo-0.4.20.jar$TEZ_JARS
```

（3）在 hive-site.xml 檔案中增加以下設定，將 Hive 的計算引擎更改為 Tez。

```xml
<property>
    <name>hive.execution.engine</name>
    <value>tez</value>
</property>
```

3. 設定 Tez

在 Hive 的 /opt/module/hive/conf 目錄下建立一個 tez-site.xml 檔案。

```
[atguigu@hadoop102 conf]$ pwd
/opt/module/hive/conf
[atguigu@hadoop102 conf]$ vim tez-site.xml
```

在檔案中增加以下內容。

```
<?xml version="1.0" encoding="UTF-8"?>
<?xml-stylesheet type="text/xsl" href="configuration.xsl"?>
<configuration>
#設定在Tez中使用的uris的jar套件路徑
<property>
 <name>tez.lib.uris</name>    <value>${fs.defaultFS}/tez/tez-
0.9.1,${fs.defaultFS}/tez/tez-0.9.1/lib</value>
</property>
#設定在Tez中使用的uris的類別路徑
<property>
 <name>tez.lib.uris.classpath</name>    <value>${fs.defaultFS}/tez/tez-
0.9.1,${fs.defaultFS}/tez/tez-0.9.1/lib</value>
</property>
#是否使用Hadoop依賴
<property>
    <name>tez.use.cluster.hadoop-libs</name>
    <value>true</value>
</property>
#設定Tez自己的歷史伺服器
<property>
    <name>tez.history.logging.service.class</name>
<value>org.apache.tez.dag.history.logging.ats.ATSHistoryLoggingService
</value>
</property>
</configuration>
```

4. 上傳 Tez 到叢集

將 /opt/module/tez-0.9.1 上傳到 HDFS 的 /tez 目錄下。

```
[atguigu@hadoop102 conf]$ hadoop fs -mkdir /tez
[atguigu@hadoop102 conf]$ hadoop fs -put /opt/module/tez-0.9.1/ /tez
[atguigu@hadoop102 conf]$ hadoop fs -ls /tez
/tez/tez-0.9.1
```

5. 測試

（1）啟動 Hive。

```
[atguigu@hadoop102 hive]$ bin/hive
```

（2）建立 LZO 表。

```
hive (default)> create table student(
id int,
name string);
```

（3）在表中插入資料。

```
hive (default)> insert into student values(1,"zhangsan");
```

（4）如果沒有顯示出錯，就表示設定成功了。

```
hive (default)> select * from student;
1       zhangsan
```

6. 小結

執行 Tez 時，有可能遇到因為 Container 使用過多記憶體而被 NodeManager 殺死處理程序的問題，如下所示。

```
Caused by: org.apache.tez.dag.api.SessionNotRunning: TezSession has
already shutdown. Application application_1546781144082_0005 failed
2 times due to AM Container for appattempt_1546781144082_0005_000002
exited with  exitCode: -103
For more detailed output, check application tracking page:http://
hadoop103:8088/cluster/app/application_1546781144082_0005Then, click
on links to logs of each attempt.
Diagnostics: Container [pid=11116,containerID=container_1546781144082_
0005_02_000001] is running beyond virtual memory limits. Current usage:
```

```
216.3 MB of 1 GB physical memory used; 2.6 GB of 2.1 GB virtual memory
used. Killing container.
```

產生的原因是 NodeManager 上執行的 Container 試圖使用過多的記憶體，而被
NodeManager 殺掉了，如下所示。

```
[摘錄] The NodeManager is killing your container. It sounds like you are
trying to use hadoop streaming which is running as a child process of the
map-reduce task. The NodeManager monitors the entire process tree of the
task and if it eats up more memory than the maximum set in mapreduce.map.
memory.mb or mapreduce.reduce.memory.mb respectively, we would expect the
Nodemanager to kill the task, otherwise your task is stealing memory
belonging to other containers, which you don't want.
```

解決方法：修改 Hadoop 的設定檔 yarn-site.xml，增加以下設定，關掉虛擬記
憶體檢查，修改後，分發設定檔，並重新啟動叢集。

```
<property>
    <name>yarn.nodemanager.vmem-check-enabled</name>
    <value>false</value>
</property>
```

6.3 資料倉儲架設——ODS 層

ODS 層為原始資料層，其保持資料原貌不進行任何修改，造成備份資料的作
用；資料採用 LZO 壓縮格式，以減少磁碟儲存空間；建立分區表，可以避免
後續對表查詢時進行全資料表掃描。在進行 ODS 層資料的匯入之前，先要建
立資料庫，用於儲存整個電子商務資料倉儲專案的所有資料資訊。

6.3.1 建立資料庫

（1）啟動 Hive。

```
[atguigu@hadoop102 hive]$ bin/hiveservices.sh start
[atguigu@hadoop102 hive]$ bin/hive
```

（2）顯示資料庫。

```
hive (default)> show databases;
```

（3）建立資料庫。

```
hive (default)> create database gmall;
```

說明：當資料庫已存在且有資料，需要強制刪除時，執行以下指令。

```
drop database gmall cascade;
```

（4）使用 gmall 資料庫，以下所有操作均在該資料庫下進行。

```
hive (default)> use gmall;
```

6.3.2 使用者行為資料

ODS 層主要載入擷取記錄檔系統落盤在 HDFS 的記錄檔資料檔案，在擷取記錄檔系統的部署過程中，我們根據記錄檔的不同類型，將記錄檔分為啟動記錄和事件記錄檔，在架設 ODS 層時，我們也將根據兩種資料類型分別進行建表匯入處理。

分析想法如下。

（1）在進行建表之前，若要建立的表已經存在，則先刪除該表，以防止指令稿出錯，影響後續執行。

（2）建立外部表。外部表即只建立表與原始資料之間的對映關係，而不改變資料的位置，在對表執行刪除操作時，只會刪除表的中繼資料，而不會刪除表的資料，相對來說更加安全，這種方式在實際工作環境中應用十分廣泛。

（3）表格的欄位為 JSON 格式的 String 類型字串。

（4）表格按照日期進行分區，方便使用者按照日期進行查詢。

（5）檔案輸入格式為 LZO 壓縮格式，由於 LZO 壓縮格式的檔案不支援 HDFS 分片，因此需要對 LZO 壓縮格式的檔案建立索引。

（6）儲存路徑根據儲存資料的不同分別指定，啟動記錄表儲存路徑為 /user/ hive/warehouse/gmall/ods/ods_start_log；事件記錄檔表儲存路徑為 /user/ hive/warehouse/gmall/ods/ods_event_log。

（7）外部表建立成功後，將 Flume 擷取落盤的檔案載入到 Hive 表中，ODS 層 即架設成功。

實際操作如下。

（1）按照上述想法建立輸入資料是 LZO 壓縮格式、輸出資料是 TEXT 儲存格 式、支援 JSON 解析的分區啟動記錄表。

```
hive (gmall)>
drop table if exists ods_start_log;
CREATE EXTERNAL TABLE ods_start_log (`line` string)
PARTITIONED BY (`dt` string)
STORED AS
  INPUTFORMAT 'com.hadoop.mapred.DeprecatedLzoTextInputFormat'
  OUTPUTFORMAT 'org.apache.hadoop.hive.ql.io.
HiveIgnoreKeyTextOutputFormat'
LOCATION '/warehouse/gmall/ods/ods_start_log';
```

（2）載入資料，指定每天資料的分區資訊為到期日的日期。

```
hive (gmall)>
load data inpath '/origin_data/gmall/log/topic_start/2020-03-10' into
table gmall.ods_start_log partition(dt='2020-03-10');
```

> ★ 注意：日期格式都設定成 YYYY-MM-DD 格式，這是 Hive 預設支援的日期 格式。

（3）檢視分區啟動記錄表是否載入成功。

```
hive (gmall)> select * from ods_start_log limit 2;
```

（4）由於 LZO 壓縮格式的檔案不支援 HDFS 分片，因此對 LZO 壓縮格式的檔 案建立索引。

```
hadoop jar /opt/module/hadoop-2.7.2/share/hadoop/common/hadoop-lzo-
0.4.20.jar com.hadoop.compression.lzo.DistributedLzoIndexer /warehouse/
gmall/ods/ods_start_log/dt=2020-03-10
```

（5）建立輸入資料是 LZO 壓縮格式、支援 JSON 解析的分區事件記錄檔表。

```
hive (gmall)>
drop table if exists ods_event_log;
CREATE EXTERNAL TABLE ods_event_log(`line` string)
PARTITIONED BY (`dt` string)
STORED AS
  INPUTFORMAT 'com.hadoop.mapred.DeprecatedLzoTextInputFormat'
  OUTPUTFORMAT 'org.apache.hadoop.hive.ql.io.
HiveIgnoreKeyTextOutputFormat'
LOCATION '/warehouse/gmall/ods/ods_event_log';
```

（6）載入資料。

```
hive (gmall)>
load data inpath '/origin_data/gmall/log/topic_event/2020-03-10' into
table gmall.ods_event_log partition(dt='2020-03-10');
```

> ✈ **注意**：日期格式都設定成 YYYY-MM-DD 格式，這是 Hive 預設支援的日期
> 格式。

（7）檢視分區事件記錄檔表是否載入成功。

```
hive (gmall)> select * from ods_event_log limit 2;
```

（8）為 LZO 壓縮格式的檔案建立索引。

```
hadoop jar /opt/module/hadoop-2.7.2/share/hadoop/common/hadoop-lzo-
0.4.20.jar com.hadoop.compression.lzo.DistributedLzoIndexer /warehouse/
gmall/ods/ods_event_log/dt=2020-03-10
```

Shell 中單引號和雙引號的區別如下。

（1）在 /home/atguigu/bin 目錄下建立一個 test.sh 檔案。

```
[atguigu@hadoop102 bin]$ vim test.sh
```

在檔案中增加以下內容。

```
#!/bin/bash
do_date=$1

echo '$do_date'
echo "$do_date"
echo "'$do_date'"
echo '"$do_date"'
echo `date`
```

（2）檢視執行結果。

```
[atguigu@hadoop102 bin]$ test.sh 2020-03-10
$do_date
2020-03-10
'2020-03-10'
"$do_date"
2020年 05月 02日 星期四 21:02:08 CST
```

（3）歸納如下。

- 單引號表示不取出變數值。
- 雙引號表示取出變數值。
- 反引號表示執行引號中的指令。
- 雙引號內部巢狀結構單引號表示取出變數值。
- 單引號內部巢狀結構雙引號表示不取出變數值。

6.3.3 ODS 層使用者行為資料匯入指令稿

將 ODS 層使用者行為資料的載入過程撰寫成指令稿，方便每日呼叫執行。

（1）在 hadoop102 的 /home/atguigu/bin 目錄下建立指令稿 ods_log.sh。

```
[atguigu@hadoop102 bin]$ vim ods_log.sh
```

在指令稿中撰寫以下內容。

```
#!/bin/bash
```

```
# 定義變數，方便後續修改
APP=gmall
hive=/opt/module/hive/bin/hive
hadoop=/opt/module/hadoop-2.7.2/bin/hadoop

# 若輸入了日期參數，則取輸入參數作為日期值；若沒有輸入日期參數，則取目前時間
的前一天作為日期值
if [ -n "$1" ] ;then
   do_date=$1
else
   do_date=`date -d "-1 day" +%F`
fi

echo "===記錄檔日期為 $do_date==="
sql="
load data inpath '/origin_data/gmall/log/topic_start/$do_date' into
table ${db}.ods_start_log partition(dt='$do_date');
load data inpath '/origin_data/gmall/log/topic_event/$do_date' into
table ${db}.ods_event_log partition(dt='$do_date');
"

$hive -e "$sql"
$hadoop jar /opt/module/hadoop-2.7.2/share/hadoop/common/hadoop-lzo-
0.4.20.jar com.hadoop.compression.lzo.DistributedLzoIndexer /
warehouse/gmall/ods/ods_start_log/dt=$do_date
$hadoop jar /opt/module/hadoop-2.7.2/share/hadoop/common/hadoop-
lzo-0.4.20.jar com.hadoop.compression.lzo.DistributedLzoIndexer /
warehouse/gmall/ods/ods_event_log/dt=$do_date
```

說明：

- [-n 變數值] 用於判斷變數的值是否為空。
- 如果變數的值不可為空，則傳回 true。
- 如果變數的值為空，則傳回 false。

（2）增加指令稿執行許可權。

```
[atguigu@hadoop102 bin]$ chmod 777 ods_log.sh
```

（3）執行指令稿，匯入資料。

```
[atguigu@hadoop102 module]$ ods_log.sh 2020-03-11
```

（4）查詢結果資料。

```
hive (gmall)>
select * from ods_log where dt='2020-03-11' limit 2;
```

6.3.4 業務資料

業務資料的 ODS 層架設與使用者行為資料的 ODS 層架設相同，都是保留原始資料，不對資料進行任何轉換處理，根據需求分析選取業務資料庫中表的必需欄位進行建表，然後將 Sqoop 匯入的原始資料載入（Load）至所建表格中。

1. 建立訂單表

```
hive (gmall)>
drop table if exists ods_order_info;
create external table ods_order_info (
    `id` string COMMENT '編號',
    `final_total_amount` decimal(10,2) COMMENT '總金額',
    `order_status` string COMMENT '訂單狀態',
    `user_id` string COMMENT '使用者id',
    `out_trade_no` string COMMENT '訂單交易編號',
    `create_time` string COMMENT '建立時間',
    `operate_time` string COMMENT '操作時間',
    `province_id` string COMMENT '省份id',
    `benefit_reduce_amount` decimal(10,2) COMMENT '優惠金額',
    `original_total_amount` decimal(10,2)  COMMENT '原價金額',
    `feight_fee` decimal(10,2)  COMMENT '運費'
) COMMENT '訂單表'
PARTITIONED BY (`dt` string)
row format delimited fields terminated by '\t'
STORED AS
  INPUTFORMAT 'com.hadoop.mapred.DeprecatedLzoTextInputFormat'
  OUTPUTFORMAT 'org.apache.hadoop.hive.ql.io.HiveIgnoreKeyTextOutputFormat'
location '/warehouse/gmall/ods/ods_order_info/';
```

2. 建立訂單詳情表

```
hive (gmall)>
drop table if exists ods_order_detail;
create external table ods_order_detail(
    `id` string COMMENT '編號',
    `order_id` string  COMMENT '訂單編號',
    `user_id` string COMMENT '使用者id',
    `sku_id` string COMMENT '商品id',
    `sku_name` string COMMENT '商品名稱',
    `order_price` decimal(10,2) COMMENT '商品價格',
    `sku_num` bigint COMMENT '商品數量',
    `create_time` string COMMENT '建立時間'
) COMMENT '訂單詳情表'
PARTITIONED BY (`dt` string)
row format delimited fields terminated by '\t'
STORED AS
  INPUTFORMAT 'com.hadoop.mapred.DeprecatedLzoTextInputFormat'
  OUTPUTFORMAT 'org.apache.hadoop.hive.ql.io.HiveIgnoreKeyTextOutputFormat'
location '/warehouse/gmall/ods/ods_order_detail/';
```

3. 建立 SKU 商品表

```
hive (gmall)>
drop table if exists ods_sku_info;
create external table ods_sku_info(
    `id` string COMMENT '商品id',
    `spu_id` string   COMMENT '標準產品單位id',
    `price` decimal(10,2) COMMENT '價格',
    `sku_name` string COMMENT '商品名稱',
    `sku_desc` string COMMENT '商品描述',
    `weight` string COMMENT '重量',
    `tm_id` string COMMENT '品牌id',
    `category3_id` string COMMENT '三級品項id',
    `create_time` string COMMENT '建立時間'
) COMMENT 'SKU商品表'
PARTITIONED BY (`dt` string)
row format delimited fields terminated by '\t'
STORED AS
```

```
  INPUTFORMAT 'com.hadoop.mapred.DeprecatedLzoTextInputFormat'
  OUTPUTFORMAT 'org.apache.hadoop.hive.ql.io.HiveIgnoreKeyTextOutputFormat'
location '/warehouse/gmall/ods/ods_sku_info/';
```

4. 建立使用者表

```
hive (gmall)>
drop table if exists ods_user_info;
create external table ods_user_info(
    `id` string COMMENT '使用者id',
    `name` string COMMENT '真實姓名',
    `birthday` string COMMENT '生日',
    `gender` string COMMENT '性別',
    `email` string COMMENT '電子郵件',
    `user_level` string COMMENT '使用者等級',
    `create_time` string COMMENT '建立時間',
    `operate_time` string COMMENT '操作時間'
) COMMENT '使用者表'
PARTITIONED BY (`dt` string)
row format delimited fields terminated by '\t'
STORED AS
  INPUTFORMAT 'com.hadoop.mapred.DeprecatedLzoTextInputFormat'
  OUTPUTFORMAT 'org.apache.hadoop.hive.ql.io.HiveIgnoreKeyTextOutputFormat'
location '/warehouse/gmall/ods/ods_user_info/';
```

5. 建立商品一級品項表

```
hive (gmall)>
drop table if exists ods_base_category1;
create external table ods_base_category1(
    `id` string COMMENT 'id',
    `name` string COMMENT '名稱'
) COMMENT '商品一級品項表'
PARTITIONED BY (`dt` string)
row format delimited fields terminated by '\t'
STORED AS
  INPUTFORMAT 'com.hadoop.mapred.DeprecatedLzoTextInputFormat'
  OUTPUTFORMAT 'org.apache.hadoop.hive.ql.io.HiveIgnoreKeyTextOutputFormat'
location '/warehouse/gmall/ods/ods_base_category1/';
```

6. 建立商品二級品項表

```
hive (gmall)>
drop table if exists ods_base_category2;
create external table ods_base_category2(
    `id` string COMMENT 'id',
    `name` string COMMENT '名稱',
    category1_id string COMMENT '一級品項id'
) COMMENT '商品二級品項表'
PARTITIONED BY (`dt` string)
row format delimited fields terminated by '\t'
STORED AS
  INPUTFORMAT 'com.hadoop.mapred.DeprecatedLzoTextInputFormat'
  OUTPUTFORMAT 'org.apache.hadoop.hive.ql.io.HiveIgnoreKeyTextOutputFormat'
location '/warehouse/gmall/ods/ods_base_category2/';
```

7. 建立商品三級品項表

```
hive (gmall)>
drop table if exists ods_base_category3;
create external table ods_base_category3(
    `id` string COMMENT 'id',
    `name` string COMMENT '名稱',
    category2_id string COMMENT '二級品項id'
) COMMENT '商品三級品項表'
PARTITIONED BY (`dt` string)
row format delimited fields terminated by '\t'
STORED AS
  INPUTFORMAT 'com.hadoop.mapred.DeprecatedLzoTextInputFormat'
  OUTPUTFORMAT 'org.apache.hadoop.hive.ql.io.HiveIgnoreKeyTextOutputFormat'
location '/warehouse/gmall/ods/ods_base_category3/';
```

8. 建立支付流水錶

```
hive (gmall)>
drop table if exists ods_payment_info;
create external table ods_payment_info(
    `id`    bigint COMMENT '編號',
    `out_trade_no`    string COMMENT '對外業務編號',
```

```
    `order_id`          string COMMENT '訂單編號',
    `user_id`           string COMMENT '使用者id',
    `alipay_trade_no`   string COMMENT '支付寶交易流水編號',
    `total_amount`      decimal(16,2) COMMENT '支付金額',
    `subject`           string COMMENT '交易內容',
    `payment_type`      string COMMENT '支付類型',
    `payment_time`      string COMMENT '支付時間'
)  COMMENT '支付流水錶'
PARTITIONED BY (`dt` string)
row format delimited fields terminated by '\t'
STORED AS
  INPUTFORMAT 'com.hadoop.mapred.DeprecatedLzoTextInputFormat'
  OUTPUTFORMAT 'org.apache.hadoop.hive.ql.io.HiveIgnoreKeyTextOutputFormat'
location '/warehouse/gmall/ods/ods_payment_info/';
```

9. 建立省份表

```
hive (gmall)>
drop table if exists ods_base_province;
create external table ods_base_province (
    `id`    bigint COMMENT '編號',
    `name`          string COMMENT '省份名稱',
    `region_id`     string COMMENT '地區id',
    `area_code`     string COMMENT '地區編碼',
    `iso_code` string COMMENT '國際編碼'
)  COMMENT '省份表'
row format delimited fields terminated by '\t'
STORED AS
  INPUTFORMAT 'com.hadoop.mapred.DeprecatedLzoTextInputFormat'
  OUTPUTFORMAT 'org.apache.hadoop.hive.ql.io.HiveIgnoreKeyTextOutputFormat'
location '/warehouse/gmall/ods/ods_base_province/';
```

10. 建立地區表

```
hive (gmall)>
drop table if exists ods_base_region;
create external table ods_base_region (
    `id`    bigint COMMENT '編號',
    `name`    string COMMENT '地區名稱'
```

```
)   COMMENT '地區表'
row format delimited fields terminated by '\t'
STORED AS
  INPUTFORMAT 'com.hadoop.mapred.DeprecatedLzoTextInputFormat'
  OUTPUTFORMAT 'org.apache.hadoop.hive.ql.io.HiveIgnoreKeyTextOutputFormat'
location '/warehouse/gmall/ods/ods_base_region/';
```

11. 建立品牌表

```
hive (gmall)>
drop table if exists ods_base_trademark;
create external table ods_base_trademark (
    `id`   bigint COMMENT 'id',
    `tm_name` string COMMENT '品牌名稱'
)   COMMENT '品牌表'
PARTITIONED BY (`dt` string)
row format delimited fields terminated by '\t'
STORED AS
  INPUTFORMAT 'com.hadoop.mapred.DeprecatedLzoTextInputFormat'
  OUTPUTFORMAT 'org.apache.hadoop.hive.ql.io.HiveIgnoreKeyTextOutputFormat'
location '/warehouse/gmall/ods/ods_base_trademark/';
```

12. 建立訂單狀態表

```
hive (gmall)>
drop table if exists ods_order_status_log;
create external table ods_order_status_log (
    `id`   bigint COMMENT '編號',
    `order_id` string COMMENT '訂單編號',
    `order_status` string COMMENT '訂單狀態',
    `operate_time` string COMMENT '操作時間'
)   COMMENT '訂單狀態表'
PARTITIONED BY (`dt` string)
row format delimited fields terminated by '\t'
STORED AS
  INPUTFORMAT 'com.hadoop.mapred.DeprecatedLzoTextInputFormat'
  OUTPUTFORMAT 'org.apache.hadoop.hive.ql.io.HiveIgnoreKeyTextOutputFormat'
location '/warehouse/gmall/ods/ods_order_status_log/';
```

13.建立 SPU 商品表

```
hive (gmall)>
drop table if exists ods_spu_info;
create external table ods_spu_info(
    `id` string COMMENT '商品id',
    `spu_name` string COMMENT '標準產品單位名稱',
    `category3_id` string COMMENT '三級品項id',
    `tm_id` string COMMENT '品牌id'
) COMMENT 'SPU商品表'
PARTITIONED BY (`dt` string)
row format delimited fields terminated by '\t'
STORED AS
  INPUTFORMAT 'com.hadoop.mapred.DeprecatedLzoTextInputFormat'
  OUTPUTFORMAT 'org.apache.hadoop.hive.ql.io.HiveIgnoreKeyTextOutputFormat'
location '/warehouse/gmall/ods/ods_spu_info/';
```

14.建立商品評價表

```
hive (gmall)>
drop table if exists ods_comment_info;
create external table ods_comment_info(
    `id` string COMMENT '編號',
    `user_id` string COMMENT '使用者id',
    `sku_id` string COMMENT '商品id',
    `spu_id` string COMMENT '標準產品單位id',
    `order_id` string COMMENT '訂單編號',
    `appraise` string COMMENT '評價',
    `create_time` string COMMENT '建立時間'
) COMMENT '商品評價表'
PARTITIONED BY (`dt` string)
row format delimited fields terminated by '\t'
STORED AS
  INPUTFORMAT 'com.hadoop.mapred.DeprecatedLzoTextInputFormat'
  OUTPUTFORMAT 'org.apache.hadoop.hive.ql.io.HiveIgnoreKeyTextOutputFormat'
location '/warehouse/gmall/ods/ods_comment_info/';
```

15.建立退款表

```
hive (gmall)>
drop table if exists ods_order_refund_info;
create external table ods_order_refund_info(
    `id` string COMMENT '編號',
    `user_id` string COMMENT '使用者id',
    `order_id` string COMMENT '訂單編號',
    `sku_id` string COMMENT '商品id',
    `refund_type` string COMMENT '退款類型',
    `refund_num` bigint COMMENT '退款件數',
    `refund_amount` decimal(16,2) COMMENT '退款金額',
    `refund_reason_type` string COMMENT '退款原因類型',
    `create_time` string COMMENT '建立時間'
) COMMENT '退款表'
PARTITIONED BY (`dt` string)
row format delimited fields terminated by '\t'
STORED AS
  INPUTFORMAT 'com.hadoop.mapred.DeprecatedLzoTextInputFormat'
  OUTPUTFORMAT 'org.apache.hadoop.hive.ql.io.HiveIgnoreKeyTextOutputFormat'
location '/warehouse/gmall/ods/ods_order_refund_info/';
```

16.建立加購表

```
hive (gmall)>
drop table if exists ods_cart_info;
create external table ods_cart_info(
    `id` string COMMENT '編號',
    `user_id` string  COMMENT '使用者id',
    `sku_id` string  COMMENT '商品id',
    `cart_price` string  COMMENT '放入購物車時的價格',
    `sku_num` string  COMMENT '數量',
    `sku_name` string  COMMENT '商品名稱 (容錯)',
    `create_time` string  COMMENT '建立時間',
    `operate_time` string COMMENT '操作時間',
    `is_ordered` string COMMENT '是否已經下單',
    `order_time` string  COMMENT '下單時間'
) COMMENT '加購表'
PARTITIONED BY (`dt` string)
```

```
row format delimited fields terminated by '\t'
STORED AS
  INPUTFORMAT 'com.hadoop.mapred.DeprecatedLzoTextInputFormat'
  OUTPUTFORMAT 'org.apache.hadoop.hive.ql.io.HiveIgnoreKeyTextOutputFormat'
location '/warehouse/gmall/ods/ods_cart_info/';
```

17. 建立商品收藏表

```
hive (gmall)>
drop table if exists ods_favor_info;
create external table ods_favor_info(
    `id` string COMMENT '編號',
    `user_id` string  COMMENT '使用者id',
    `sku_id` string  COMMENT '商品id',
    `spu_id` string  COMMENT '標準產品單位id',
    `is_cancel` string  COMMENT '是否取消',
    `create_time` string  COMMENT '收藏時間',
    `cancel_time` string  COMMENT '取消時間'
) COMMENT '商品收藏表'
PARTITIONED BY (`dt` string)
row format delimited fields terminated by '\t'
STORED AS
  INPUTFORMAT 'com.hadoop.mapred.DeprecatedLzoTextInputFormat'
  OUTPUTFORMAT 'org.apache.hadoop.hive.ql.io.HiveIgnoreKeyTextOutputFormat'
location '/warehouse/gmall/ods/ods_favor_info/';
```

18. 建立優惠券領用表

```
hive (gmall)>
drop table if exists ods_coupon_use;
create external table ods_coupon_use(
    `id` string COMMENT '編號',
    `coupon_id` string  COMMENT '優惠券id',
    `user_id` string  COMMENT '使用者id',
    `order_id` string  COMMENT '訂單編號',
    `coupon_status` string  COMMENT '優惠券狀態',
    `get_time` string  COMMENT '領取時間',
    `using_time` string  COMMENT '使用(下單)時間',
    `used_time` string  COMMENT '使用(支付)時間'
```

```
) COMMENT '優惠券領用表'
PARTITIONED BY (`dt` string)
row format delimited fields terminated by '\t'
STORED AS
  INPUTFORMAT 'com.hadoop.mapred.DeprecatedLzoTextInputFormat'
  OUTPUTFORMAT 'org.apache.hadoop.hive.ql.io.HiveIgnoreKeyTextOutputFormat'
location '/warehouse/gmall/ods/ods_coupon_use/';
```

19.建立優惠券表

```
hive (gmall)>
drop table if exists ods_coupon_info;
create external table ods_coupon_info(
  `id` string COMMENT '優惠券編號',
  `coupon_name` string COMMENT '優惠券名稱',
  `coupon_type` string COMMENT '優惠券類型 1 現金券 2 折扣券 3 滿減券 4
滿件打折券',
  `condition_amount` string COMMENT '滿減金額',
  `condition_num` string COMMENT '滿減件數',
  `activity_id` string COMMENT '活動編號',
  `benefit_amount` string COMMENT '優惠金額',
  `benefit_discount` string COMMENT '優惠折扣',
  `create_time` string COMMENT '建立時間',
  `range_type` string COMMENT '範圍類型1商品2品項3品牌',
  `spu_id` string COMMENT '標準產品單位id',
  `tm_id` string COMMENT '品牌id',
  `category3_id` string COMMENT '三級品項id',
  `limit_num` string COMMENT '最多領用次數',
  `operate_time` string COMMENT '操作時間',
  `expire_time` string COMMENT '過期時間'
) COMMENT '優惠券表'
PARTITIONED BY (`dt` string)
row format delimited fields terminated by '\t'
STORED AS
  INPUTFORMAT 'com.hadoop.mapred.DeprecatedLzoTextInputFormat'
  OUTPUTFORMAT 'org.apache.hadoop.hive.ql.io.HiveIgnoreKeyTextOutputFormat'
location '/warehouse/gmall/ods/ods_coupon_info/';
```

20. 建立活動表

```
hive (gmall)>
drop table if exists ods_activity_info;
create external table ods_activity info(
    `id` string COMMENT '活動id',
    `activity_name` string  COMMENT '活動名稱',
    `activity_type` string  COMMENT '活動類型',
    `start_time` string  COMMENT '開始時間',
    `end_time` string  COMMENT '結束時間',
    `create_time` string  COMMENT '建立時間'
) COMMENT '活動表'
PARTITIONED BY (`dt` string)
row format delimited fields terminated by '\t'
STORED AS
  INPUTFORMAT 'com.hadoop.mapred.DeprecatedLzoTextInputFormat'
  OUTPUTFORMAT 'org.apache.hadoop.hive.ql.io.HiveIgnoreKeyTextOutputFormat'
location '/warehouse/gmall/ods/ods_activity_info/';
```

21. 建立活動訂單連結表

```
hive (gmall)>
drop table if exists ods_activity_order;
create external table ods_activity_order(
    `id` string COMMENT '編號',
    `activity_id` string  COMMENT '活動id',
    `order_id` string  COMMENT '訂單編號',
    `create_time` string  COMMENT '建立時間'
) COMMENT '活動訂單連結表'
PARTITIONED BY (`dt` string)
row format delimited fields terminated by '\t'
STORED AS
  INPUTFORMAT 'com.hadoop.mapred.DeprecatedLzoTextInputFormat'
  OUTPUTFORMAT 'org.apache.hadoop.hive.ql.io.HiveIgnoreKeyTextOutputFormat'
location '/warehouse/gmall/ods/ods_activity_order/';
```

22.建立優惠規則表

```
hive (gmall)>
drop table if exists ods_activity_rule;
create external table ods_activity_rule(
    `id` string COMMENT '編號',
    `activity_id` string  COMMENT '活動id',
    `condition_amount` string  COMMENT '滿減金額',
    `condition_num` string  COMMENT '滿減件數',
    `benefit_amount` string  COMMENT '優惠金額',
    `benefit_discount` string  COMMENT '優惠折扣',
    `benefit_level` string  COMMENT '優惠等級'
) COMMENT '優惠規則表'
PARTITIONED BY (`dt` string)
row format delimited fields terminated by '\t'
STORED AS
  INPUTFORMAT 'com.hadoop.mapred.DeprecatedLzoTextInputFormat'
  OUTPUTFORMAT 'org.apache.hadoop.hive.ql.io.HiveIgnoreKeyTextOutputFormat'
location '/warehouse/gmall/ods/ods_activity_rule/';
```

23.建立編碼字典表

```
hive (gmall)>
drop table if exists ods_base_dic;
create external table ods_base_dic(
    `dic_code` string COMMENT '編號',
    `dic_name` string  COMMENT '編碼名稱',
    `parent_code` string  COMMENT '父編碼',
    `create_time` string  COMMENT '建立時間',
    `operate_time` string  COMMENT '修改時間'
) COMMENT '編碼字典表'
PARTITIONED BY (`dt` string)
row format delimited fields terminated by '\t'
STORED AS
  INPUTFORMAT 'com.hadoop.mapred.DeprecatedLzoTextInputFormat'
  OUTPUTFORMAT 'org.apache.hadoop.hive.ql.io.HiveIgnoreKeyTextOutputFormat'
location '/warehouse/gmall/ods/ods_base_dic/';
```

24.建立參與活動商品表

```
hive (gmall)>
drop table if exists ods_activity_sku;
create external table ods_activity_sku (
    `id` string COMMENT '編號',
    `activity_id` string  COMMENT '活動id',
    `sku_id` string  COMMENT '商品id',
    `create_time` string  COMMENT '建立時間'
) COMMENT '參與活動商品表'
PARTITIONED BY (`dt` string)
row format delimited fields terminated by '\t'
STORED AS
  INPUTFORMAT 'com.hadoop.mapred.DeprecatedLzoTextInputFormat'
  OUTPUTFORMAT 'org.apache.hadoop.hive.ql.io.HiveIgnoreKeyTextOutputFormat'
location '/warehouse/gmall/ods/ods_activity_sku/';
```

6.3.5 ODS 層業務資料匯入指令稿

將 ODS 層業務資料的載入過程撰寫成指令稿，方便每日呼叫執行。

（1）在 /home/atguigu/bin 目錄下建立指令稿 hdfs_to_ods_db.sh。

```
[atguigu@hadoop102 bin]$ vim hdfs_to_ods_db.sh
```

在指令稿中撰寫以下內容。

```
#!/bin/bash

APP=gmall
hive=/opt/module/hive/bin/hive

#若輸入了日期參數，則取輸入參數作為日期值；若沒有輸入日期參數，則取目前時間的
前一天作為日期值
if [ -n "$2" ] ;then
    do_date=$2
else
    do_date=`date -d "-1 day" +%F`
fi
```

```
sql1="
load data inpath '/origin_data/$APP/db/order_info/$do_date' OVERWRITE
into table ${APP}.ods_order_info partition(dt='$do_date');

load data inpath '/origin_data/$APP/db/order_detail/$do_date' OVERWRITE
into table ${APP}.ods_order_detail partition(dt='$do_date');

load data inpath '/origin_data/$APP/db/sku_info/$do_date' OVERWRITE into
table ${APP}.ods_sku_info partition(dt='$do_date');

load data inpath '/origin_data/$APP/db/user_info/$do_date' OVERWRITE into
table ${APP}.ods_user_info partition(dt='$do_date');

load data inpath '/origin_data/$APP/db/payment_info/$do_date' OVERWRITE
into table ${APP}.ods_payment_info partition(dt='$do_date');

load data inpath '/origin_data/$APP/db/base_category1/$do_date' OVERWRITE
into table ${APP}.ods_base_category1 partition(dt='$do_date');

load data inpath '/origin_data/$APP/db/base_category2/$do_date' OVERWRITE
into table ${APP}.ods_base_category2 partition(dt='$do_date');

load data inpath '/origin_data/$APP/db/base_category3/$do_date' OVERWRITE
into table ${APP}.ods_base_category3 partition(dt='$do_date');

load data inpath '/origin_data/$APP/db/base_trademark/$do_date' OVERWRITE
into table ${APP}.ods_base_trademark partition(dt='$do_date');

load data inpath '/origin_data/$APP/db/activity_info/$do_date' OVERWRITE
into table ${APP}.ods_activity_info partition(dt='$do_date');

load data inpath '/origin_data/$APP/db/activity_order/$do_date' OVERWRITE
into table ${APP}.ods_activity_order partition(dt='$do_date');

load data inpath '/origin_data/$APP/db/cart_info/$do_date' OVERWRITE into
table ${APP}.ods_cart_info partition(dt='$do_date');

load data inpath '/origin_data/$APP/db/comment_info/$do_date' OVERWRITE
```

```
into table ${APP}.ods_comment_info partition(dt='$do_date');

load data inpath '/origin_data/$APP/db/coupon_info/$do_date' OVERWRITE
into table ${APP}.ods_coupon_info partition(dt='$do_date');

load data inpath '/origin_data/$APP/db/coupon_use/$do_date' OVERWRITE
into table ${APP}.ods_coupon_use partition(dt='$do_date');

load data inpath '/origin_data/$APP/db/favor_info/$do_date' OVERWRITE
into table ${APP}.ods_favor_info partition(dt='$do_date');

load data inpath '/origin_data/$APP/db/order_refund_info/$do_date' OVERWRITE
into table ${APP}.ods_order_refund_info partition(dt='$do_date');

load data inpath '/origin_data/$APP/db/order_status_log/$do_date'
OVERWRITE into table ${APP}.ods_order_status_log partition(dt='$do_date');

load data inpath '/origin_data/$APP/db/spu_info/$do_date' OVERWRITE into
table ${APP}.ods_spu_info partition(dt='$do_date');

load data inpath '/origin_data/$APP/db/activity_rule/$do_date' OVERWRITE
into table ${APP}.ods_activity_rule partition(dt='$do_date');

load data inpath '/origin_data/$APP/db/base_dic/$do_date' OVERWRITE into
table ${APP}.ods_base_dic partition(dt='$do_date');
load data inpath '/origin_data/$APP/db/activity_sku/$do_date' OVERWRITE
into table ${APP}.ods_activity_sku partition(dt='$do_date');
"

sql2="
load data inpath '/origin_data/$APP/db/base_province/$do_date' OVERWRITE
into table ${APP}.ods_base_province;

load data inpath '/origin_data/$APP/db/base_region/$do_date' OVERWRITE
into table ${APP}.ods_base_region;
"
case $1 in
"first"){
    $hive -e "$sql1"
```

```
    $hive -e "$sql2"
};;
"all"){
    $hive -e "$sql1"
};;
esac
```

（2）增加指令稿執行許可權。

```
[atguigu@hadoop102 bin]$ chmod 777 hdfs_to_ods_db.sh
```

（3）初次執行指令稿，傳入 first，匯入 2020-03-10 的資料。

```
[atguigu@hadoop102 bin]$ hdfs_to_ods_db.sh first 2020-03-10
```

（4）再次執行指令稿，傳入 all，匯入 2020-03-11 的資料。

```
[atguigu@hadoop102 bin]$ hdfs_to_ods_db.sh all 2020-03-11
```

（5）測試資料是否匯入成功。

```
hive (gmall)> select * from ods_order_detail where dt='2020-03-11' limit 10;
```

6.4 資料倉儲架設——DWD 層

對 ODS 層的資料進行判空過濾，然後對商品品項表進行維度退化（降維），使用 Parquet 格式進行儲存，並儲存為 LZO 壓縮格式，以減少儲存空間的佔用。

6.4.1 使用者行為啟動記錄表解析

在第 4 章中，透過記錄檔產生程式產生的啟動記錄資料範例如下。

```
{
    "mid":"995",
    "uid":"995",
    "vc":"10",
    "vn":"1.3.4",
    "l":"en",
```

```
    "sr":"B",
    "os":"8.1.2",
    "ar":"MX",
    "md":"HTC-2",
    "ba":"HTC",
        "sv":"V2.0.6",
    "g":"43R2SEQX@gmail.com",
    "hw":"640*960",
    "t":"1561472502444",
    "nw":"4G",
    "ln":"-99.3",
    "la":"20.4",
    "entry":"2",
    "open_ad_type":"2",
    "action":"1",
    "loading_time":"2",
    "detail":"",
    "extend1":"",
    "en":"start",
}
```

記錄檔資料為 JSON 格式，Hive 內建了 JSON 字串解析工具，進一步可以獲
得字串內欄位的對應資訊，根據範例資料中的欄位資訊，可確定啟動記錄表
中所包含的欄位。啟動記錄表為分區表，以日期為分區，方便使用者進行分
區查詢。將資料儲存為 Parquet 格式，並儲存為 LZO 壓縮格式，可大幅減少
儲存空間的佔用。

1. 建立啟動記錄表

在 gmall 資料庫中執行以下建表敘述。

```
hive (gmall)>
drop table if exists dwd_start_log;
CREATE EXTERNAL TABLE dwd_start_log( --欄位為範例資料中出現的欄位
    `mid_id` string,
    `user_id` string,
    `version_code` string,
    `version_name` string,
```

```
    `lang` string,
    `source` string,
    `os` string,
    `area` string,
    `model` string,
    `brand` string,
    `sdk_version` string,
    `gmail` string,
    `height_width` string,
    `app_time` string,
    `network` string,
    `lng` string,
    `lat` string,
    `entry` string,
    `open_ad_type` string,
    `action` string,
    `loading_time` string,
    `detail` string,
    `extend1` string
)
PARTITIONED BY (dt string) --以日期為分區
stored as  parquet  --儲存格式為Parquet
location '/warehouse/gmall/dwd/dwd_start_log/' --指定儲存目錄
tblproperties ("parquet.compression"="lzo") --指定壓縮格式為LZO
;
```

2. 向啟動記錄表中匯入資料

（1）使用 insert 敘述向已經建立的啟動記錄表中匯入資料，其中，欄位資訊由 Hive 內建的 JSON 字串解析函數 get_json_object() 解析獲得。get_json_object() 函數的第一個參數填寫 JSON 物件變數，第二個參數使用 $ 表示 JSON 變數標識。

```
hive (gmall)>
insert overwrite table dwd_start_log -PARTITION (dt='2020-03-10') --指定分區
select
    --使用Hive內建的JSON字串解析函數解析JSON字串，取得欄位值
    get_json_object(line,'$.mid') mid_id,
```

```
    get_json_object(line,'$.uid') user_id,
    get_json_object(line,'$.vc') version_code,
    get_json_object(line,'$.vn') version_name,
    get_json_object(line,'$.l') lang,
    get_json_object(line,'$.sr') source,
    get_json_object(line,'$.os') os,
    get_json_object(line,'$.ar') area,
    get_json_object(line,'$.md') model,
    get_json_object(line,'$.ba') brand,
    get_json_object(line,'$.sv') sdk_version,
    get_json_object(line,'$.g') gmail,
    get_json_object(line,'$.hw') height_width,
    get_json_object(line,'$.t') app_time,
    get_json_object(line,'$.nw') network,
    get_json_object(line,'$.ln') lng,
    get_json_object(line,'$.la') lat,
    get_json_object(line,'$.entry') entry,
    get_json_object(line,'$.open_ad_type') open_ad_type,
    get_json_object(line,'$.action') action,
    get_json_object(line,'$.loading_time') loading_time,
    get_json_object(line,'$.detail') detail,
    get_json_object(line,'$.extend1') extend1
from ods_start_log  --資料來源為ODS層的啟動記錄表
where dt='2020-03-10'; --指定資料日期
```

（2）測試。

```
hive (gmall)> select * from dwd_start_log limit 2;
```

3. DWD 層啟動記錄表載入資料指令稿

（1）在 hadoop102 的 /home/atguigu/bin 目錄下建立指令稿 dwd_start_log.sh。

```
[atguigu@hadoop102 bin]$ vim dwd_start_log.sh
```

在指令稿中撰寫以下內容。

```
#!/bin/bash
# 定義變數，方便後續修改
APP=gmall
```

```
hive=/opt/module/hive/bin/hive

#如果輸入了日期參數，則取輸入參數作為日期值；如果沒有輸入日期參數，則取目前時
間的前一天作為日期值
if [ -n "$1" ] ;then
 do_date=$1
else
 do_date=`date -d "-1 day" +%F`
fi

sql="
insert overwrite table "$APP".dwd_start_log
PARTITION (dt='$do_date')
select
    get_json_object(line,'$.mid') mid_id,
    get_json_object(line,'$.uid') user_id,
    get_json_object(line,'$.vc') version_code,
    get_json_object(line,'$.vn') version_name,
    get_json_object(line,'$.l') lang,
    get_json_object(line,'$.sr') source,
    get_json_object(line,'$.os') os,
    get_json_object(line,'$.ar') area,
    get_json_object(line,'$.md') model,
    get_json_object(line,'$.ba') brand,
    get_json_object(line,'$.sv') sdk_version,
    get_json_object(line,'$.g') gmail,
    get_json_object(line,'$.hw') height_width,
    get_json_object(line,'$.t') app_time,
    get_json_object(line,'$.nw') network,
    get_json_object(line,'$.ln') lng,
    get_json_object(line,'$.la') lat,
    get_json_object(line,'$.entry') entry,
    get_json_object(line,'$.open_ad_type') open_ad_type,
    get_json_object(line,'$.action') action,
    get_json_object(line,'$.loading_time') loading_time,
    get_json_object(line,'$.detail') detail,
    get_json_object(line,'$.extend1') extend1
from "$APP".ods_start_log
where dt='$do_date';
```

```
"
$hive -e "$sql"
```

（2）增加指令稿執行許可權。

```
[atguigu@hadoop102 bin]$ chmod 777 dwd_start_log.sh
```

（3）執行指令稿，匯入資料。

```
[atguigu@hadoop102 module]$ dwd_start_log.sh 2020-03-11
```

（4）查詢結果資料。

```
hive (gmall)>
select * from dwd_start_log where dt='2020-03-11' limit 2;
```

6.4.2　使用者行為事件表拆分

我們對 ODS 層的事件表進行解析的基本想法如圖 6-18 和圖 6-19 所示，透過自訂 UDF 函數和自訂 UDTF 函數將事件表解析成 dwd_base_event_log 表，再將 dwd_base_event_log 表拆解成分類事件表。

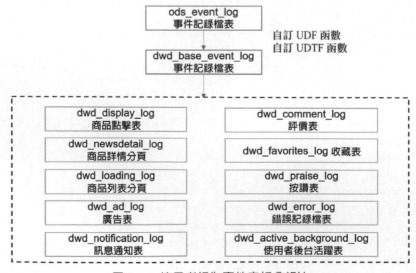

圖 6-18　使用者行為事件表拆分想法

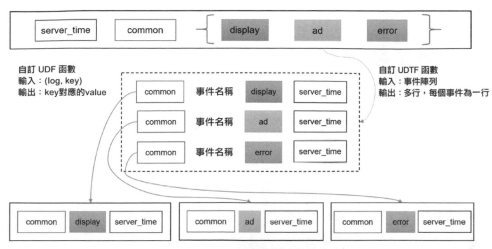

圖 6-19 自訂函數解析記錄檔想法

如圖 6-20 所示為 DWD 層使用者行為事件表資料解析想法。

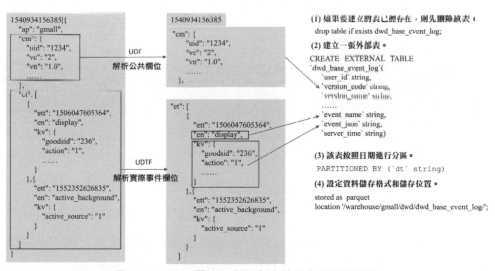

圖 6-20 DWD 層使用者行為事件表資料解析想法

1. 建立基礎明細表

基礎明細表用於儲存從 ODS 層原始表中轉換過來的明細資料。

建立事件記錄檔基礎明細表。

```
hive (gmall)>
drop table if exists dwd_base_event_log;
CREATE EXTERNAL TABLE dwd_base_event_log(
    `mid_id` string,
    `user_id` string,
    `version_code` string,
    `version_name` string,
    `lang` string,
    `source` string,
    `os` string,
    `area` string,
    `model` string,
    `brand` string,
    `sdk_version` string,
    `gmail` string,
    `height_width` string,
    `app_time` string,
    `network` string,
    `lng` string,
    `lat` string,
    `event_name` string,
    `event_json` string,
    `server_time` string
)
PARTITIONED BY (`dt` string)
stored as parquet
location '/warehouse/gmall/dwd/dwd_base_event_log/'
TBLPROPERTIES('parquet.compression'='lzo');
```

其中，event_name 和 event_json 分別對應事件名稱和整個事件。這個地方將原始記錄檔一對多的形式拆分出來了，操作的時候我們需要將原始記錄檔進行解析，將用到 UDF 和 UDTF。

UDF 是什麼？
UDF 的全稱為 User-Defined Function，意思為使用者定義函數，它的作用是什麼呢？有的時候，使用者的需求無法透過 Hive 提供的內建函數來實現，這時候透過撰寫 UDF，使用者可以方便地定義自己需要的處理邏輯，並在查詢中使用它們。

UDTF 是什麼？

Hive 中有三種 UDF：（普通）UDF、使用者定義聚集函數（User-Defined Aggregate Function，UDAF）和使用者定義表產生函數（User-Defined Table-generating Function，UDTF）。UDTF 操作作用於單一資料行，並且將產生多個資料行作為輸出。

2. 自訂 UDF 函數（解析公共欄位）

自訂 UDF 函數，解析公共欄位，想法如圖 6-21 所示。

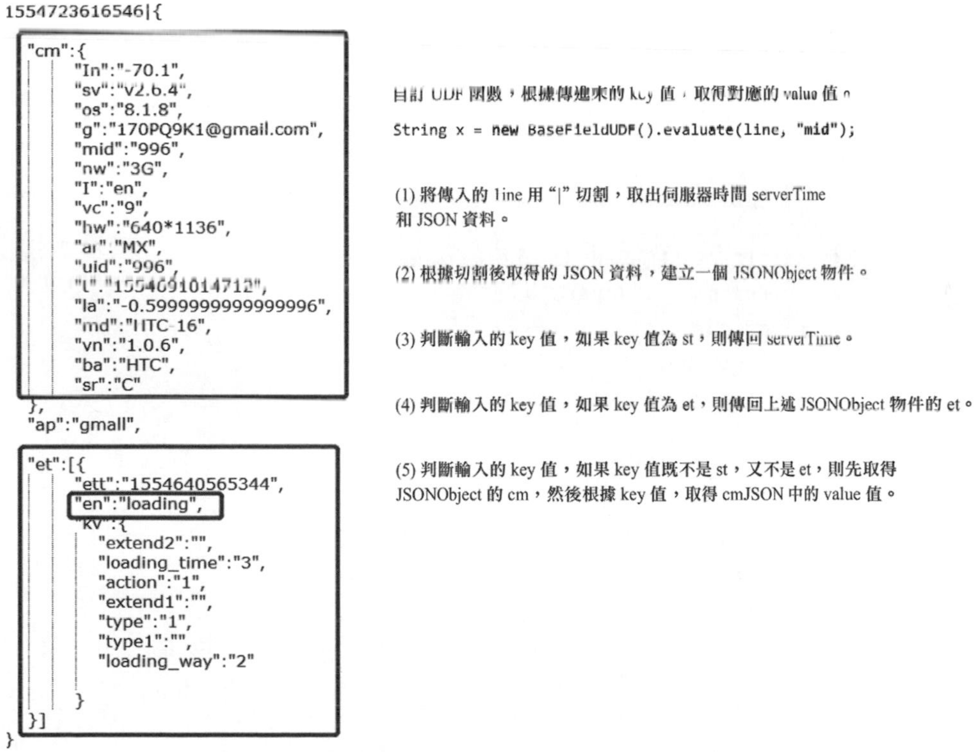

```
1554723616546|{
    "cm":{
        "ln":"-70.1",
        "sv":"v2.6.4",
        "os":"8.1.8",
        "g":"170PQ9K1@gmail.com",
        "mid":"996",
        "nw":"3G",
        "l":"en",
        "vc":"9",
        "hw":"640*1136",
        "ar":"MX",
        "uid":"996",
        "t":"1554691014712",
        "la":"-0.5999999999999996",
        "md":"HTC-16",
        "vn":"1.0.6",
        "ba":"HTC",
        "sr":"C"
    },
    "ap":"gmall",
    "et":[{
        "ett":"1554640565344",
        "en":"loading",
        "kv":{
            "extend2":"",
            "loading_time":"3",
            "action":"1",
            "extend1":"",
            "type":"1",
            "type1":"",
            "loading_way":"2"
        }
    }]
}
```

自訂 UDF 函數，根據傳進來的 key 值，取得對應的 value 值。

String x = new BaseFieldUDF().evaluate(line, "mid");

(1) 將傳入的 line 用 "|" 切割，取出伺服器時間 serverTime 和 JSON 資料。

(2) 根據切割後取得的 JSON 資料，建立一個 JSONObject 物件。

(3) 判斷輸入的 key 值，如果 key 值為 st，則傳回 serverTime。

(4) 判斷輸入的 key 值，如果 key 值為 et，則傳回上述 JSONObject 物件的 et。

(5) 判斷輸入的 key 值，如果 key 值既不是 st，又不是 et，則先取得 JSONObject 的 cm，然後根據 key 值，取得 cmJSON 中的 value 值。

圖 6-21 自訂 UDF 函數解析公共欄位的想法

（1）建立一個 maven 專案：hivefunction。

（2）建立套件名稱：com.atguigu.udf。

（3）在 pom.xml 檔案中增加以下內容。

```xml
<properties>
    <project.build.sourceEncoding>UTF8</project.build.sourceEncoding>
    <hive.version>1.2.1</hive.version>
</properties>

<dependencies>
    <!--增加Hive依賴-->
    <dependency>
        <groupId>org.apache.hive</groupId>
        <artifactId>hive-exec</artifactId>
        <version>${hive.version}</version>
    </dependency>
</dependencies>

<build>
    <plugins>
        <plugin>
            <artifactId>maven-compiler-plugin</artifactId>
            <version>2.3.2</version>
            <configuration>
                <source>1.8</source>
                <target>1.8</target>
            </configuration>
        </plugin>
        <plugin>
            <artifactId>maven-assembly-plugin</artifactId>
            <configuration>
                <descriptorRefs>
                    <descriptorRef>jar-with-dependencies</descriptorRef>
                </descriptorRefs>
            </configuration>
            <executions>
                <execution>
                    <id>make-assembly</id>
                    <phase>package</phase>
```

```
                    <goals>
                        <goal>single</goal>
                    </goals>
                </execution>
            </executions>
        </plugin>
    </plugins>
</build>
```

（4）自訂 UDF 函數，解析公共欄位。

```java
package com.atguigu.udf;
import org.apache.commons.lang.StringUtils;
import org.apache.hadoop.hive.ql.exec.UDF;
import org.json.JSONException;
import org.json.JSONObject;

public class BaseFieldUDF extends UDF {

    public String evaluate(String line, String key) throws JSONException {

        // 1按"\\|"對記錄檔line進行切割
        String[] log = line.split("\\|");

        // 2 合法性驗證
        if (log.length != 2 || StringUtils.isBlank(log[1])) {
            return "";
        }

        // 3 開始處理JSON
        JSONObject baseJson = new JSONObject(log[1].trim());

        String result = "";

        // 4 根據傳進來的key值，尋找對應的value值
        if ("et".equals(key)) {
            if (baseJson.has("et")) {
                result = baseJson.getString("et");
            }
```

```
        } else if ("st".equals(key)) {
            result = log[0].trim();
        } else {
            JSONObject cm = baseJson.getJSONObject("cm");
            if (cm.has(key)) {
                result = cm.getString(key);
            }
        }


        return result;
}
 public static void main(String[] args) {

        String line = "1541217850324|{\"cm\":{\"mid\":\"m7856\",\"uid\":
\"u8739\",\"ln\":\"-74.8\",\"sv\":\"V2.2.2\",\"os\":\"8.1.3\",\"g\":
\"P7XC9126@gmail.com\",\"nw\":\"3G\",\"l\":\"es\",\"vc\":\"6\",\"hw\":
\"640*960\",\"ar\":\"MX\",\"t\":\"1541204134250\",\"la\":\"-31.7\",
\"md\":\"huawei-17\",\"vn\":\"1.1.2\",\"sr\":\"O\",\"ba\":\"Huawei\"},
\"ap\":\"weather\",\"et\":[{\"ett\":\"1541146624055\",\"en\":\"display\",
\"kv\":{\"goodsid\":\"n4195\",\"copyright\":\"ESPN\",\"content_provider\":
\"CNN\",\"extend2\":\"5\",\"action\":\"2\",\"extend1\":\"2\",\"place\":
\"3\",\"showtype\":\"2\",\"category\":\"72\",\"newstype\":\"5\"}},
{\"ett\":\"1541213331817\",\"en\":\"loading\",\"kv\":{\"extend2\":\"\",
\"loading_time\":\"15\",\"action\":\"3\",\"extend1\":\"\",\"type1\":\"\",
\"type\":\"3\",\"loading_way\":\"1\"}},{\"ett\":\"1541126195645\",\"en\":
\"ad\",\"kv\":{\"entry\":\"3\",\"show_style\":\"0\",\"action\":\"2\",
\"detail\":\"325\",\"source\":\"4\",\"behavior\":\"2\",\"content\":\"1\",
\"newstype\":\"5\"}},{\"ett\":\"1541202678812\",\"en\":\"notification\",
\"kv\":{\"ap_time\":\"1541184614380\",\"action\":\"3\",\"type\":\"4\",
\"content\":\"\"}},{\"ett\":\"1541194686688\",\"en\":\"active_background\",
\"kv\":{\"active_source\":\"3\"}}]}";
        String x = new BaseFieldUDF().evaluate(line, "mid,uid,vc,vn,l,
sr,os,ar,md,ba,sv,g,hw,nw,ln,la,t");
        System.out.println(x);
    }
}
```

★ **注意**：main() 函數主要用於模擬資料測試。

3. 自訂 UDTF 函數（解析實際事件欄位）

自訂 UDTF 函數，解析實際事件欄位，取得事件名稱，想法如圖 6-22 所示。

圖 6-22 自訂 UDTF 函數取得事件名稱的想法

（1）建立套件名稱：com.atguigu.udtf。

（2）在 com.atguigu.udtf 套件下建立類別名稱：EventJsonUDTF。

（3）EventJsonUDTF 繼承 GenericUDTF 類別，並重新定義該抽象類別的相關
方法。

```
package com.atguigu.udtf;

import org.apache.commons.lang.StringUtils;
import org.apache.hadoop.hive.ql.exec.UDFArgumentException;
import org.apache.hadoop.hive.ql.metadata.HiveException;
import org.apache.hadoop.hive.ql.udf.generic.GenericUDTF;
import org.apache.hadoop.hive.serde2.objectinspector.ObjectInspector;
import org.apache.hadoop.hive.serde2.objectinspector.ObjectInspectorFactory;
import org.apache.hadoop.hive.serde2.objectinspector.StructObjectInspector;
import org.apache.hadoop.hive.serde2.objectinspector.primitive.
PrimitiveObjectInspectorFactory;
```

```java
import org.json.JSONArray;
import org.json.JSONException;

import java.util.ArrayList;

public class EventJsonUDTF extends GenericUDTF {

    //在該方法中，我們將指定輸出參數的名稱和參數類型
  @Override
    public StructObjectInspector initialize(ObjectInspector[] argOIs)
throws UDFArgumentException {

        ArrayList<String> fieldNames = new ArrayList<String>();
        ArrayList<ObjectInspector> fieldOIs = new ArrayList
<ObjectInspector>();

        fieldNames.add("event_name");
        fieldOIs.add(PrimitiveObjectInspectorFactory.
javaStringObjectInspector);
        fieldNames.add("event_json");
        fieldOIs.add(PrimitiveObjectInspectorFactory.
javaStringObjectInspector);

        return ObjectInspectorFactory.getStandardStructObjectInspector(fieldNames,
fieldOIs);
    }

    //輸入1筆記錄，輸出許多條結果
  @Override
    public void process(Object[] objects) throws HiveException {

        // 取得傳入的et
    String input = objects[0].toString();

        // 如果傳進來的資料為空，則直接傳回，過濾掉該資料
    if (StringUtils.isBlank(input)) {
            return;
        } else {
```

```java
        try {
            // 取得事件的個數（ad/facoriters）
    JSONArray ja = new JSONArray(input);

            if (ja == null)
                return;

            // 循環檢查每個事件
    for (int i = 0; i < ja.length(); i++) {
                String[] result = new String[2];

                try {
                    // 取出每個事件的名稱（ad/facoriters）
        result[0] = ja.getJSONObject(i).getString("en");

                    // 取出每個事件整體
        result[1] = ja.getString(i);
                } catch (JSONException e) {
                    continue;
                }

                // 將結果傳回
        forward(result);
            }
        } catch (JSONException e) {
            e.printStackTrace();
        }
    }
  }

    //當沒有記錄處理的時候，該方法會被呼叫，用來清理程式或產生額外的輸出
    @Override
    public void close() throws HiveException {

    }
}
```

（4）包裝自訂函數，如圖 6-23 所示。

圖 6-23 包裝自訂函數

（5）將 hivefunction-1.0-SNAPSHOT 上傳到 HDFS 的 /user/hive/jars 目錄下。

（6）建立永久函數，與建立好的自訂函數類別進行連結。

```
hive (gmall)>
create function base_analizer as 'com.atguigu.udf.BaseFieldUDF' using jar
'hdfs://hadoop102:9000/user/hive/jars/hivefunction-1.0-SNAPSHOT.jar';

create function flat_analizer as 'com.atguigu.udtf.EventJsonUDTF' using
jar 'hdfs://hadoop102:9000/user/hive/jars/hivefunction-1.0-SNAPSHOT.jar';
```

4. 解析事件記錄檔基礎明細表

（1）解析事件記錄檔基礎明細表。

```
hive (gmall)>
insert overwrite table dwd_base_event_log partition(dt='2020-03-10')
select
    base_analizer(line,'mid') as mid_id,
    base_analizer(line,'uid') as user_id,
    base_analizer(line,'vc') as version_code,
    base_analizer(line,'vn') as version_name,
    base_analizer(line,'l') as lang,
    base_analizer(line,'sr') as source,
    base_analizer(line,'os') as os,
    base_analizer(line,'ar') as area,
    base_analizer(line,'md') as model,
    base_analizer(line,'ba') as brand,
    base_analizer(line,'sv') as sdk_version,
    base_analizer(line,'g') as gmail,
    base_analizer(line,'hw') as height_width,
    base_analizer(line,'t') as app_time,
    base_analizer(line,'nw') as network,
```

```
    base_analizer(line,'ln') as lng,
    base_analizer(line,'la') as lat,
    event_name,
    event_json,
    base_analizer(line,'st') as server_time
from ods_event_log lateral view flat_analizer(base_analizer(line,'et'))
tmp_flat as event_name,event_json
where dt='2020-03-10' and base_analizer(line,'et')<>'';
```

（2）測試。

```
hive (gmall)> select * from dwd_base_event_log limit 2;
```

5. DWD 層資料解析指令稿

（1）在 hadoop102 的 /home/atguigu/bin 目錄下建立指令稿 dwd_base_log.sh。

```
[atguigu@hadoop102 bin]$ vim dwd_base_log.sh
```

在指令稿中撰寫以下內容。

```
#!/bin/bash

# 定義變數，方便後續修改
APP=gmall
hive=/opt/module/hive/bin/hive

# 如果輸入了日期參數，則取輸入參數作為日期值；如果沒有輸入日期參數，則取目前
時間的前一天作為日期值
if [ -n "$1" ] ;then
 do_date=$1
else
 do_date=`date -d "-1 day" +%F`
fi

sql="
insert overwrite table "$APP".dwd_base_event_log partition(dt='$do_date')
select
    "$APP".base_analizer(line,'mid') as mid_id,
    "$APP".base_analizer(line,'uid') as user_id,
```

```
    "$APP".base_analizer(line,'vc') as version_code,
    "$APP".base_analizer(line,'vn') as version_name,
    "$APP".base_analizer(line,'l') as lang,
    "$APP".base_analizer(line,'sr') as source,
    "$APP".base_analizer(line,'os') as os,
    "$APP".base_analizer(line,'ar') as area,
    "$APP".base_analizer(line,'md') as model,
    "$APP".base_analizer(line,'ba') as brand,
    "$APP".base_analizer(line,'sv') as sdk_version,
    "$APP".base_analizer(line,'g') as gmail,
    "$APP".base_analizer(line,'hw') as height_width,
    "$APP".base_analizer(line,'t') as app_time,
    "$APP".base_analizer(line,'nw') as network,
    "$APP".base_analizer(line,'ln') as lng,
    "$APP".base_analizer(line,'la') as lat,
    event_name,
    event_json,
    "$APP".base_analizer(line,'st') as server_time
from "$APP".ods_event_log lateral view "$APP".flat_analizer("$APP".base_
analizer(line,'et')) tem_flat as event_name,event_json
where dt='$do_date'  and "$APP".base_analizer(line,'et')<>'';
"

$hive -e "$sql"
```

（2）增加指令稿執行許可權。

```
[atguigu@hadoop102 bin]$ chmod 777 dwd_base_log.sh
```

（3）執行指令稿，匯入資料。

```
[atguigu@hadoop102 module]$ dwd_base_log.sh 2020-03-11
```

（4）查詢結果資料。

```
hive (gmall)>
select * from dwd_base_event_log where dt='2020-03-11' limit 2;
```

6.4.3 使用者行為事件表解析

獲得 DWD 層的事件記錄檔基礎明細表後,可將各實際事件表從事件記錄檔基礎明細表的 event_json 欄位中解析出來。根據不同的事件名稱欄位 event_name,我們需要分別建立商品點擊表、商品詳情分頁、商品列表分頁、廣告表、訊息通知表、使用者後台活躍表、評價表、收藏表、按讚表和錯誤記錄檔表,並根據事件名稱欄位對事件記錄檔基礎明細表進行查詢,然後將查詢結果一一插入建立好的實際事件表中,各實際事件表的建表敘述相似,只有 event_json 解析出的欄位資訊不同,所以不再進行單獨說明,實際操作步驟如下。

1. 商品點擊表

1)建表敘述

```
hive (gmall)>
drop table if exists dwd_display_log;
CREATE EXTERNAL TABLE dwd_display_log(
    `mid_id` string,
    `user_id` string,
    `version_code` string,
    `version_name` string,
    `lang` string,
    `source` string,
    `os` string,
    `area` string,
    `model` string,
    `brand` string,
    `sdk_version` string,
    `gmail` string,
    `height_width` string,
    `app_time` string,
    `network` string,
    `lng` string,
    `lat` string,
    `action` string,
    `goodsid` string,
```

```
    `place` string,
    `extend1` string,
    `category` string,
    `server_time` string
)
PARTITIONED BY (dt string)
location '/warehouse/gmall/dwd/dwd_display_log/';
```

2）匯入資料

```
hive (gmall)>
insert overwrite table dwd_display_log
PARTITION (dt='2020-03-10')
select
    mid_id,
    user_id,
    version_code,
    version_name,
    lang,
    source,
    os,
    area,
    model,
    brand,
    sdk_version,
    gmail,
    height_width,
    app_time,
    network,
    lng,
    lat,
    get_json_object(event_json,'$.kv.action') action,
    get_json_object(event_json,'$.kv.goodsid') goodsid,
    get_json_object(event_json,'$.kv.place') place,
    get_json_object(event_json,'$.kv.extend1') extend1,
    get_json_object(event_json,'$.kv.category') category,
    server_time
from dwd_base_event_log
where dt='2020-03-10' and event_name='display';
```

3）測試

```
hive (gmall)> select * from dwd_display_log limit 2;
```

2. 商品詳情分頁

1）建表敘述

```
hive (gmall)>
drop table if exists dwd_newsdetail_log;
CREATE EXTERNAL TABLE dwd_newsdetail_log(
    `mid_id` string,
    `user_id` string,
    `version_code` string,
    `version_name` string,
    `lang` string,
    `source` string,
    `os` string,
    `area` string,
    `model` string,
    `brand` string,
    `sdk_version` string,
    `gmail` string,
    `height_width` string,
    `app_time` string,
    `network` string,
    `lng` string,
    `lat` string,
    `entry` string,
    `action` string,
    `goodsid` string,
    `showtype` string,
    `news_staytime` string,
    `loading_time` string,
    `type1` string,
    `category` string,
    `server_time` string
)
PARTITIONED BY (dt string)
location '/warehouse/gmall/dwd/dwd_newsdetail_log/';
```

2）匯入資料

```
hive (gmall)>
insert overwrite table dwd_newsdetail_log PARTITION (dt='2020-03-10')
select
    mid_id,
    user_id,
    version_code,
    version_name,
    lang,
    source,
    os,
    area,
    model,
    brand,
    sdk_version,
    gmail,
    height_width,
    app_time,
    network,
    lng,
    lat,
    get_json_object(event_json,'$.kv.entry') entry,
    get_json_object(event_json,'$.kv.action') action,
    get_json_object(event_json,'$.kv.goodsid') goodsid,
    get_json_object(event_json,'$.kv.showtype') showtype,
    get_json_object(event_json,'$.kv.news_staytime') news_staytime,
    get_json_object(event_json,'$.kv.loading_time') loading_time,
    get_json_object(event_json,'$.kv.type1') type1,
    get_json_object(event_json,'$.kv.category') category,
    server_time
from dwd_base_event_log
where dt='2020-03-10' and event_name='newsdetail';
```

3）測試

```
hive (gmall)> select * from dwd_newsdetail_log limit 2;
```

3. 商品列表分頁

1）建表敘述

```
hive (gmall)>
drop table if exists dwd_loading_log;
CREATE EXTERNAL TABLE dwd_loading_log(
    `mid_id` string,
    `user_id` string,
    `version_code` string,
    `version_name` string,
    `lang` string,
    `source` string,
    `os` string,
    `area` string,
    `model` string,
    `brand` string,
    `sdk_version` string,
    `gmail` string,
    `height_width` string,
    `app_time` string,
    `network` string,
    `lng` string,
    `lat` string,
    `action` string,
    `loading_time` string,
    `loading_way` string,
    `extend1` string,
    `extend2` string,
    `type` string,
    `type1` string,
    `server_time` string
)
PARTITIONED BY (dt string)
location '/warehouse/gmall/dwd/dwd_loading_log/';
```

2）匯入資料

```
hive (gmall)>
```

```
insert overwrite table dwd_loading_log PARTITION (dt='2020-03-10')
select
    mid_id,
    user_id,
    version_code,
    version_name,
    lang,
    source,
    os,
    area,
    model,
    brand,
    sdk_version,
    gmail,
    height_width,
    app_time,
    network,
    lng,
    lat,
    get_json_object(event_json,'$.kv.action') action,
    get_json_object(event_json,'$.kv.loading_time') loading_time,
    get_json_object(event_json,'$.kv.loading_way') loading_way,
    get_json_object(event_json,'$.kv.extend1') extend1,
    get_json_object(event_json,'$.kv.extend2') extend2,
    get_json_object(event_json,'$.kv.type') type,
    get_json_object(event_json,'$.kv.type1') type1,
    server_time
from dwd_base_event_log
where dt='2020-03-10' and event_name='loading';
```

3）測試

```
hive (gmall)> select * from dwd_loading_log limit 2;
```

4. 廣告表

1）建表敘述

```
hive (gmall)>
```

```
drop table if exists dwd_ad_log;
CREATE EXTERNAL TABLE dwd_ad_log(
    `mid_id` string,
    `user_id` string,
    `version_code` string,
    `version_name` string,
    `lang` string,
    `source` string,
    `os` string,
    `area` string,
    `model` string,
    `brand` string,
    `sdk_version` string,
    `gmail` string,
    `height_width` string,
    `app_time` string,
    `network` string,
    `lng` string,
    `lat` string,
    `entry` string,
    `action` string,
    `content` string,
    `detail` string,
    `ad_source` string,
    `behavior` string,
    `newstype` string,
    `show_style` string,
    `server_time` string
)
PARTITIONED BY (dt string)
location '/warehouse/gmall/dwd/dwd_ad_log/';
```

2）匯入資料

```
hive (gmall)>

insert overwrite table dwd_ad_log PARTITION (dt='2020-03-10')
select
    mid_id,
```

```
    user_id,
    version_code,
    version_name,
    lang,
    source,
    os,
    area,
    model,
    brand,
    sdk_version,
    gmail,
    height_width,
    app_time,
    network,
    lng,
    lat,
    get_json_object(event_json,'$.kv.entry') entry,
    get_json_object(event_json,'$.kv.action') action,
    get_json_object(event_json,'$.kv.content') content,
    get_json_object(event_json,'$.kv.detail') detail,
    get_json_object(event_json,'$.kv.source') ad_source,
    get_json_object(event_json,'$.kv.behavior') behavior,
    get_json_object(event_json,'$.kv.newstype') newstype,
    get_json_object(event_json,'$.kv.show_style') show_style,
    server_time
from dwd_base_event_log
where dt='2020-03-10' and event_name='ad';
```

3）測試

```
hive (gmall)> select * from dwd_ad_log limit 2;
```

5. 訊息通知表

1）建表敘述

```
hive (gmall)>
drop table if exists dwd_notification_log;
CREATE EXTERNAL TABLE dwd_notification_log(
```

```
    `mid_id` string,
    `user_id` string,
    `version_code` string,
    `version_name` string,
    `lang` string,
    `source` string,
    `os` string,
    `area` string,
    `model` string,
    `brand` string,
    `sdk_version` string,
    `gmail` string,
    `height_width` string,
    `app_time` string,
    `network` string,
    `lng` string,
    `lat` string,
    `action` string,
    `noti_type` string,
    `ap_time` string,
    `content` string,
    `server_time` string
)
PARTITIONED BY (dt string)
location '/warehouse/gmall/dwd/dwd_notification_log/';
```

2）匯入資料

```
hive (gmall)>

insert overwrite table dwd_notification_log
PARTITION (dt='2020-03-10')
select
    mid_id,
    user_id,
    version_code,
    version_name,
    lang,
    source,
```

```
        os,
        area,
        model,
        brand,
        sdk_version,
        gmail,
        height_width,
        app_time,
        network,
        lng,
        lat,
        get_json_object(event_json,'$.kv.action') action,
        get_json_object(event_json,'$.kv.noti_type') noti_type,
        get_json_object(event_json,'$.kv.ap_time') ap_time,
        get_json_object(event_json,'$.kv.content') content,
        server_time
from dwd_base_event_log
where dt='2020-03-10' and event_name='notification';
```

3）測試

```
hive (gmall)> select * from dwd_notification_log limit 2;
```

6. 使用者後台活躍表

1）建表敘述

```
hive (gmall)>
drop table if exists dwd_active_background_log;
CREATE EXTERNAL TABLE dwd_active_background_log(
    `mid_id` string,
    `user_id` string,
    `version_code` string,
    `version_name` string,
    `lang` string,
    `source` string,
    `os` string,
    `area` string,
    `model` string,
    `brand` string,
```

```
    `sdk_version` string,
    `gmail` string,
    `height_width` string,
    `app_time` string,
    `network` string,
    `lng` string,
    `lat` string,
    `active_source` string,
    `server_time` string
)
PARTITIONED BY (dt string)
location '/warehouse/gmall/dwd/dwd_background_log/';
```

2）匯入資料

```
hive (gmall)>

insert overwrite table dwd_active_background_log
PARTITION (dt='2020-03-10')
select
    mid_id,
    user_id,
    version_code,
    version_name,
    lang,
    source,
    os,
    area,
    model,
    brand,
    sdk_version,
    gmail,
    height_width,
    app_time,
    network,
    lng,
    lat,
    get_json_object(event_json,'$.kv.active_source') active_source,
    server_time
```

```
from dwd_base_event_log
where dt='2020-03-10' and event_name='active_background';
```

3）測試

```
hive (gmall)> select * from dwd_active_background_log limit 2;
```

7. 評價表

1）建表敘述

```
hive (gmall)>
drop table if exists dwd_comment_log;
CREATE EXTERNAL TABLE dwd_comment_log(
    `mid_id` string,
    `user_id` string,
    `version_code` string,
    `version_name` string,
    `lang` string,
    `source` string,
    `os` string,
    `area` string,
    `model` string,
    `brand` string,
    `sdk_version` string,
    `gmail` string,
    `height_width` string,
    `app_time` string,
    `network` string,
    `lng` string,
    `lat` string,
    `comment_id` int,
    `userid` int,
    `p_comment_id` int,
    `content` string,
    `addtime` string,
    `other_id` int,
    `praise_count` int,
    `reply_count` int,
    `server_time` string
```

```
)
PARTITIONED BY (dt string)
location '/warehouse/gmall/dwd/dwd_comment_log/';
```

2）匯入資料

```
hive (gmall)>

insert overwrite table dwd_comment_log
PARTITION (dt='2020-03-10')
select
    mid_id,
    user_id,
    version_code,
    version_name,
    lang,
    source,
    os,
    area,
    model,
    brand,
    sdk_version,
    gmail,
    height_width,
    app_time,
    network,
    lng,
    lat,
    get_json_object(event_json,'$.kv.comment_id') comment_id,
    get_json_object(event_json,'$.kv.userid') userid,
    get_json_object(event_json,'$.kv.p_comment_id') p_comment_id,
    get_json_object(event_json,'$.kv.content') content,
    get_json_object(event_json,'$.kv.addtime') addtime,
    get_json_object(event_json,'$.kv.other_id') other_id,
    get_json_object(event_json,'$.kv.praise_count') praise_count,
    get_json_object(event_json,'$.kv.reply_count') reply_count,
    server_time
from dwd_base_event_log
where dt='2020-03-10' and event_name='comment';
```

3）測試

```
hive (gmall)> select * from dwd_comment_log limit 2;
```

8. 收藏表

1）建表敘述

```
hive (gmall)>
drop table if exists dwd_favorites_log;
CREATE EXTERNAL TABLE dwd_favorites_log(
    `mid_id` string,
    `user_id` string,
    `version_code` string,
    `version_name` string,
    `lang` string,
    `source` string,
    `os` string,
    `area` string,
    `model` string,
    `brand` string,
    `sdk_version` string,
    `gmail` string,
    `height_width` string,
    `app_time` string,
    `network` string,
    `lng` string,
    `lat` string,
    `id` int,
    `course_id` int,
    `userid` int,
    `add_time` string,
    `server_time` string
)
PARTITIONED BY (dt string)
location '/warehouse/gmall/dwd/dwd_favorites_log/';
```

2）匯入資料

```
hive (gmall)>
```

```
insert overwrite table dwd_favorites_log
PARTITION (dt='2020-03-10')
select
    mid_id,
    user_id,
    version_code,
    version_name,
    lang,
    source,
    os,
    area,
    model,
    brand,
    sdk_version,
    gmail,
    height_width,
    app_time,
    network,
    lng,
    lat,
    get_json_object(event_json,'$.kv.id') id,
    get_json_object(event_json,'$.kv.course_id') course_id,
    get_json_object(event_json,'$.kv.userid') userid,
    get_json_object(event_json,'$.kv.add_time') add_time,
    server_time
from dwd_base_event_log
where dt='2020-03-10' and event_name='favorites';
```

3）測試

```
hive (gmall)> select * from dwd_favorites_log limit 2;
```

9. 按讚表

1）建表敘述

```
hive (gmall)>
drop table if exists dwd_praise_log;
```

```
CREATE EXTERNAL TABLE dwd_praise_log(
    `mid_id` string,
    `user_id` string,
    `version_code` string,
    `version_name` string,
    `lang` string,
    `source` string,
    `os` string,
    `area` string,
    `model` string,
    `brand` string,
    `sdk_version` string,
    `gmail` string,
    `height_width` string,
    `app_time` string,
    `network` string,
    `lng` string,
    `lat` string,
    `id` string,
    `userid` string,
    `target_id` string,
    `type` string,
    `add_time` string,
    `server_time` string
)
PARTITIONED BY (dt string)
location '/warehouse/gmall/dwd/dwd_praise_log/';
```

2）匯入資料

```
hive (gmall)>

insert overwrite table dwd_praise_log
PARTITION (dt='2020-03-10')
select
    mid_id,
    user_id,
    version_code,
    version_name,
```

```
    lang,
    source,
    os,
    area,
    model,
    brand,
    sdk_version,
    gmail,
    height_width,
    app_time,
    network,
    lng,
    lat,
    get_json_object(event_json,'$.kv.id') id,
    get_json_object(event_json,'$.kv.userid') userid,
    get_json_object(event_json,'$.kv.target_id') target_id,
    get_json_object(event_json,'$.kv.type') type,
    get_json_object(event_json,'$.kv.add_time') add_time,
    server_time
from dwd_base_event_log
where dt='2020-03-10' and event_name='praise';
```

3）測試

```
hive (gmall)> select * from dwd_praise_log limit 2;
```

10.錯誤記錄檔表

1）建表敘述

```
hive (gmall)>
drop table if exists dwd_error_log;
CREATE EXTERNAL TABLE dwd_error_log(
    `mid_id` string,
    `user_id` string,
    `version_code` string,
    `version_name` string,
    `lang` string,
    `source` string,
```

```
    `os` string,
    `area` string,
    `model` string,
    `brand` string,
    `sdk_version` string,
    `gmail` string,
    `height_width` string,
    `app_time` string,
    `network` string,
    `lng` string,
    `lat` string,
    `errorBrief` string,
    `errorDetail` string,
    `server_time` string
)
PARTITIONED BY (dt string)
location '/warehouse/gmall/dwd/dwd_error_log/';
```

2）匯入資料

```
hive (gmall)>

insert overwrite table dwd_error_log
PARTITION (dt='2020-03-10')
select
    mid_id,
    user_id,
    version_code,
    version_name,
    lang,
    source,
    os,
    area,
    model,
    brand,
    sdk_version,
    gmail,
    height_width,
    app_time,
```

```
    network,
    lng,
    lat,
    get_json_object(event_json,'$.kv.errorBrief') errorBrief,
    get_json_object(event_json,'$.kv.errorDetail') errorDetail,
    server_time
from dwd_base_event_log
where dt='2020-03-10' and event_name='error';
```

3）測試

```
hive (gmall)> select * from dwd_error_log limit 2;
```

11. DWD 層事件表載入資料指令稿

將載入及解析資料的過程撰寫成指令稿，方便每日呼叫執行。

（1）在 hadoop102 的 /home/atguigu/bin 目錄下建立指令稿 ods_to_dwd_event_
log.sh。

```
[atguigu@hadoop102 bin]$ vim ods_to_dwd_event_log.sh
```

在指令稿中撰寫以下內容。

```
#!/bin/bash

# 定義變數，方便後續修改
APP=gmall
hive=/opt/module/hive/bin/hive

# 如果輸入了日期參數，則取輸入參數作為日期值；如果沒有輸入日期參數，則取目前
時間的前一天作為日期值
if [ -n "$1" ] ;then
    do_date=$1
else
    do_date=`date -d "-1 day" +%F`
fi

sql="
insert overwrite table "$APP".dwd_display_log
```

```
PARTITION (dt='$do_date')
select
    mid_id,
    user_id,
    version_code,
    version_name,
    lang,
    source,
    os,
    area,
    model,
    brand,
    sdk_version,
    gmail,
    height_width,
    app_time,
    network,
    lng,
    lat,
    get_json_object(event_json,'$.kv.action') action,
    get_json_object(event_json,'$.kv.goodsid') goodsid,
    get_json_object(event_json,'$.kv.place') place,
    get_json_object(event_json,'$.kv.extend1') extend1,
    get_json_object(event_json,'$.kv.category') category,
    server_time
from "$APP".dwd_base_event_log
where dt='$do_date' and event_name='display';

insert overwrite table "$APP".dwd_newsdetail_log
PARTITION (dt='$do_date')
select
    mid_id,
    user_id,
    version_code,
    version_name,
    lang,
    source,
    os,
    area,
```

```
    model,
    brand,
    sdk_version,
    gmail,
    height_width,
    app_time,
    network,
    lng,
    lat,
    get_json_object(event_json,'$.kv.entry') entry,
    get_json_object(event_json,'$.kv.action') action,
    get_json_object(event_json,'$.kv.goodsid') goodsid,
    get_json_object(event_json,'$.kv.showtype') showtype,
    get_json_object(event_json,'$.kv.news_staytime') news_staytime,
    get_json_object(event_json,'$.kv.loading_time') loading_time,
    get_json_object(event_json,'$.kv.type1') type1,
    get_json_object(event_json,'$.kv.category') category,
    server_time
from "$APP".dwd_base_event_log
where dt='$do_date' and event_name='newsdetail';

insert overwrite table "$APP".dwd_loading_log
PARTITION (dt='$do_date')
select
    mid_id,
    user_id,
    version_code,
    version_name,
    lang,
    source,
    os,
    area,
    model,
    brand,
    sdk_version,
    gmail,
    height_width,
    app_time,
    network,
```

```
    lng,
    lat,
    get_json_object(event_json,'$.kv.action') action,
    get_json_object(event_json,'$.kv.loading_time') loading_time,
    get_json_object(event_json,'$.kv.loading_way') loading_way,
    get_json_object(event_json,'$.kv.extend1') extend1,
    get_json_object(event_json,'$.kv.extend2') extend2,
    get_json_object(event_json,'$.kv.type') type,
    get_json_object(event_json,'$.kv.type1') type1,
    server_time
from "$APP".dwd_base_event_log
where dt='$do_date' and event_name='loading';

insert overwrite table "$APP".dwd_ad_log
PARTITION (dt='$do_date')
select
    mid_id,
    user_id,
    version_code,
    version_name,
    lang,
    source,
    os,
    area,
    model,
    brand,
    sdk_version,
    gmail,
    height_width,
    app_time,
    network,
    lng,
    lat,
    get_json_object(event_json,'$.kv.entry') entry,
    get_json_object(event_json,'$.kv.action') action,
    get_json_object(event_json,'$.kv.contentType') contentType,
    get_json_object(event_json,'$.kv.displayMills') displayMills,
    get_json_object(event_json,'$.kv.itemId') itemId,
    get_json_object(event_json,'$.kv.activityId') activityId,
```

```
    server_time
from "$APP".dwd_base_event_log
where dt='$do_date' and event_name='ad';


insert overwrite table "$APP".dwd_notification_log
PARTITION (dt='$do_date')
select
    mid_id,
    user_id,
    version_code,
    version_name,
    lang,
    source,
    os,
    area,
    model,
    brand,
    sdk_version,
    gmail,
    height_width,
    app_time,
    network,
    lng,
    lat,
    get_json_object(event_json,'$.kv.action') action,
    get_json_object(event_json,'$.kv.noti_type') noti_type,
    get_json_object(event_json,'$.kv.ap_time') ap_time,
    get_json_object(event_json,'$.kv.content') content,
    server_time
from "$APP".dwd_base_event_log
where dt='$do_date' and event_name='notification';


insert overwrite table "$APP".dwd_active_background_log
PARTITION (dt='$do_date')
select
    mid_id,
    user_id,
    version_code,
    version_name,
```

```
    lang,
    source,
    os,
    area,
    model,
    brand,
    sdk_version,
    gmail,
    height_width,
    app_time,
    network,
    lng,
    lat,
    get_json_object(event_json,'$.kv.active_source') active_source,
    server_time
from "$APP".dwd_base_event_log
where dt='$do_date' and event_name='active_background';

insert overwrite table "$APP".dwd_comment_log
PARTITION (dt='$do_date')
select
    mid_id,
    user_id,
    version_code,
    version_name,
    lang,
    source,
    os,
    area,
    model,
    brand,
    sdk_version,
    gmail,
    height_width,
    app_time,
    network,
    lng,
    lat,
    get_json_object(event_json,'$.kv.comment_id') comment_id,
```

```
    get_json_object(event_json,'$.kv.userid') userid,
    get_json_object(event_json,'$.kv.p_comment_id') p_comment_id,
    get_json_object(event_json,'$.kv.content') content,
    get_json_object(event_json,'$.kv.addtime') addtime,
    get_json_object(event_json,'$.kv.other_id') other_id,
    get_json_object(event_json,'$.kv.praise_count') praise_count,
    get_json_object(event_json,'$.kv.reply_count') reply_count,
    server_time
from "$APP".dwd_base_event_log
where dt='$do_date' and event_name='comment';

insert overwrite table "$APP".dwd_favorites_log
PARTITION (dt='$do_date')
select
    mid_id,
    user_id,
    version_code,
    version_name,
    lang,
    source,
    os,
    area,
    model,
    brand,
    sdk_version,
    gmail,
    height_width,
    app_time,
    network,
    lng,
    lat,
    get_json_object(event_json,'$.kv.id') id,
    get_json_object(event_json,'$.kv.course_id') course_id,
    get_json_object(event_json,'$.kv.userid') userid,
    get_json_object(event_json,'$.kv.add_time') add_time,
    server_time
from "$APP".dwd_base_event_log
where dt='$do_date' and event_name='favorites';
```

```
insert overwrite table "$APP".dwd_praise_log
PARTITION (dt='$do_date')
select
    mid_id,
    user_id,
    version_code,
    version_name,
    lang,
    source,
    os,
    area,
    model,
    brand,
    sdk_version,
    gmail,
    height_width,
    app_time,
    network,
    lng,
    lat,
    get_json_object(event_json,'$.kv.id') id,
    get_json_object(event_json,'$.kv.userid') userid,
    get_json_object(event_json,'$.kv.target_id') target_id,
    get_json_object(event_json,'$.kv.type') type,
    get_json_object(event_json,'$.kv.add_time') add_time,
    server_time
from "$APP".dwd_base_event_log
where dt='$do_date' and event_name='praise';

insert overwrite table "$APP".dwd_error_log
PARTITION (dt='$do_date')
select
    mid_id,
    user_id,
    version_code,
    version_name,
    lang,
    source,
    os,
```

```
    area,
    model,
    brand,
    sdk_version,
    gmail,
    height_width,
    app_time,
    network,
    lng,
    lat,
    get_json_object(event_json,'$.kv.errorBrief') errorBrief,
    get_json_object(event_json,'$.kv.errorDetail') errorDetail,
    server_time
from "$APP".dwd_base_event_log
where dt='$do_date' and event_name='error';
"

$hive -e "$sql"
```

（2）增加指令稿執行許可權。

```
[atguigu@hadoop102 bin]$ chmod 777 ods_to_dwd_event_log.sh
```

（3）執行指令稿，匯入資料。

```
[atguigu@hadoop102 module]$ ods_to_dwd_event_log.sh 2020-03-11
```

（4）查詢結果資料。

```
hive (gmall)>
select * from dwd_comment_log where dt='2020-03-11' limit 2;
```

6.4.4 業務資料維度資料表解析

關於業務資料，DWD 層的架設主要需要注意維度的退化，ODS 層的業務資料有二十多張表，形成了比較複雜的關係模型，這種情況下想要取得一些細節維度的資訊，通常需要進行多表 join 才能獲得，為了使查詢更加方便，也為了避免進行大量的表 join 計算，將關係模型進行適度的維度退化很有必要。

資料倉儲建模基本想法如圖 6-24 所示，從圖 6-24 中可以看出，建置業務資料 DWD 層的想法是：將許多維度按照類型和關係退化為 7 張主要的維度資料表。

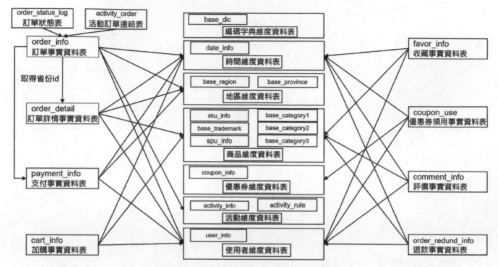

圖 6-24 資料倉儲建模基本想法

接下來對進行維度退化後的幾張主要的維度資料表説明。

1. 商品維度資料表（全量表）

1）建表敘述

```
hive (gmall)>
DROP TABLE IF EXISTS `dwd_dim_sku_info`;
CREATE EXTERNAL TABLE `dwd_dim_sku_info` (
    `id` string COMMENT '商品id',
    `spu_id` string COMMENT '標準產品單位id',
    `price` double COMMENT '商品價格',
    `sku_name` string COMMENT '商品名稱',
    `sku_desc` string COMMENT '商品描述',
    `weight` double COMMENT '重量',
    `tm_id` string COMMENT '品牌id',
    `tm_name` string COMMENT '品牌名稱',
```

```
    `category3_id` string COMMENT '三級品項id',
    `category2_id` string COMMENT '二級品項id',
    `category1_id` string COMMENT '一級品項id',
    `category3_name` string COMMENT '三級品項名稱',
    `category2_name` string COMMENT '二級品項名稱',
    `category1_name` string COMMENT '一級品項名稱',
    `spu_name` string COMMENT '標準產品單位名稱',
    `create_time` string COMMENT '建立時間'
) COMMENT '商品維度資料表'
PARTITIONED BY (`dt` string)
stored as parquet
location '/warehouse/gmall/dwd/dwd_dim_sku_info/'
tblproperties ("parquet.compression"="lzo");
```

2）匯入資料

```
hive (gmall)>
insert overwrite table dwd_dim_sku_info partition(dt='2020-03-10')
select
    sku.id,
    sku.spu_id,
    sku.price,
    sku.sku_name,
    sku.sku_desc,
    sku.weight,
    sku.tm_id,
    ob.tm_name,
    sku.category3_id,
    c2.id category2_id,
    c1.id category1_id,
    c3.name category3_name,
    c2.name category2_name,
    c1.name category1_name,
    spu.spu_name,
    sku.create_time
from
(
    select * from ods_sku_info where dt='2020-03-10'
)sku
```

```
join
(
    select * from ods_base_trademark where dt='2020-03-10'
)ob on sku.tm_id=ob.tm_id
join
(
    select * from ods_spu_info where dt='2020-03-10'
)spu on spu.id = sku.spu_id
join
(
    select * from ods_base_category3 where dt='2020-03-10'
)c3 on sku.category3_id=c3.id
join
(
    select * from ods_base_category2 where dt='2020-03-10'
)c2 on c3.category2_id=c2.id
join
(
    select * from ods_base_category1 where dt='2020-03-10'
)c1 on c2.category1_id=c1.id;
```

3）查詢結果資料

```
hive (gmall)> select * from dwd_dim_sku_info where dt='2020-03-10';
```

2. 優惠券維度資料表（全量）

將 ODS 層 ods_coupon_info 表中的資料匯入 DWD 層優惠券維度資料表中，在匯入過程中可以進行適當的清洗。

1）建表敘述

```
hive (gmall)>
drop table if exists dwd_dim_coupon_info;
create external table dwd_dim_coupon_info(
    `id` string COMMENT '優惠券編號',
    `coupon_name` string COMMENT '優惠券名稱',
    `coupon_type` string COMMENT '優惠券類型 1 現金券 2 折扣券 3 滿減券 4
滿件打折券',
```

```
    `condition_amount` string COMMENT '滿額數',
    `condition_num` string COMMENT '滿件數',
    `activity_id` string COMMENT '活動編號',
    `benefit_amount` string COMMENT '滿減金額',
    `benefit_discount` string COMMENT '折扣',
    `create_time` string COMMENT '建立時間',
    `range_type` string COMMENT '範圍類型 1商品 2品項 3品牌',
    `spu_id` string COMMENT '標準產品單位id',
    `tm_id` string COMMENT '品牌id',
    `category3_id` string COMMENT '品項id',
    `limit_num` string COMMENT '最多領用次數',
    `operate_time`  string COMMENT '操作時間',
    `expire_time`  string COMMENT '過期時間'
) COMMENT '優惠券維度資料表'
PARTITIONED BY (`dt` string)
row format delimited fields terminated by '\t'
stored as parquet
location '/warehouse/gmall/dwd/dwd_dim_coupon_info/';
tblproperties ("parquet.compression"="lzo");
```

2）匯入資料

```
hive (gmall)>
insert overwrite table dwd_dim_coupon_info partition(dt='2020-03-10')
select
    id,
    coupon_name,
    coupon_type,
    condition_amount,
    condition_num,
    activity_id,
    benefit_amount,
    benefit_discount,
    create_time,
    range_type,
    spu_id,
    tm_id,
    category3_id,
```

```
    limit_num,
    operate_time,
    expire_time
from ods_coupon_info
where dt='2020-03-10';
```

3）查詢結果資料

```
hive (gmall)> select * from dwd_dim_coupon_info where dt='2020-03-10';
```

3. 活動維度資料表（全量）

1）建表敘述

```
hive (gmall)>
drop table if exists dwd_dim_activity_info;
create external table dwd_dim_activity_info(
    `id` string COMMENT '編號',
    `activity_name` string  COMMENT '活動名稱',
    `activity_type` string  COMMENT '活動類型',
    `condition_amount` string  COMMENT '滿減金額',
    `condition_num` string  COMMENT '滿減件數',
    `benefit_amount` string  COMMENT '優惠金額',
    `benefit_discount` string  COMMENT '優惠折扣',
    `benefit_level` string  COMMENT '優惠等級',
    `start_time` string  COMMENT '開始時間',
    `end_time` string  COMMENT '結束時間',
    `create_time` string  COMMENT '建立時間'
) COMMENT '活動維度資料表'
PARTITIONED BY (`dt` string)
row format delimited fields terminated by '\t'
stored as parquet
location '/warehouse/gmall/dwd/dwd_dim_activity_info/';
tblproperties ("parquet.compression"="lzo");
```

2）匯入資料

```
hive (gmall)>
insert overwrite table dwd_dim_activity_info partition(dt='2020-03-10')
```

```
select
    info.id,
    info.activity_name,
    info.activity_type,
    rule.condition_amount,
    rule.condition_num,
    rule.benefit_amount,
    rule.benefit_discount,
    rule.benefit_level,
    info.start_time,
    info.end_time,
    info.create_time
from
(
    select * from ods_activity_info where dt='2020-03-10'
)info
left join
(
    select * from ods_activity_rule where dt='2020-03-10'
)rule on info.id = rule.activity_id;
```

3）查詢結果資料

```
hive (gmall)> select * from dwd_dim_activity_info where dt='2020-03-10';
```

4. 地區維度資料表（特殊）

1）建表敘述

```
hive (gmall)>
DROP TABLE IF EXISTS `dwd_dim_base_province`;
CREATE EXTERNAL TABLE `dwd_dim_base_province` (
    `id` string COMMENT 'id',
    `province_name` string COMMENT '省市名稱',
    `area_code` string COMMENT '地區編碼',
    `iso_code` string COMMENT 'ISO編碼',
    `region_id` string COMMENT '地區id',
    `region_name` string COMMENT '地區名稱'
) COMMENT '地區維度資料表'
```

```
stored as parquet
location '/warehouse/gmall/dwd/dwd_dim_base_province/';
tblproperties ("parquet.compression"="lzo");
```

2）匯入資料

```
hive (gmall)>
insert overwrite table dwd_dim_base_province
select
    bp.id,
    bp.name,
    bp.area_code,
    bp.iso_code,
    bp.region_id,
    br.region_name
from ods_base_province bp
join ods_base_region br
on bp.region_id=br.id;
```

3）查詢結果資料

```
hive (gmall)> select * from dwd_dim_base_province;
```

5. 時間維度資料表（特殊）

時間維度資料表比較特殊，時間的維度是不會發生改變的，所以只需匯入一份固定的資料即可。

（1）建表敘述。

```
hive (gmall)>
DROP TABLE IF EXISTS `dwd_dim_date_info`;
CREATE EXTERNAL TABLE `dwd_dim_date_info`(
    `date_id` string COMMENT '日',
    `week_id` int COMMENT '周',
    `week_day` int COMMENT '周的第幾天',
    `day` int COMMENT '每月的第幾天',
    `month` int COMMENT '第幾月',
    `quarter` int COMMENT '第幾季',
```

```
    `year` int COMMENT '年',
    `is_workday` int COMMENT '是否是週末',
    `holiday_id` int COMMENT '是否是節假日'
)COMMENT '時間維度資料表'
row format delimited fields terminated by '\t'
stored as parquet
location '/warehouse/gmall/dwd/dwd_dim_date_info/';
tblproperties ("parquet.compression"="lzo");
```

（2）將 date_info.txt 檔案上傳到 hadoop102 的 /opt/module/db_log/ 目錄下。

（3）匯入資料。

```
hive (gmall)>
load data local inpath '/opt/module/db_log/date_info.txt' into table
dwd_dim_date_info;
```

（4）查詢結果資料。

```
hive (gmall)> select * from dwd_dim_date_info;
```

6.4.5 業務資料事實資料表解析

DWD 層中事實資料表的建立，則需要根據各張表的特點進行不同的處理。本節主要對 DWD 層業務資料事實資料表的建立和資料匯入說明。

1. 訂單詳情事實資料表（交易型事實資料表）

如表 6-15 所示，訂單詳情事實資料表與時間、使用者、地區、商品四個維度有關，其中，與時間、使用者、商品的連結分別透過 dt、user_id、sku_id 欄位建立，與地區的連結則需要透過與 ODS 層的訂單詳情表進行 join 獲得。

表 6-15 訂單詳情事實資料表相關維度

	時間	使用者	地區	商品	優惠券	活動	編碼	度量值
訂單詳情	√	√	√	√				件數 / 金額

1）建表敘述

```
hive (gmall)>
drop table if exists dwd_fact_order_detail;
create external table dwd_fact_order_detail (
    `id` string COMMENT 'id',
    `order_id` string COMMENT '訂單編號',
    `province_id` string COMMENT '省份id',
    `user_id` string COMMENT '使用者id',
    `sku_id` string COMMENT '商品id',
    `create_time` string COMMENT '建立時間',
    `total_amount` decimal(16,2) COMMENT '總金額',
    `sku_num` bigint COMMENT '商品數量'
)COMMENT '訂單詳情事實資料表'
PARTITIONED BY (`dt` string)
stored as parquet
location '/warehouse/gmall/dwd/dwd_fact_order_detail/'
tblproperties ("parquet.compression"="lzo")
;
```

2）匯入資料

```
insert overwrite table dwd_fact_order_detail partition(dt='2020-03-10')
select
    od.id,
    od.order_id,
    oi.province_id,
    od.user_id,
    od.sku_id,
    od.create_time,
    od.order_price*od.sku_num,
    od.sku_num
from
(
    select * from ods_order_detail where dt='2020-03-10'
) od
join
(
    select * from ods_order_info where dt='2020-03-10'
```

```
) oi
on od.order_id=oi.id;
```

3）查詢結果資料

```
hive (gmall)> select * from dwd_fact_order_detail where dt='2020-03-10';
```

2. 支付事實資料表（交易型事實資料表）

如表 6-16 所示，支付事實資料表與時間、使用者、地區三個維度有關，其中，與時間、使用者的連結分別透過 dt、user_id 欄位建立，與地區的連結則需要透過與 ODS 層的訂單詳情表進行 join 獲得。

表 6-16　支付事實資料表相關維度

	時間	使用者	地區	商品	優惠券	活動	編碼	度量值
支付	√	√	√					次數 / 金額

1）建表敘述

```
hive (gmall)>
drop table if exists dwd_fact_payment_info;
create external table dwd_fact_payment_info (
    `id` string COMMENT '',
    `out_trade_no` string COMMENT '對外業務編號',
    `order_id` string COMMENT '訂單編號',
    `user_id` string COMMENT '使用者編號',
    `alipay_trade_no` string COMMENT '支付寶交易流水編號',
    `payment_amount`    decimal(16,2) COMMENT '支付金額',
    `subject`          string COMMENT '交易內容',
    `payment_type` string COMMENT '支付類型',
    `payment_time` string COMMENT '支付時間',
    `province_id` string COMMENT '省份id'
) COMMENT'支付事實資料表'
PARTITIONED BY (`dt` string)
stored as parquet
location '/warehouse/gmall/dwd/dwd_fact_payment_info/'
tblproperties ("parquet.compression"="lzo");
```

2）匯入資料

```
hive (gmall)>
insert overwrite table dwd_fact_payment_info partition(dt='2020-03-10')
select
    pi.id,
    pi.out_trade_no,
    pi.order_id,
    pi.user_id,
    pi.alipay_trade_no,
    pi.total_amount,
    pi.subject,
    pi.payment_type,
    pi.payment_time,
    oi.province_id
from
(
    select * from ods_payment_info where dt='2020-03-10'
)pi
join
(
    select id, province_id from ods_order_info where dt='2020-03-10'
)oi
on pi.order_id = oi.id;
```

3）查詢結果資料

```
hive (gmall)> select * from dwd_fact_payment_info where dt='2020-03-10';
```

3. 退款事實資料表（交易型事實資料表）

如表 6-17 所示，退款事實資料表與時間、使用者、商品三個維度有關，ODS 層的退單表已經具有所有連結欄位，所以無須從其他表格獲得連結，直接將原 ODS 層退單表中的資料匯入即可。

表 6-17 退款事實資料表相關維度

	時間	使用者	地區	商品	優惠券	活動	編碼	度量值
退款	√	√		√				件數 / 金額

1）建表敘述

```
hive (gmall)>
drop table if exists dwd_fact_order_refund_info;
create external table dwd_fact_order_refund_info(
    `id` string COMMENT '編號',
    `user_id` string COMMENT '使用者id',
    `order_id` string COMMENT '訂單編號',
    `sku_id` string COMMENT '商品id',
    `refund_type` string COMMENT '退款類型',
    `refund_num` bigint COMMENT '退款件數',
    `refund_amount` decimal(16,2) COMMENT '退款金額',
    `refund_reason_type` string COMMENT '退款原因類型',
    `create_time` string COMMENT '退款時間'
) COMMENT '退款事實資料表'
PARTITIONED BY (`dt` string)
stored as parquet
row format delimited fields terminated by '\t'
location '/warehouse/gmall/dwd/dwd_fact_order_refund_info/'
tblproperties ("parquet.compression"="lzo");
```

2）匯入資料

```
hive (gmall)>
insert overwrite table dwd_fact_order_refund_info partition(dt='2020-03-10')
select
    id,
    user_id,
    order_id,
    sku_id,
    refund_type,
    refund_num,
    refund_amount,
    refund_reason_type,
    create_time
from ods_order_refund_info
where dt='2020-03-10';
```

3）查詢結果資料

```
hive (gmall)> select * from dwd_fact_order_refund_info where dt='2020-03-10';
```

4. 評價事實資料表（交易型事實資料表）

如表 6-18 所示，評價事實資料表與時間、使用者、商品三個維度有關，ODS 層的商品評價表已經具有所有連結欄位，所以無須從其他表格中獲得連結，直接將 ODS 層商品評價表中的資料匯入即可。

表 6-18 評價事實資料表相關維度

	時間	使用者	地區	商品	優惠券	活動	編碼	度量值
評價	√	√		√				筆數

1）建表敘述

```
hive (gmall)>
drop table if exists dwd_fact_comment_info;
create external table dwd_fact_comment_info(
    `id` string COMMENT '編號',
    `user_id` string COMMENT '使用者id',
    `sku_id` string COMMENT '商品id',
    `spu_id` string COMMENT '標準產品單位id',
    `order_id` string COMMENT '訂單編號',
    `appraise` string COMMENT '評價',
    `create_time` string COMMENT '評價時間'
) COMMENT '評價事實資料表'
PARTITIONED BY (`dt` string)
stored as parquet
row format delimited fields terminated by '\t'
location '/warehouse/gmall/dwd/dwd_fact_comment_info/'
tblproperties ("parquet.compression"="lzo");
```

2）匯入資料

```
hive (gmall)>
insert overwrite table dwd_fact_comment_info partition(dt='2020-03-10')
select
    id,
```

```
    user_id,
    sku_id,
    spu_id,
    order_id,
    appraise,
    create_time
from ods_comment_info
where dt='2020-03-10';
```

3）查詢結果資料

```
hive (gmall)> select * from dwd_fact_comment_info where dt='2020-03-10';
```

5. 加購事實資料表（週期型快照事實資料表，每日快照）

由於購物車中的資料經常會發生變化，所以不適合採用每日增量同步策略匯入資料。我們採用的策略是每天做一次快照，進行全量資料匯入。這樣做的劣勢是儲存的資料量會比較大。

由於週期型快照事實資料表儲存的資料比較注重時效性，儲存時間過於久遠的資料存在的意義不大，所以可以定時刪除以前的資料來釋放記憶體。

如表 6-19 所示，加購事實資料表與時間、使用者、商品三個維度有關。

表 6-19 加購事實資料表相關維度

	時間	使用者	地區	商品	優惠券	活動	編碼	度量值
加購	√	√		√				件數 / 金額

1）建表敘述

```
hive (gmall)>
drop table if exists dwd_fact_cart_info;
create external table dwd_fact_cart_info(
    `id` string COMMENT '編號',
    `user_id` string  COMMENT '使用者id',
    `sku_id` string  COMMENT '商品id',
    `cart_price` string  COMMENT '放入購物車時的價格',
    `sku_num` string  COMMENT '數量',
```

```
    `sku_name` string   COMMENT '商品名稱 (容錯)',
    `create_time` string   COMMENT '建立時間',
    `operate_time` string COMMENT '操作時間',
    `is_ordered` string COMMENT '是否已經下單，1為已下單;0為未下單',
    `order_time` string   COMMENT '下單時間'
) COMMENT '加購事實資料表'
PARTITIONED BY (`dt` string)
stored as parquet
row format delimited fields terminated by '\t'
location '/warehouse/gmall/dwd/dwd_fact_cart_info/'
tblproperties ("parquet.compression"="lzo");
```

2）匯入資料

```
hive (gmall)>
insert overwrite table dwd_fact_cart_info partition(dt='2020-03-10')
select
    id,
    user_id,
    sku_id,
    cart_price,
    sku_num,
    sku_name,
    create_time,
    operate_time,
    is_ordered,
    order_time
from ods_cart_info
where dt='2020-03-10';
```

3）查詢結果資料

```
hive (gmall)> select * from dwd_fact_cart_info where dt='2020-03-10';
```

6. 收藏事實資料表（週期型快照事實資料表，每日快照）

收藏事實資料表採用的同步策略與加購事實資料表相同。

如表 6-20 所示，收藏事實資料表與時間、使用者、商品三個維度有關。

表 6-20 收藏事實資料表相關維度

	時間	使用者	地區	商品	優惠券	活動	編碼	度量值
收藏	√	√		√				個數

1）建表敘述

```
hive (gmall)>
drop table if exists dwd_fact_favor_info;
create external table dwd_fact_favor_info(
    `id` string COMMENT '編號',
    `user_id` string  COMMENT '使用者id',
    `sku_id` string  COMMENT '商品id',
    `spu_id` string  COMMENT '標準產品單位id',
    `is_cancel` string  COMMENT '是否取消',
    `create_time` string  COMMENT '收藏時間',
    `cancel_time` string  COMMENT '取消時間'
) COMMENT '收藏事實資料表'
PARTITIONED BY (`dt` string)
stored as parquet
row format delimited fields terminated by '\t'
location '/warehouse/gmall/dwd/dwd_fact_favor_info/'
tblproperties ("parquet.compression"="lzo");
```

2）匯入資料

```
hive (gmall)>
insert overwrite table dwd_fact_favor_info partition(dt='2020-03-10')
select
    id,
    user_id,
    sku_id,
    spu_id,
    is_cancel,
    create_time,
    cancel_time
from ods_favor_info
where dt='2020-03-10';
```

3）查詢結果資料

```
hive (gmall)> select * from dwd_fact_favor_info where dt='2020-03-10';
```

7. 優惠券領用事實資料表（累積型快照事實資料表）

如表 6-21 所示，優惠券領用事實資料表與時間、使用者、優惠券三個維度有關。

表 6-21　優惠券領用事實資料表相關維度

	時間	使用者	地區	商品	優惠券	活動	編碼	度量值
優惠券領用	√	√			√			個數

優惠券的使用有一定的生命週期：領取優惠券→使用優惠券下單→優惠券參與支付。所以優惠券領用事實資料表符合累積型快照事實資料表的特徵，即將優惠券的領用、下單使用、支付使用三個時間節點進行快照記錄。

累積型快照事實資料表可以用來統計優惠券領取次數、使用優惠券下單的次數及優惠券參與支付的次數等資料。

1）建表敘述

```
hive (gmall)>
drop table if exists dwd_fact_coupon_use;
create external table dwd_fact_coupon_use(
    `id` string COMMENT '編號',
    `coupon_id` string  COMMENT '優惠券id',
    `user_id` string  COMMENT 'userid',
    `order_id` string  COMMENT '訂單id',
    `coupon_status` string  COMMENT '優惠券狀態',
    `get_time` string  COMMENT '領取時間',
    `using_time` string  COMMENT '使用(下單)時間',
    `used_time` string  COMMENT '使用(支付)時間'
) COMMENT '優惠券領用事實資料表'
PARTITIONED BY (`dt` string)
stored as parquet
row format delimited fields terminated by '\t'
location '/warehouse/gmall/dwd/dwd_fact_coupon_use/'
tblproperties ("parquet.compression"="lzo");
```

> ★ **注意**：dt 是按照優惠券領取時間 get_time 進行分區的。

2）匯入資料

優惠券領用事實資料表匯入資料的想法如圖 6-25、圖 6-26 和圖 6-27 所示。

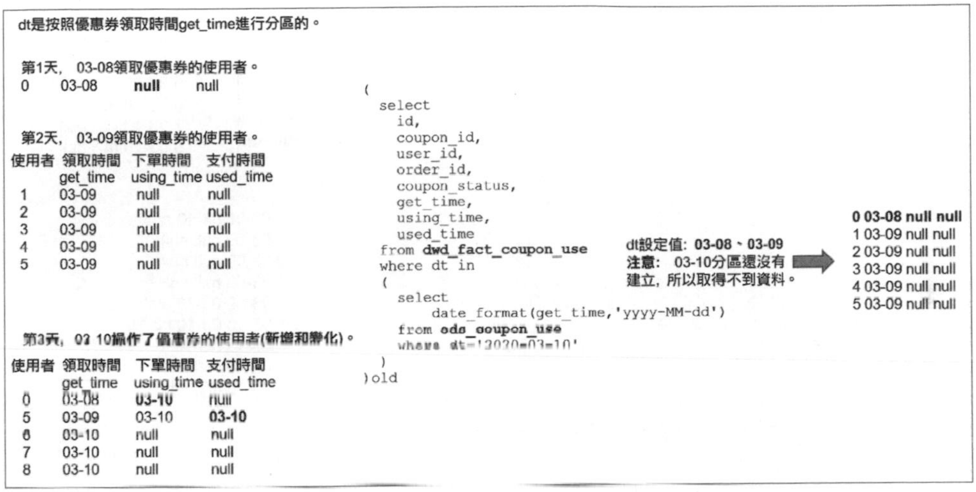

```
dt是按照優惠券領取時間get_time進行分區的。

第1天， 03-08領取優惠券的使用者。
0    03-08    null    null                          (
                                                     select
                                                       id,
第2天， 03-09領取優惠券的使用者。                          coupon_id,
使用者 領取時間  下單時間    支付時間                       user_id,
      get_time using_time used_time                   order_id,
1     03-09    null    null                           coupon_status,
2     03-09    null    null                           get_time,
3     03-09    null    null                           using_time,
4     03-09    null    null                           used_time
5     03-09    null    null                         from dwd_fact_coupon_use
                                                     where dt in
                                                     (
                                                       select
                                                         date_format(get_time,'yyyy-MM-dd')
第3天， 03-10操作了優惠券的使用者(新增和變化)。              from ods_coupon_use
使用者 領取時間  下單時間    支付時間                       where dt='2020-03-10'
      get_time using_time used_time                   )
0     03-08    03-10    null                         )old
5     03-09    03-10    03-10
6     03-10    null    null
7     03-10    null    null
8     03-10    null    null
```

dt設定值：**03-08、03-09**
注意： 03-10分區還沒有
建立，所以取得不到資料。 ➡

```
0 03-08 null null
1 03-09 null null
2 03-09 null null
3 03-09 null null
4 03-09 null null
5 03-09 null null
```

圖 6-25 優惠券領用事實資料表匯入資料的想法（1）

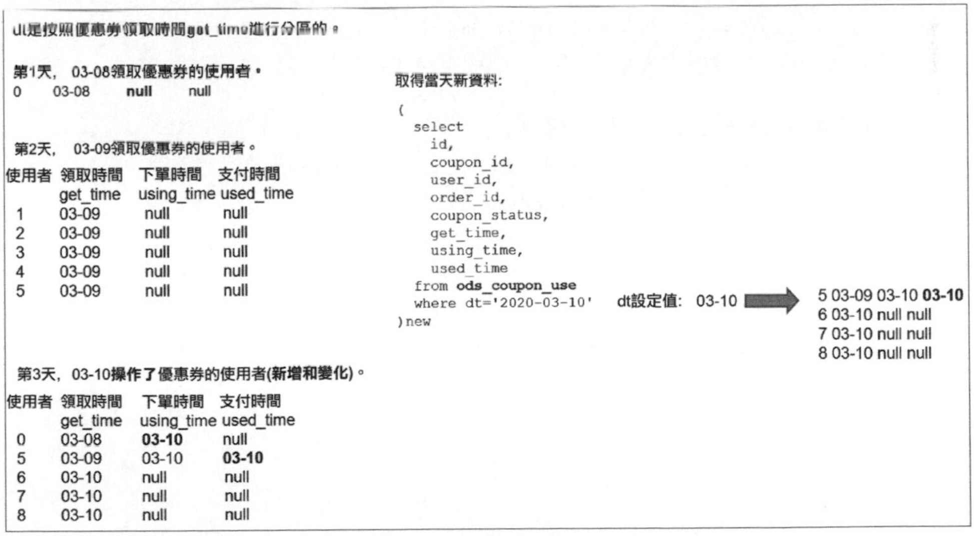

```
dt是按照優惠券領取時間get_time進行分區的。

第1天， 03-08領取優惠券的使用者。                    取得當天新資料：
0    03-08    null    null                          (
                                                     select
                                                       id,
第2天， 03-09領取優惠券的使用者。                          coupon_id,
使用者 領取時間  下單時間    支付時間                       user_id,
      get_time using_time used_time                   order_id,
1     03-09    null    null                           coupon_status,
2     03-09    null    null                           get_time,
3     03-09    null    null                           using_time,
4     03-09    null    null                           used_time
5     03-09    null    null                         from ods_coupon_use
                                                     where dt='2020-03-10'
                                                     )new
```

dt設定值： 03-10 ➡

```
5 03-09 03-10 03-10
6 03-10 null null
7 03-10 null null
8 03-10 null null
```

```
第3天， 03-10操作了優惠券的使用者(新增和變化)。
使用者 領取時間  下單時間    支付時間
      get_time using_time used_time
0     03-08    03-10    null
5     03-09    03-10    03-10
6     03-10    null    null
7     03-10    null    null
8     03-10    null    null
```

圖 6-26 優惠券領用事實資料表匯入資料的想法（2）

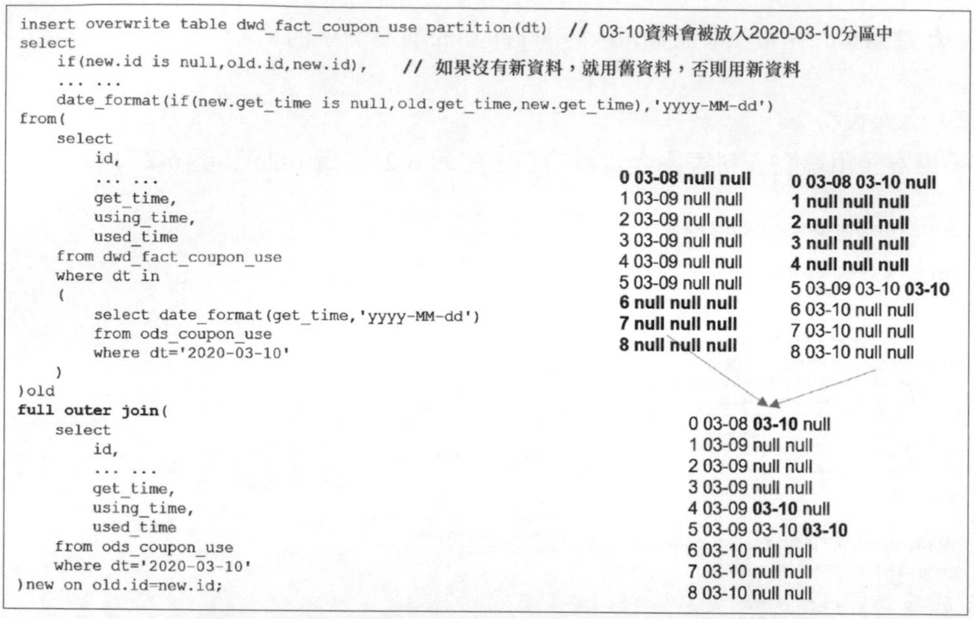

```
insert overwrite table dwd_fact_coupon_use partition(dt)     // 03-10資料會被放入2020-03-10分區中
select
    if(new.id is null,old.id,new.id),      // 如果沒有新資料，就用舊資料，否則用新資料
    ... ...
    date_format(if(new.get_time is null,old.get_time,new.get_time),'yyyy-MM-dd')
from(
    select
        id,
        ... ...
        get_time,
        using_time,
        used_time
    from dwd_fact_coupon_use
    where dt in
    (
        select date_format(get_time,'yyyy-MM-dd')
        from ods_coupon_use
        where dt='2020-03-10'
    )
)old
full outer join(
    select
        id,
        ... ...
        get_time,
        using_time,
        used_time
    from ods_coupon_use
    where dt='2020-03-10'
)new on old.id=new.id;
```

0 03-08 null null	0 03-08 03-10 null	
1 03-09 null null	1 null null null	
2 03-09 null null	2 null null null	
3 03-09 null null	3 null null null	
4 03-09 null null	4 null null null	
5 03-09 null null	5 03-09 03-10 03-10	
6 null null null	6 03-10 null null	
7 null null null	7 03-10 null null	
8 null null null	8 03-10 null null	

0 03-08 **03-10** null
1 03-09 null null
2 03-09 null null
3 03-09 null null
4 03-09 **03-10** null
5 03-09 03-10 **03-10**
6 03-10 null null
7 03-10 null null
8 03-10 null null

圖 6-27 優惠券領用事實資料表匯入資料的想法（3）

程式如下。

```
hive (gmall)>
set hive.exec.dynamic.partition.mode=nonstrict;
insert overwrite table dwd_fact_coupon_use partition(dt)
select
    if(new.id is null,old.id,new.id),
    if(new.coupon_id is null,old.coupon_id,new.coupon_id),
    if(new.user_id is null,old.user_id,new.user_id),
    if(new.order_id is null,old.order_id,new.order_id),
    if(new.coupon_status is null,old.coupon_status,new.coupon_status),
    if(new.get_time is null,old.get_time,new.get_time),
    if(new.using_time is null,old.using_time,new.using_time),
    if(new.used_time is null,old.used_time,new.used_time),
    date_format(if(new.get_time is null,old.get_time,new.get_time),
'yyyy-MM-dd')
from
(
    select
```

```
        id,
        coupon_id,
        user_id,
        order_id,
        coupon_status,
        get_time,
        using_time,
        used_time
    from dwd_fact_coupon_use
    where dt in
    (
        select
            date_format(get_time,'yyyy-MM-dd')
        from ods_coupon_use
        where dt='2020-03-10'
    )
)old
full outer join
(
    select
        id,
        coupon_id,
        user_id,
        order_id,
        coupon_status,
        get_time,
        using_time,
        used_time
    from ods_coupon_use
    where dt='2020-03-10'
)new
on old.id=new.id;
```

3）查詢結果資料

```
hive (gmall)> select * from dwd_fact_coupon_use where dt='2020-03-10';
```

8. 訂單事實資料表（累積型快照事實資料表）

1）資料匯入過程相關的函數

（1）concat() 函數。concat() 函數用於連接字串，在連接字串時，只要其中一個字串是 NULL，結果就傳回 NULL。

```
hive> select concat('a','b');
ab

hive> select concat('a','b',null);
NULL
```

（2）concat_ws() 函數。concat_ws() 函數同樣用於連接字串，在連接字串時，只要有一個字串不是 NULL，結果就不會傳回 NULL。concat_ws() 函數需要指定分隔符號。

```
hive> select concat_ws('-','a','b');
a-b

hive> select concat_ws('-','a','b',null);
a-b

hive> select concat_ws('','a','b',null);
ab
```

（3）str_to_map() 函數。

■ 語法描述。

str_to_map(VARCHAR text, VARCHAR listDelimiter, VARCHAR keyValueDelimiter)。

■ 功能描述。

使用 listDelimiter 將 text 分隔成 key-value 對，然後使用 keyValueDelimiter 分隔每個 key-value 對，並組裝成 MAP 傳回。預設 listDelimiter 為 ","，keyValueDelimiter 為 "="。

■ 案例。

```
str_to_map('1001=2020-03-10,1002=2020-03-10',  ','  ,  '=')
```

輸出：

```
{"1001":"2020-03-10","1002":"2020-03-10"}
```

2）建表敘述

如表 6-22 所示，訂單事實資料表與時間、使用者、地區、活動四個維度有關。

表 6-22　訂單事實資料表相關維度

	時間	使用者	地區	商品	優惠券	活動	編碼	度量值
訂單	√	√	√			√		件數 / 金額

訂單從建立到完成同樣具有一定的生命週期，這個生命週期為建立→支付→取消→完成→退款→退款完成。

由於 ODS 層的訂單表只有建立時間和操作時間兩個狀態，不能表達所有時間節點，所以需要連結訂單狀態表。訂單事實資料表中增加了活動 id，所以需要連結活動訂單表。

```
hive (gmall)>
drop table if exists dwd_fact_order_info;
create external table dwd_fact_order_info (
    `id` string COMMENT '訂單編號',
    `order_status` string COMMENT '訂單狀態',
    `user_id` string COMMENT '使用者id',
    `out_trade_no` string COMMENT '支付流水號',
    `create_time` string COMMENT '建立時間(未支付狀態)',
    `payment_time` string COMMENT '支付時間(已支付狀態)',
    `cancel_time` string COMMENT '取消時間(已取消狀態)',
    `finish_time` string COMMENT '完成時間(已完成狀態)',
    `refund_time` string COMMENT '退款時間(退款中狀態)',
    `refund_finish_time` string COMMENT '退款完成時間(退款完成狀態)',
    `province_id` string COMMENT '省份id',
    `activity_id` string COMMENT '活動id',
    `original_total_amount` string COMMENT '原價金額',
    `benefit_reduce_amount` string COMMENT '優惠金額',
    `feight_fee` string COMMENT '運費',
```

```
    `final_total_amount` decimal(10,2) COMMENT '訂單金額'
) COMMENT '訂單事實資料表'
PARTITIONED BY (`dt` string)
stored as parquet
location '/warehouse/gmall/dwd/dwd_fact_order_info/'
tblproperties ("parquet.compression"="lzo");
```

3）匯入資料的想法

匯入資料的想法如圖 6-28 和圖 6-29 所示。

```
hive (gmall)>
set hive.exec.dynamic.partition.mode=nonstrict;
insert overwrite table dwd_fact_order_info partition(dt)
select
    if(new.id is null,old.id,new.id),
    if(new.order_status is null,old.order_status,new.order_status),
    if(new.user_id is null,old.user_id,new.user_id),
    if(new.out_trade_no is null,old.out_trade_no,new.out_trade_no),
    if(new.tms['1001'] is null,old.create_time,new.tms['1001']),--1001 對應未支付狀態
    if(new.tms['1002'] is null,old.payment_time,new.tms['1002']),
    if(new.tms['1003'] is null,old.cancel_time,new.tms['1003']),
    if(new.tms['1004'] is null,old.finish_time,new.tms['1004']),
    if(new.tms['1005'] is null,old.refund_time,new.tms['1005']),
    if(new.tms['1006'] is null,old.refund_finish_time,new.tms['1006']),
    if(new.province_id is null,old.province_id,new.province_id),
    if(new.activity_id is null,old.activity_id,new.activity_id),
    if(new.original_total_amount is null,old.original_total_amount,new.original_total_amount),
    if(new.benefit_reduce_amount is
null,old.benefit_reduce_amount,new.benefit_reduce_amount),
    if(new.feight_fee is null,old.feight_fee,new.feight_fee),
    if(new.final_total_amount is null,old.final_total_amount,new.final_total_amount),
    date_format(if(new.tms['1001'] is null,old.create_time,new.tms['1001']),'yyyy-MM-dd')
from (
    select
        *
    from dwd_fact_order_info
    where dt
    in  (
            select
                date_format(create_time,'%Y-%m-%d')
            from ods_order_info
            where dt='2020-03-10'
        )
)old

full outer join
(
    select
        info.id,
        info.order_status,
        info.user_id,
        info.out_trade_no,
        info.province_id,
        act.activity_id,
        log.tms,
        info.original_total_amount,
        info.benefit_reduce_amount,
        info.feight_fee,
        info.final_total_amount
    from
    (
        select
            order_id,
            str_to_map(concat_ws(',',collect_set(concat(order_status,'=',operate_time)))
,',','=') tms
        from ods_order_status_log
        where dt='2020-03-10'
        group by order_id
    )log
    join (
        select * from ods_order_info where dt='2020-03-10'
    )info on log.order_id=info.id
    left join (
        select * from ods_activity_order where dt='2020-03-10'
    )act   on log.order_id=act.order_id
)new
on old.id=new.id;
```

圖 6-28 訂單事實資料表匯入資料的想法（1）

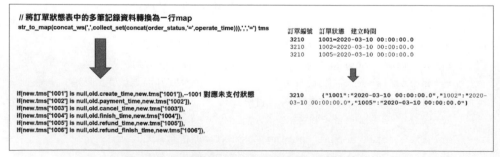

圖 6-29 訂單事實資料表匯入資料的想法（2）

4）常用函數在本次數據匯入中的使用

```
hive (gmall)> select order_id, concat(order_status,'=', operate_time)
from ods_order_status_log where dt='2020-03-10';

3210    1001=2020-03-10 00:00:00.0
3211    1001=2020-03-10 00:00:00.0
3212    1001=2020-03-10 00:00:00.0
3210    1002=2020-03-10 00:00:00.0
3211    1002=2020-03-10 00:00:00.0
3212    1002=2020-03-10 00:00:00.0
3210    1005=2020-03-10 00:00:00.0
3211    1004=2020-03-10 00:00:00.0
3212    1004=2020-03-10 00:00:00.0

hive (gmall)> select order_id, collect_set(concat(order_status,'-',
operate_time)) from ods_order_status_log where dt='2020-03-10' group by \
order_id;

3210    ["1001=2020-03-10 00:00:00.0","1002=2020-03-10 00:00:00.0",
"1005=2020-03-10 00:00:00.0"]
3211    ["1001=2020-03-10 00:00:00.0","1002=2020-03-10 00:00:00.0",
"1004=2020-03-10 00:00:00.0"]
3212    ["1001=2020-03-10 00:00:00.0","1002=2020-03-10 00:00:00.0",
"1004=2020-03-10 00:00:00.0"]

hive (gmall)>
select order_id, concat_ws(',', collect_set(concat(order_status,'=',
operate_time))) from ods_order_status_log where dt='2020-03-10' group by
order_id;

3210    1001=2020-03-10 00:00:00.0,1002=2020-03-10 00:00:00.0,1005=2020-
03-10 00:00:00.0
3211    1001=2020-03-10 00:00:00.0,1002=2020-03-10 00:00:00.0,1004=2020-
03-10 00:00:00.0
3212    1001=2020-03-10 00:00:00.0,1002=2020-03-10 00:00:00.0,1004=2020-
03-10 00:00:00.0

hive (gmall)>
```

```
select order_id, str_to_map(concat_ws(',',collect_set(concat(order_status,
'=',
operate_time))), ',' , '=') from ods_order_status_log where dt='2020-
03-10' group by order_id;

3210    {"1001":"2020-03-10 00:00:00.0","1002":"2020-03-10 00:00:00.0",
"1005":"2020-03-10 00:00:00.0"}
3211    {"1001":"2020-03-10 00:00:00.0","1002":"2020-03-10 00:00:00.0",
"1004":"2020-03-10 00:00:00.0"}
3212    {"1001":"2020-03-10 00:00:00.0","1002":"2020-03-10 00:00:00.0",
"1004":"2020-03-10 00:00:00.0"}
```

5）匯入資料

```
hive (gmall)>
set hive.exec.dynamic.partition.mode=nonstrict;
insert overwrite table dwd_fact_order_info partition(dt)
select
    if(new.id is null,old.id,new.id),
    if(new.order_status is null,old.order_status,new.order_status),
    if(new.user_id is null,old.user_id,new.user_id),
    if(new.out_trade_no is null,old.out_trade_no,new.out_trade_no),
    --1001對應未支付狀態
    if(new.tms['1001'] is null,old.create_time,new.tms['1001']),
    if(new.tms['1002'] is null,old.payment_time,new.tms['1002']),
    if(new.tms['1003'] is null,old.cancel_time,new.tms['1003']),
    if(new.tms['1004'] is null,old.finish_time,new.tms['1004']),
    if(new.tms['1005'] is null,old.refund_time,new.tms['1005']),
    if(new.tms['1006'] is null,old.refund_finish_time,new.tms['1006']),
    if(new.province_id is null,old.province_id,new.province_id),
    if(new.activity_id is null,old.activity_id,new.activity_id),
    if(new.original_total_amount is null,old.original_total_amount,
new.original_total_amount),
    if(new.benefit_reduce_amount is null,old.benefit_reduce_amount,
new.benefit_reduce_amount),
    if(new.feight_fee is null,old.feight_fee,new.feight_fee),
    if(new.final_total_amount is null,old.final_total_amount,
new.final_total_amount),
    date_format(if(new.tms['1001'] is null,old.create_time,
```

```
new.tms['1001']),'yyyy-MM-dd')
from
(
    select
        id,
        order_status,
        user_id,
        out_trade_no,
        create_time,
        payment_time,
        cancel_time,
        finish_time,
        refund_time,
        refund_finish_time,
        province_id,
        activity_id,
        original_total_amount,
        benefit_reduce_amount,
        feight_fee,
        final_total_amount
    from dwd_fact_order_info
    where dt
    in
    (
        select
          date_format(create_time,'yyyy-MM-dd')
        from ods_order_info
        where dt='2020-03-10'
    )
)old
full outer join
(
    select
        info.id,
        info.order_status,
        info.user_id,
        info.out_trade_no,
        info.province_id,
        act.activity_id,
```

```
        log.tms,
        info.original_total_amount,
        info.benefit_reduce_amount,
        info.feight_fee,
        info.final_total_amount
    from
    (
        select
            order_id,
            str_to_map(concat_ws(',',collect_set(concat(order_status,
'=',operate_time))),',','=') tms
        from ods_order_status_log
        where dt='2020-03-10'
        group by order_id
    )log
    join
    (
        select * from ods_order_info where dt='2020-03-10'
    )info
    on log.order_id=info.id
    left join
    (
        select * from ods_activity_order where dt='2020-03-10'
    )act
    on log.order_id=act.order_id
)new
on old.id=new.id;
```

6）查詢結果資料

```
hive (gmall)> select * from dwd_fact_order_info where dt='2020-03-10';
```

6.4.6 拉鏈表建置之使用者維度資料表

什麼是拉鏈表？

拉鏈表是維護歷史狀態及最新狀態資料的一種表，用於記錄每筆資訊的生命週期，一旦一筆資訊的生命週期結束，就重新開始記錄一條新的資訊，並把目前日期放入生效開始日期，如表 6-23 所示。

表 6-23　使用者狀態拉鏈表

使用者 id	手機號碼	生效開始日期	生效結束日期
1	136****9090	2019-01-01	2019-05-01
1	137****8989	2019-05-02	2019-07-02
1	182****7878	2019-07-03	2019-09-05
1	155****1234	2019-09-06	9999-99-99

如果目前資訊至今有效，則在生效結束日期中填入一個極大值（如 9999-99-99）。

如表 6-23 所示的使用者狀態拉鏈表，舉例來説，使用者想得到 2019-08-01 所有的使用者狀態資訊，則可以透過「生效開始日期≤ 2019-08-01 ≤生效結束日期」計算獲得。

拉鏈表適用於以下場景，資料量比較大，且資料部分欄位會發生變化，變化的比例不大且頻率不高，若採用每日全量同步策略匯入資料，則會佔用大量記憶體且會儲存很多不變的資訊。在此種情況下使用拉鏈表，既能反映資料的歷史狀態，又能大幅地節省儲存空間。

舉例來説，使用者資訊會發生變化，但是變化比例不大。如果使用者數量具有一定規模，則按照每日全量同步策略儲存，效率會很低。

使用者表中的資料每日有可能新增，也有可能修改，但修改頻率並不高，屬於緩慢變化維度，所以此處採用拉鏈表儲存使用者維度數據。

如何製作拉鏈表？

如圖 6-30 所示，將使用者當日全部資料和 MySQL 中當日變化資料連接在一起，形成一張新的使用者拉鍊臨時表，使用使用者拉鍊臨時表中的資料覆蓋舊的使用者拉鏈表中的資料，即可解決 Hive 中資料不能更新的問題。

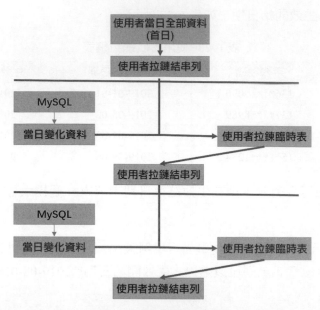

圖 6-30 拉鏈表製作想法

拉鏈表製作過程如圖 6-31 所示。

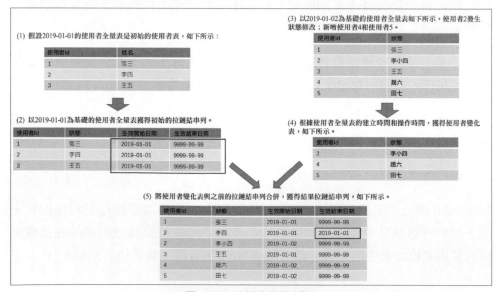

圖 6-31 拉鏈表製作過程

（1）步驟 1：初始化拉鏈表（第一次獨立執行）。

① 建立使用者拉鏈表。

```
hive (gmall)>
drop table if exists dwd_dim_user_info_his;
create external table dwd_dim_user_info_his(
    `id` string COMMENT '使用者id',
    `name` string COMMENT '姓名',
    `birthday` string COMMENT '生日',
    `gender` string COMMENT '性別',
    `email` string COMMENT '電子郵件',
    `user_level` string COMMENT '使用者等級',
    `create_time` string COMMENT '建立時間',
    `operate_time` string COMMENT '操作時間',
    `start_date` string COMMENT '生效開始日期',
    `end_date` string COMMENT '生效結束日期'
) COMMENT '使用者拉鏈表'
stored as parquet
location '/warehouse/gmall/dwd/dwd_dim_user_info_his/'
tblproperties ("parquet.compression"="lzo");
```

② 初始化使用者拉鏈表。

```
hive (gmall)>
insert overwrite table dwd_dim_user_info_his
select
    id,
    name,
    birthday,
    gender,
    email,
    user_level,
    create_time,
    operate_time,
    '2020-03-10',
    '9999-99-99'
from ods_user_info oi
where oi.dt='2020-03-10';
```

（2）步驟 2：製作當日變動資料表，包含新增資料和變動資料。
① 獲得當日變動資料表的想法如下。

- 表內最好有建立時間和變動時間。如果沒有，則可以利用第三方工具，如 canal，監控 MySQL 的即時變化，並進行記錄，這種方式比較麻煩。
- 逐行比較前後兩天的資料，檢查全部可能變化的欄位是否相同。
- 要求業務資料庫提供變動流水。

② 因為 ods_user_info 表本身匯入進來就是新增變動明細的表，表中有建立時間欄位和更改時間欄位，所以不用處理，可直接透過查詢篩選這兩個欄位，取得新增和變動資料。

- 透過 Sqoop 把 2020-03-11 的所有資料匯入。

```
[atguigu@hadoop102 ~]$ mysqlTohdfs.sh all 2020-03-11
```

- 將 ODS 層資料匯入。

```
[atguigu@hadoop102 ~]$ hdfs_to_ods_db.sh all 2020-03-11
```

（3）步驟 3：先合併變動資訊，再追加新增資訊，插入臨時表中。
① 建立使用者拉鍊臨時表，與使用者拉鏈表欄位完全相同，注意使用者拉鍊臨時表以 _tmp 結尾。

```
hive (gmall)>
drop table if exists dwd_dim_user_info_his_tmp;
create external table dwd_dim_user_info_his_tmp(
    `id` string COMMENT '使用者id',
    `name` string COMMENT '姓名',
    `birthday` string COMMENT '生日',
    `gender` string COMMENT '性別',
    `email` string COMMENT '電子郵件',
    `user_level` string COMMENT '使用者等級',
    `create_time` string COMMENT '建立時間',
    `operate_time` string COMMENT '操作時間',
    `start_date`  string COMMENT '生效開始日期',
    `end_date`  string COMMENT '生效結束日期'
```

```
) COMMENT '使用者拉鍊臨時表'
stored as parquet
location '/warehouse/gmall/dwd/dwd_dim_user_info_his_tmp/'
tblproperties ("parquet.compression"="lzo");
```

② 匯入資料。

ods_user_info 表中的 operate_time 欄位為當天日期的資料即為當時修改資料，
在增加當時修改資料的同時，應該修改原使用者拉鏈表中狀態已經發生改變
的資料，將生效結束日期修改為前一日，可以透過將原使用者拉鏈表與當日
新增及當時修改資料按照 id 進行連接，若新增及當時修改資料中不存在對目
前狀態的修改，則保留原生效結束日期；若新增及當時修改資料中存在對目
前狀態的修改，則將生效結束日期修改為前一日。

```
hive (gmall)>
insert overwrite table dwd_dim_user_info_his_tmp
select * from
(
    select
        id,
        name,
        birthday,
        gender,
        email,
        user_level,
        create_time,
        operate_time,
        '2020-03-11' start_date,
        '9999-99-99' end_date
    from ods_user_info where dt='2020-03-11'

    union all
    select
        uh.id,
        uh.name,
        uh.birthday,
        uh.gender,
        uh.email,
```

```
        uh.user_level,
        uh.create_time,
        uh.operate_time,
        uh.start_date,
        if(ui.id is not null  and uh.end_date='9999-99-99',
date_add(ui.dt,-1), uh.end_date) end_date
    from dwd_dim_user_info_his uh left join
    (
        select
            *
        from ods_user_info
        where dt='2020-03-11'
    ) ui on uh.id=ui.id
)his
order by his.id, start_date;
```

（4）步驟 4：使用使用者拉鍊臨時表中的資料覆蓋使用者拉鏈表中的資料。

① 匯入資料。

```
hive (gmall)>
insert overwrite table dwd_dim_user_info_his
select * from dwd_dim_user_info_his_tmp;
```

② 查詢結果資料。

```
hive (gmall)> select id, start_date, end_date from dwd_dim_user_info_his;
```

6.4.7 DWD 層資料匯入指令稿

將 DWD 層的資料匯入過程撰寫成指令稿，方便每日呼叫執行。

（1）在 /home/atguigu/bin 目錄下建立指令稿 ods_to_dwd_db.sh。

```
[atguigu@hadoop102 bin]$ vim ods_to_dwd_db.sh
```

在指令稿中撰寫以下內容。

```
#!/bin/bash

APP=gmall
```

```
hive=/opt/module/hive/bin/hive

# 如果輸入了日期參數,則取輸入參數作為日期值;如果沒有輸入日期參數,則取目前
時間的前一天作為日期值
if [ -n "$2" ] ;then
    do_date=$2
else
    do_date=`date -d "-1 day" +%F`
fi

sql1="
set hive.exec.dynamic.partition.mode=nonstrict;

insert overwrite table ${APP}.dwd_dim_sku_info partition(dt='$do_date')
select
    sku.id,
    sku.spu_id,
    sku.price,
    sku.sku_name,
    sku.sku_desc,
    sku.weight,
    sku.tm_id,
    ob.tm_name,
    sku.category3_id,
    c2.id category2_id,
    c1.id category1_id,
    c3.name category3_name,
    c2.name category2_name,
    c1.name category1_name,
    spu.spu_name,
    sku.create_time
from
(
    select * from ${APP}.ods_sku_info where dt='$do_date'
)sku
join
(
    select * from ${APP}.ods_base_trademark where dt='$do_date'
)ob on sku.tm_id=ob.tm_id
```

```
join
(
    select * from ${APP}.ods_spu_info where dt='$do_date'
)spu on spu.id = sku.spu_id
join
(
    select * from ${APP}.ods_base_category3 where dt='$do_date'
)c3 on sku.category3_id=c3.id
join
(
    select * from ${APP}.ods_base_category2 where dt='$do_date'
)c2 on c3.category2_id=c2.id
join
(
    select * from ${APP}.ods_base_category1 where dt='$do_date'
)c1 on c2.category1_id=c1.id;

insert overwrite table ${APP}.dwd_dim_coupon_info partition(dt=
'$do_date')
select
    id,
    coupon_name,
    coupon_type,
    condition_amount,
    condition_num,
    activity_id,
    benefit_amount,
    benefit_discount,
    create_time,
    range_type,
    spu_id,
    tm_id,
    category3_id,
    limit_num,
    operate_time,
    expire_time
from ${APP}.ods_coupon_info
where dt='$do_date';
```

```
insert overwrite table ${APP}.dwd_dim_activity_info partition(dt=
'$do_date')
select
    info.id,
    info.activity_name,
    info.activity_type,
    rule.condition_amount,
    rule.condition_num,
    rule.benefit_amount,
    rule.benefit_discount,
    rule.benefit_level,
    info.start_time,
    info.end_time,
    info.create_time
from
(
    select * from ${APP}.ods_activity_info where dt='$do_date'
)info
left join
(
    select * from ${APP}.ods_activity_rule where dt='$do_date'
)rule on info.id = rule.activity_id;

insert overwrite table ${APP}.dwd_fact_order_detail partition(dt=
'$do_date')
select
    od.id,
    od.order_id,
    od.user_id,
    od.sku_id,
    od.sku_name,
    od.order_price,
    od.sku_num,
    od.create_time,
    oi.province_id,
    od.order_price*od.sku_num
from
(
    select * from ${APP}.ods_order_detail where dt='$do_date'
```

```
) od
join
(
    select * from ${APP}.ods_order_info where dt='$do_date'
) oi
on od.order_id=oi.id;

insert overwrite table ${APP}.dwd_fact_payment_info partition(dt=
'$do_date')
select
    pi.id,
    pi.out_trade_no,
    pi.order_id,
    pi.user_id,
    pi.alipay_trade_no,
    pi.total_amount,
    pi.subject,
    pi.payment_type,
    pi.payment_time,
    oi.province_id
from
(
    select * from ${APP}.ods_payment_info where dt='$do_date'
)pi
join
(
    select id, province_id from ${APP}.ods_order_info where dt='$do_date'
)oi
on pi.order_id = oi.id;

insert overwrite table ${APP}.dwd_fact_order_refund_info partition(dt='$do_
date')
select
    id,
    user_id,
    order_id,
    sku_id,
    refund_type,
    refund_num,
```

```
    refund_amount,
    refund_reason_type,
    create_time
from ${APP}.ods_order_refund_info
where dt='$do_date';


insert overwrite table ${APP}.dwd_fact_comment_info partition(dt='$do_date')
select
    id,
    user_id,
    sku_id,
    spu_id,
    order_id,
    appraise,
    create_time
from ${APP}.ods_comment_info
where dt='$do_date';


insert overwrite table ${APP}.dwd_fact_cart_info partition(dt='$do_date')
select
    id,
    user_id,
    sku_id,
    cart_price,
    sku_num,
    sku_name,
    create_time,
    operate_time,
    is_ordered,
    order_time
from ${APP}.ods_cart_info
where dt='$do_date';


insert overwrite table ${APP}.dwd_fact_favor_info partition(dt='$do_date')
select
    id,
    user_id,
    sku_id,
    spu_id,
```

```
    is_cancel,
    create_time,
    cancel_time
from ${APP}.ods_favor_info
where dt='$do_date';

insert overwrite table ${APP}.dwd_fact_coupon_use partition(dt)
select
    if(new.id is null,old.id,new.id),
    if(new.coupon_id is null,old.coupon_id,new.coupon_id),
    if(new.user_id is null,old.user_id,new.user_id),
    if(new.order_id is null,old.order_id,new.order_id),
    if(new.coupon_status is null,old.coupon_status,new.coupon_status),
    if(new.get_time is null,old.get_time,new.get_time),
    if(new.using_time is null,old.using_time,new.using_time),
    if(new.used_time is null,old.used_time,new.used_time),
    date_format(if(new.get_time is null,old.get_time,new.get_time),
'yyyy-MM-dd')
from
(
    select
        id,
        coupon_id,
        user_id,
        order_id,
        coupon_status,
        get_time,
        using_time,
        used_time
    from ${APP}.dwd_fact_coupon_use
    where dt in
    (
        select
            date_format(get_time,'yyyy-MM-dd')
        from ${APP}.ods_coupon_use
        where dt='$do_date'
    )
)old
full outer join
```

```
(
    select
        id,
        coupon_id,
        user_id,
        order_id,
        coupon_status,
        get_time,
        using_time,
        used_time
    from ${APP}.ods_coupon_use
    where dt='$do_date'
)new
on old.id=new.id;

insert overwrite table ${APP}.dwd_fact_order_info partition(dt)
select
    if(new.id is null,old.id,new.id),
    if(new.order_status is null,old.order_status,new.order_status),
    if(new.user_id is null,old.user_id,new.user_id),
    if(new.out_trade_no is null,old.out_trade_no,new.out_trade_no),
    --1001對應未支付狀態
    if(new.tms['1001'] is null,old.create_time,new.tms['1001']),
    if(new.tms['1002'] is null,old.payment_time,new.tms['1002']),
    if(new.tms['1003'] is null,old.cancel_time,new.tms['1003']),
    if(new.tms['1004'] is null,old.finish_time,new.tms['1004']),
    if(new.tms['1005'] is null,old.refund_time,new.tms['1005']),
    if(new.tms['1006'] is null,old.refund_finish_time,new.tms['1006']),
    if(new.province_id is null,old.province_id,new.province_id),
    if(new.activity_id is null,old.activity_id,new.activity_id),
    if(new.original_total_amount is null,old.original_total_amount,
new.original_total_amount),
    if(new.benefit_reduce_amount is null,old.benefit_reduce_amount,
new.benefit_reduce_amount),
    if(new.feight_fee is null,old.feight_fee,new.feight_fee),
    if(new.final_total_amount is null,old.final_total_amount,
new.final_total_amount),
    date_format(if(new.tms['1001'] is null,old.create_time,
```

```
new.tms['1001']),'yyyy-MM-dd')
from
(
    select
        id,
        order_status,
        user_id,
        out_trade_no,
        create_time,
        payment_time,
        cancel_time,
        finish_time,
        refund_time,
        refund_finish_time,
        province_id,
        activity_id,
        original_total_amount,
        benefit_reduce_amount,
        feight_fee,
        final_total_amount
    from ${APP}.dwd_fact_order_info
    where dt
    in
    (
        select
          date_format(create_time,'yyyy-MM-dd')
        from ${APP}.ods_order_info
        where dt='$do_date'
    )
)old
full outer join
(
    select
        info.id,
        info.order_status,
        info.user_id,
        info.out_trade_no,
        info.province_id,
        act.activity_id,
```

```
        log.tms,
        info.original_total_amount,
        info.benefit_reduce_amount,
        info.feight_fee,
        info.final_total_amount
    from
    (
        select
            order_id,
            str_to_map(concat_ws(',',collect_set(concat(order_status,
'=',operate_time))),',','=') tms
        from ${APP}.ods_order_status_log
        where dt='$do_date'
        group by order_id
    )log
    join
    (
        select * from ${APP}.ods_order_info where dt='$do_date'
    )info
    on log.order_id=info.id
    left join
    (
        select * from ${APP}.ods_activity_order where dt='$do_date'
    )act
    on log.order_id=act.order_id
)new
on old.id=new.id;

insert overwrite table ${APP}.dwd_dim_user_info_his_tmp
select * from
(
    select
        id,
        name,
        birthday,
        gender,
        email,
        user_level,
        create_time,
```

```
            operate_time,
        '$do_date' start_date,
        '9999-99-99' end_date
    from ${APP}.ods_user_info where dt='$do_date'

    union all
    select
        uh.id,
        uh.name,
        uh.birthday,
        uh.gender,
        uh.email,
        uh.user_level,
        uh.create_time,
        uh.operate_time,
        uh.start_date,
        if(ui.id is not null  and uh.end_date='9999-99-99',
date_add(ui.dt,-1), uh.end_date) end_date
    from ${APP}.dwd_dim_user_info_his uh left join
    (
        select
            *
        from ${APP}.ods_user_info
        where dt='$do_date'
    ) ui on uh.id=ui.id
)his
order by his.id, start_date;

insert overwrite table ${APP}.dwd_dim_user_info_his select * from ${APP}.
dwd_dim_user_info_his_tmp;
"

sql2="
insert overwrite table ${APP}.dwd_dim_base_province
select
    bp.id,
    bp.name,
    bp.area_code,
    bp.iso_code,
```

```
    bp.region_id,
    br.region_name
from ${APP}.ods_base_province bp
join ${APP}.ods_base_region br
on bp.region_id=br.id;
"

case $1 in
"first"){
    $hive -e "$sql1"
    $hive -e "$sql2"
};;
"all"){
    $hive -e "$sql1"
};;
esac
```

（2）增加指令稿執行許可權。

```
[atguigu@hadoop102 bin]$ chmod 777 ods_to_dwd_db.sh
```

（3）執行指令稿，匯入資料。

```
[atguigu@hadoop102 bin]$ ods_to_dwd_db.sh all 2020-03-11
```

（4）查詢結果資料。

```
hive (gmall)>
select * from dwd_fact_order_info where dt='2020-03-11';
select * from dwd_fact_order_detail where dt='2020-03-11';
select * from dwd_fact_comment_info where dt='2020-03-11';
select * from dwd_fact_order_refund_info where dt='2020-03-11';
```

6.5 資料倉儲架設——DWS 層

DWS 層採用寬表化方法，建置公共指標資料。其站在不同主題的角度，將資料進行整理和聚合，獲得每天每個主題的相關資料。

6.5.1 系統函數

本節在 DWS 層的架設中，需要用到的重要函數如下所示。

1. collect_set() 函數

（1）建立原資料表。

```
hive (gmall)>
drop table if exists stud;
create table stud (name string, area string, course string, score int);
```

（2）向原資料表中插入資料。

```
hive (gmall)>
insert into table stud values('zhang3','bj','math',88);
insert into table stud values('li4','bj','math',99);
insert into table stud values('wang5','sh','chinese',92);
insert into table stud values('zhao6','sh','chinese',54);
insert into table stud values('tian7','bj','chinese',91);
```

（3）查詢表中的資料。

```
hive (gmall)> select * from stud;
stud.name      stud.area      stud.course      stud.score
zhang3         bj             math             88
li4            bj             math             99
wang5          sh             chinese          92
zhao6          sh             chinese          54
tian7          bj             chinese          91
```

（4）把同一分組中不同行的資料聚合成一個集合。

```
hive (gmall)> select course, collect_set(area), avg(score) from stud
group by course;
chinese ["sh","bj"]     79.0
math    ["bj"]          93.5
```

（5）使用索引取得聚合結果的某一個值。

```
hive (gmall)> select course, collect_set(area)[0], avg(score) from stud
```

```
group by course;
chinese sh      79.0
math    bj      93.5
```

2. nvl() 函數

基本語法：nvl(運算式 1, 運算式 2)。

如果運算式 1 為空值，則 nvl() 函數傳回運算式 2 的值，否則傳回運算式 1 的值。nvl() 函數的作用是把一個空值（null）轉換成一個實際的值。其運算式的資料類型可以是數字型、字元型和日期型。需要注意的是，運算式 1 和運算式 2 的資料類型必須相同。

3. 日期處理函數

1）date_format() 函數（根據格式整理日期）

```
hive (gmall)> select date_format('2020-03-10','yyyy-MM');
2020-03
```

2）date_add() 函數（加減日期）

```
hive (gmall)> select date_add('2020-03-10',-1);
2020-03-09
hive (gmall)> select date_add('2020-03-10',1);
2020-03-11
```

3）next_day() 函數

（1）取得目前日期的下一個星期一。

```
hive (gmall)> select next_day('2020-03-12','MO');
2020-03-16
```

（2）取得目前周的星期一。

```
hive (gmall)> select date_add(next_day('2020-03-12','MO'),-7);
2020-03-11
```

4）last_day() 函數（取得當月最後一天的日期）

```
hive (gmall)> select last_day('2020-03-10');
2020-03-31
```

6.5.2 使用者行為資料聚合

出於對後續每日活躍裝置、每週活躍裝置、每日新增裝置等需求的考慮，我們利用使用者行為 DWD 層的啟動記錄表，按照裝置 id 進行聚合，獲得 DWS 層的裝置行為表。

在聚合的過程中，為避免細節資料的遺失，將聚合後的欄位使用 concat_ws() 函數進行連接，若後期在需求開發過程中需要用到這些細節資料，則可以透過使用爆炸函數 explode() 再次取得。

每日裝置行為表主要按照裝置唯一標識 mid_id 進行分組統計。

1）建表敘述

```
hive (gmall)>
drop table if exists dws_uv_detail_daycount;
create external table dws_uv_detail_daycount
(
    `mid_id` string COMMENT '裝置唯一標識',
    `user_id` string COMMENT '使用者標識',
    `version_code` string COMMENT '程式版本編號',
    `version_name` string COMMENT '程式版本名',
    `lang` string COMMENT '系統語言',
    `source` string COMMENT '通路號',
    `os` string COMMENT 'Android版本',
    `area` string COMMENT '區域',
    `model` string COMMENT '手機型號',
    `brand` string COMMENT '手機品牌',
    `sdk_version` string COMMENT 'sdkVersion',
    `gmail` string COMMENT 'gmail',
    `height_width` string COMMENT '螢幕長寬',
    `app_time` string COMMENT '用戶端記錄檔產生時的時間',
```

```
    `network` string COMMENT '網路模式',
    `lng` string COMMENT '經度',
    `lat` string COMMENT '緯度',
    `login_count` bigint COMMENT '活躍次數'
) COMMENT'每日裝置行為表'
partitioned by(dt string)
stored as parquet
location '/warehouse/gmall/dws/dws_uv_detail_daycount';
```

2）匯入資料

```
hive (gmall)>
insert overwrite table dws_uv_detail_daycount partition(dt='2020-03-10')
select
    mid_id,
    concat_ws('|', collect_set(user_id)) user_id,
    concat_ws('|', collect_set(version_code)) version_code,
    concat_ws('|', collect_set(version_name)) version_name,
    concat_ws('|', collect_set(lang))lang,
    concat_ws('|', collect_set(source)) source,
    concat_ws('|', collect_set(os)) os,
    concat_ws('|', collect_set(area)) area,
    concat_ws('|', collect_set(model)) model,
    concat_ws('|', collect_set(brand)) brand,
    concat_ws('|', collect_set(sdk_version)) sdk_version,
    concat_ws('|', collect_set(gmail)) gmail,
    concat_ws('|', collect_set(height_width)) height_width,
    concat_ws('|', collect_set(app_time)) app_time,
    concat_ws('|', collect_set(network)) network,
    concat_ws('|', collect_set(lng)) lng,
    concat_ws('|', collect_set(lat)) lat,
    count(*) login_count
from dwd_start_log
where dt='2020-03-10'
group by mid_id;
```

3）查詢結果資料

```
hive (gmall)> select * from dws_uv_detail_daycount where dt='2020-03-10';
```

6.5.3 業務資料聚合

DWS 層的寬表欄位是站在不同維度的角度去看事實資料表的，特別注意事實
資料表的度量值。我們在業務資料 DWD 層主要建置的維度資料表如圖 6-32
所示，其中，編碼字典維度資料表、時間維度資料表和地區維度資料表是特
殊表（不發生變化）。我們主要關注其餘 4 張維度資料表，分別是使用者維度
資料表、商品維度資料表、優惠券維度資料表和活動維度資料表，透過與之
連結的事實資料表，獲得不同事實資料表的度量值。

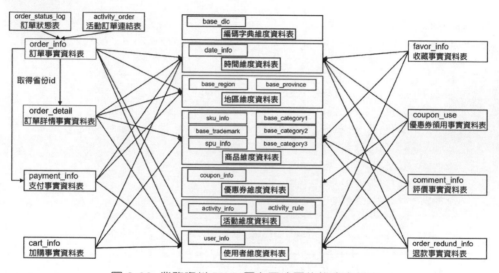

圖 6-32 業務資料 DWD 層主要建置的維度資料表

接下來我們將按照以上想法分別建置 DWS 層的業務資料寬表。

1. 每日會員行為

每日會員行為表主要以會員為中心，關注會員的行為，以及該行為對應的度
量值。

1）建表敘述

```
hive (gmall)>
drop table if exists dws_user_action_daycount;
```

```
create external table dws_user_action_daycount
(
    user_id string comment '使用者id',
    login_count bigint comment '登入次數',
    cart_count bigint comment '加入購物車次數',
    cart_amount double comment '加入購物車金額',
    order_count bigint comment '下單次數',
    order_amount    decimal(16,2)  comment '下單金額',
    payment_count    bigint        comment '支付次數',
    payment_amount  decimal(16,2) comment '支付金額',
    order_stats array<struct<sku_id:string,sku_num:bigint,
order_count:bigint,
order_amount:decimal(20,2)>> comment '下單明細統計'
) COMMENT '每日會員行為表'
PARTITIONED BY (`dt` string)
stored as parquet
location '/warehouse/gmall/dws/dws_user_action_daycount/'
```

2）匯入資料

```
hive (gmall)>
with
tmp_login as --當日登入統計，統計每個會員的當日登入次數
(
    select
        user_id,
        count(*) login_count --登入次數
    from dwd_start_log
    where dt='2020-03-10'
    and user_id is not null
    group by user_id
),
tmp_cart as --當日加入購物車統計，統計每個會員的當日加入購物車情況
(
    select
        user_id,
        count(*) cart_count, --加入購物車次數
        sum(cart_price*sku_num) cart_amount --加入購物車金額
    from dwd_fact_cart_info
```

```
    where dt='2020-03-10'
and user_id is not null
and date_format(create_time,'yyyy-MM-dd')='2020-03-10'
    group by user_id
),
tmp_order as --當日下單統計,統計每個會員的當日下單情況
(
    select
        user_id,
        count(*) order_count, --下單次數
        sum(final_total_amount) order_amount --下單金額
    from dwd_fact_order_info
    where dt='2020-03-10'
    group by user_id
) ,
tmp_payment as --當日支付統計,統計每個會員的當日支付情況
(
    select
        user_id,
        count(*) payment_count, --支付次數
        sum(payment_amount) payment_amount --支付金額
    from dwd_fact_payment_info
    where dt='2020-03-10'
    group by user_id
),
tmp_order_detail as --當日訂單詳情統計,統計每個會員的當日下單詳細資訊
(
    select
        user_id,
        --結構為struct陣列,每個陣列元素對應該會員當日下單的一件商品,包含
sku_id、sku_num
        --(下單例數)、order_count(下單次數)、order_amount(下單金額)
        collect_set(named_struct('sku_id',sku_id,'sku_num',sku_num,
'order_count',order_count,'order_amount',order_amount)) order_stats
    from
    (
        select
            user_id,
            sku_id,
```

```
            sum(sku_num) sku_num,
            count(*) order_count,
            cast(sum(total_amount) as decimal(20,2)) order_amount
        from dwd_fact_order_detail
        where dt='2020-03-10'
        group by user_id,sku_id
    )tmp
    group by user_id
)

insert overwrite table dws_user_action_daycount partition(dt='2020-03-10')
select
    coalesce(tmp_login.user_id,tmp_cart.user_id,tmp_order.user_id,
tmp_payment.
user_id,tmp_order_detail.user_id),
    login_count,
    nvl(cart_count,0),
    nvl(cart_amount,0),
    nvl(order_count,0),
    nvl(order_amount,0),
    nvl(payment_count,0),
    nvl(payment_amount,0),
    order_stats
from tmp_login
full outer join tmp_cart on tmp_login.user_id=tmp_cart.user_id
full outer join tmp_order on tmp_login.user_id=tmp_order.user_id
full outer join tmp_payment on tmp_login.user_id=tmp_payment.user_id
full outer join tmp_order_detail on tmp_login.user_id=tmp_order_detail.
user_id
```

3）查詢結果資料

```
hive (gmall)> select * from dws_user_action_daycount where dt='2020-03-10';
```

2. 每日商品行為

每日商品行為表以商品為中心，透過與商品維度有關的事實資料表獲得與商品相關的不同維度的度量值。

1）建表敘述

```
hive (gmall)>
drop table if exists dws_sku_action_daycount;
create external table dws_sku_action_daycount
(
    sku_id string comment '商品id',
    order_count bigint comment '被下單次數',
    order_num bigint comment '被下單例數',
    order_amount decimal(16,2) comment '被下單金額',
    payment_count bigint  comment '被支付次數',
    payment_num bigint comment '被支付件數',
    payment_amount decimal(16,2) comment '被支付金額',
    refund_count bigint  comment '被退款次數',
    refund_num bigint comment '被退款件數',
    refund_amount  decimal(16,2) comment '被退款金額',
    cart_count bigint comment '被加入購物車次數',
    cart_num bigint comment '被加入購物車件數',
    favor_count bigint comment '被收藏次數',
    appraise_good_count bigint comment '好評數',
    appraise_mid_count bigint comment '中評數',
    appraise_bad_count bigint comment '差評數',
    appraise_default_count bigint comment '預設評價數'
) COMMENT '每日商品行為表'
PARTITIONED BY (`dt` string)
stored as parquet
location '/warehouse/gmall/dws/dws_sku_action_daycount/'
```

2）匯入資料

注意：如果是 23:59 下單，則支付日期跨天，需要從訂單詳情中取出支付時間
是今天，下單時間是昨天或今天的訂單。

```
hive (gmall)>
with
tmp_order as --下單情況統計，統計每件商品（SKU）當日被下單的情況
(
    select
```

```
        sku_id,
        count(*) order_count, --被下單次數
        sum(sku_num) order_num, --被下單例數
        sum(total_amount) order_amount --被下單金額
    from dwd_fact_order_detail
    where dt='2020-03-10'
    group by sku_id
),
tmp_payment as --支付統計，統計每件商品（SKU）當日被支付的情況
(
    select
        sku_id,
        count(*) payment_count, --被支付次數
        sum(sku_num) payment_num, --被支付件數
        sum(total_amount) payment_amount --被支付金額
    from dwd_fact_order_detail
    where dt='2020-03-10'
    and order_id in
    (
        select
            id
        from dwd_fact_order_info
        where (dt='2020-03-10'
        or dt=date_add('2020-03-10',-1))
        and date_format(payment_time,'yyyy-MM-dd')='2020-03-10'
    )
    group by sku_id
),
tmp_refund as --退款統計
(
    select
        sku_id,
        count(*) refund_count, --被退款次數
        sum(refund_num) refund_num, --被退款件數
        sum(refund_amount) refund_amount --被退款金額
    from dwd_fact_order_refund_info
    where dt='2020-03-10'
    group by sku_id
),
```

```
tmp_cart as --加入購物車統計
(
    select
        sku_id,
        count(*) cart_count, --被加入購物車次數
        sum(sku_num) cart_num --被加入購物車件數
    from dwd_fact_cart_info
    where dt='2020-03-10'
    and date_format(create_time,'yyyy-MM-dd')='2020-03-10'
    group by sku_id
),
tmp_favor as --收藏統計
(
    select
        sku_id,
        count(*) favor_count --被收藏次數
    from dwd_fact_favor_info
    where dt='2020-03-10'
    and date_format(create_time,'yyyy-MM-dd')='2020-03-10'
    group by sku_id
),
tmp_appraise as
(
select
    sku_id,
    sum(if(appraise='1201',1,0)) appraise_good_count,
    sum(if(appraise='1202',1,0)) appraise_mid_count,
    sum(if(appraise='1203',1,0)) appraise_bad_count,
    sum(if(appraise='1204',1,0)) appraise_default_count
from dwd_fact_comment_info
where dt='2020-03-10'
group by sku_id
)

insert overwrite table dws_sku_action_daycount partition(dt='2020-03-10')
select
    sku_id,
    sum(order_count),
    sum(order_num),
```

```
    sum(order_amount),
    sum(payment_count),
    sum(payment_num),
    sum(payment_amount),
    sum(refund_count),
    sum(refund_num),
    sum(refund_amount),
    sum(cart_count),
    sum(cart_num),
    sum(favor_count),
    sum(appraise_good_count),
    sum(appraise_mid_count),
    sum(appraise_bad_count),
    sum(appraise_default_count)
from
(
    select
        sku_id,
        order_count,
        order_num,
        order_amount,
        0 payment_count,
        0 payment_num,
        0 payment_amount,
        0 refund_count,
        0 refund_num,
        0 refund_amount,
        0 cart_count,
        0 cart_num,
        0 favor_count,
        0 appraise_good_count,
        0 appraise_mid_count,
        0 appraise_bad_count,
        0 appraise_default_count
    from tmp_order
    union all
    select
        sku_id,
        0 order_count,
```

```
        0 order_num,
        0 order_amount,
        payment_count,
        payment_num,
        payment_amount,
        0 refund_count,
        0 refund_num,
        0 refund_amount,
        0 cart_count,
        0 cart_num,
        0 favor_count,
        0 appraise_good_count,
        0 appraise_mid_count,
        0 appraise_bad_count,
        0 appraise_default_count
    from tmp_payment
    union all
    select
        sku_id,
        0 order_count,
        0 order_num,
        0 order_amount,
        0 payment_count,
        0 payment_num,
        0 payment_amount,
        refund_count,
        refund_num,
        refund_amount,
        0 cart_count,
        0 cart_num,
        0 favor_count,
        0 appraise_good_count,
        0 appraise_mid_count,
        0 appraise_bad_count,
        0 appraise_default_count
    from tmp_refund
    union all
    select
        sku_id,
```

```
    0 order_count,
    0 order_num,
    0 order_amount,
    0 payment_count,
    0 payment_num,
    0 payment_amount,
    0 refund_count,
    0 refund_num,
    0 refund_amount,
    cart_count,
    cart_num,
    0 favor_count,
    0 appraise_good_count,
    0 appraise_mid_count,
    0 appraise_bad_count,
    0 appraise_default_count
from tmp_cart
union all
select
    sku_id,
    0 order_count,
    0 order_num,
    0 order_amount,
    0 payment_count,
    0 payment_num,
    0 payment_amount,
    0 refund_count,
    0 refund_num,
    0 refund_amount,
    0 cart_count,
    0 cart_num,
    favor_count,
    0 appraise_good_count,
    0 appraise_mid_count,
    0 appraise_bad_count,
    0 appraise_default_count
from tmp_favor
union all
select
```

```
        sku_id,
        0 order_count,
        0 order_num,
        0 order_amount,
        0 payment_count,
        0 payment_num,
        0 payment_amount,
        0 refund_count,
        0 refund_num,
        0 refund_amount,
        0 cart_count,
        0 cart_num,
        0 favor_count,
        appraise_good_count,
        appraise_mid_count,
        appraise_bad_count,
        appraise_default_count
    from tmp_appraise
)tmp
group by sku_id;
```

3）查詢結果資料

```
hive (gmall)> select * from dws_sku_action_daycount where dt='2020-03-10';
```

3. 每日優惠券統計

每日優惠券統計表以優惠券為中心，統計優惠券的相關行為數，透過優惠券
領用事實資料表與優惠券維度資料表左連接，獲得優惠券的基本資訊及領用
行為的相關資料。

1）建表敘述

```
hive (gmall)>
drop table if exists dws_coupon_use_daycount;
create external table dws_coupon_use_daycount
(
    `coupon_id` string  COMMENT '優惠券id',
    `coupon_name` string COMMENT '優惠券名稱',
```

```
    `coupon_type` string COMMENT '優惠券類型 1 現金券 2 折扣券 3 滿減券 4
滿件打折券',
    `condition_amount` string COMMENT '滿額數',
    `condition_num` string COMMENT '滿件數',
    `activity_id` string COMMENT '活動編號',
    `benefit_amount` string COMMENT '滿減金額',
    `benefit_discount` string COMMENT '折扣',
    `create_time` string COMMENT '建立時間',
    `range_type` string COMMENT '範圍類型 1商品 2品項 3品牌',
    `spu_id` string COMMENT '標準產品單位id',
    `tm_id` string COMMENT '品牌id',
    `category3_id` string COMMENT '品項id',
    `limit_num` string COMMENT '最多領用次數',
    `get_count` bigint COMMENT '領用次數',
    `using_count` bigint COMMENT '使用(下單)次數',
    `used_count` bigint COMMENT '使用(支付)次數'
) COMMENT '每日優惠券統計表'
PARTITIONED BY (`dt` string)
stored as parquet
location '/warehouse/gmall/dws/dws_coupon_use_daycount/'
```

2）匯入資料

```
hive (gmall)>
insert overwrite table dws_coupon_use_daycount partition(dt='2020-03-10')
select
    cu.coupon_id,
    ci.coupon_name,
    ci.coupon_type,
    ci.condition_amount,
    ci.condition_num,
    ci.activity_id,
    ci.benefit_amount,
    ci.benefit_discount,
    ci.create_time,
    ci.range_type,
    ci.spu_id,
    ci.tm_id,
```

```
    ci.category3_id,
    ci.limit_num,
    cu.get_count,
    cu.using_count,
    cu.used_count
from
(
    select
        coupon_id,
        sum(if(date_format(get_time,'yyyy-MM-dd')='2020-03-10',1,0))
get_count,
        sum(if(date_format(using_time,'yyyy-MM-dd')='2020-03-10',1,0))
using_count,
        sum(if(date_format(used_time,'yyyy-MM-dd')='2020-03-10',1,0))
used_count
    from dwd_fact_coupon_use
    where dt='2020-03-10'
    group by coupon_id
)cu
left join
(
    select
        *
    from dwd_dim_coupon_info
    where dt='2020-03-10'
)ci on cu.coupon_id=ci.id;
```

3）查詢結果資料

```
hive (gmall)> select * from dws_coupon_use_daycount where dt='2020-03-10';
```

4. 每日活動統計

每日活動統計表主要以活動為中心，透過訂單事實資料表 dwd_fact_order_info 獲得與活動有關的度量值。

1）建表敘述

```
hive (gmall)>
```

```
drop table if exists dws_activity_info_daycount;
create external table dws_activity_info_daycount(
    `id` string COMMENT '編號',
    `activity_name` string  COMMENT '活動名稱',
    `activity_type` string  COMMENT '活動類型',
    `start_time` string  COMMENT '開始時間',
    `end_time` string  COMMENT '結束時間',
    `create_time` string  COMMENT '建立時間',
    `order_count` bigint COMMENT '下單次數',
    `payment_count` bigint COMMENT '支付次數'
) COMMENT '每日活動統計表'
PARTITIONED BY (`dt` string)
row format delimited fields terminated by '\t'
location '/warehouse/gmall/dws/dws_activity_info_daycount/'
```

2）匯入資料

```
hive (gmall)>
insert overwrite table dws activity info daycount partition(dt='2020-03-10')
select
    oi.activity_id,
    ai.activity_name,
    ai.activity_type,
    ai.start_time,
    ai.end_time,
    ai.create_time,
    oi.order_count,
    oi.payment_count
from
(
    select
        activity_id,
        sum(if(date_format(create_time,'yyyy-MM-dd')='2020-03-10',1,0))
order_count,
        sum(if(date_format(payment_time,'yyyy-MM-dd')='2020-03-10',1,0))
payment_count
    from dwd_fact_order_info
    where (dt='2020-03-10' or dt=date_add('2020-03-10',-1))
    and activity_id is not null
```

```
    group by activity_id
)oi
join
(
    select
        *
    from dwd_dim_activity_info
    where dt='2020-03-10'
)ai
on oi.activity_id=ai.id;
```

3）查詢結果資料

```
hive (gmall)> select * from dws_activity_info_daycount where dt='2020-03-10';
```

6.5.4 DWS 層資料匯入指令稿

將 DWS 層的資料匯入過程撰寫成資料匯入指令稿。

（1）在 /home/atguigu/bin 目錄下建立指令稿 dwd_to_dws.sh。

```
[atguigu@hadoop102 bin]$ vim dwd_to_dws.sh
```

在指令稿中撰寫以下內容。

```
#!/bin/bash

APP=gmall
hive=/opt/module/hive/bin/hive

# 如果輸入了日期參數，則取輸入參數作為日期值；如果沒有輸入日期參數，則取目前
時間的前一天作為日期值
if [ -n "$1" ] ;then
    do_date=$1
else
    do_date=`date -d "-1 day" +%F`
fi

sql="
insert overwrite table ${APP}.dws_uv_detail_daycount partition(dt='$do_date')
```

```
select
    mid_id,
    concat_ws('|', collect_set(user_id)) user_id,
    concat_ws('|', collect_set(version_code)) version_code,
    concat_ws('|', collect_set(version_name)) version_name,
    concat_ws('|', collect_set(lang))lang,
    concat_ws('|', collect_set(source)) source,
    concat_ws('|', collect_set(os)) os,
    concat_ws('|', collect_set(area)) area,
    concat_ws('|', collect_set(model)) model,
    concat_ws('|', collect_set(brand)) brand,
    concat_ws('|', collect_set(sdk_version)) sdk_version,
    concat_ws('|', collect_set(gmail)) gmail,
    concat_ws('|', collect_set(height_width)) height_width,
    concat_ws('|', collect_set(app_time)) app_time,
    concat_ws('|', collect_set(network)) network,
    concat_ws('|', collect_set(lng)) lng,
    concat_ws('|', collect_set(lat)) lat,
    count(*) login_count
from ${APP}.dwd_start_log
where dt='$do_date'
group by mid_id;

with
tmp_login as
(
    select
        user_id,
        count(*) login_count
    from ${APP}.dwd_start_log
    where dt='$do_date'
    and user_id is not null
    group by user_id
),
tmp_cart as
(
    select
        user_id,
        count(*) cart_count,
```

```
            sum(cart_price*sku_num) cart_amount
    from ${APP}.dwd_fact_cart_info
    where dt='$do_date'
    and user_id is not null
    and date_format(create_time,'yyyy-MM-dd')='$do_date'
    group by user_id
),
tmp_order as
(
    select
        user_id,
        count(*) order_count,
        sum(final_total_amount) order_amount
    from ${APP}.dwd_fact_order_info
    where dt='$do_date'
    group by user_id
) ,
tmp_payment as
(
    select
        user_id,
        count(*) payment_count,
        sum(payment_amount) payment_amount
    from ${APP}.dwd_fact_payment_info
    where dt='$do_date'
    group by user_id
),
tmp_order_detail as
(
    select
        user_id,
        collect_set(named_struct('sku_id',sku_id,'sku_num',sku_num,
'order_count',order_count,'order_amount',order_amount)) order_stats
    from
    (
        select
            user_id,
            sku_id,
            sum(sku_num) sku_num,
```

```
            count(*) order_count,
            cast(sum(total_amount) as decimal(20,2)) order_amount
        from ${APP}.dwd_fact_order_detail
        where dt='$do_date'
        group by user_id,sku_id
    )tmp
    group by user_id
)

insert overwrite table ${APP}.dws_user_action_daycount partition(dt=
'$do_date')
select
    coalesce(tmp_login.user_id,tmp_cart.user_id,tmp_order.user_id,
tmp_payment.user_id,tmp_order_detail.user_id),
    login_count,
    nvl(cart_count,0),
    nvl(cart_amount,0),
    nvl(order_count,0),
    nvl(order_amount,0),
    nvl(payment_count,0),
    nvl(payment_amount,0),
    order_stats
from tmp_login
full outer join tmp_cart on tmp_login.user_id=tmp_cart.user_id
full outer join tmp_order on tmp_login.user_id=tmp_order.user_id
full outer join tmp_payment on tmp_login.user_id=tmp_payment.user_id
full outer join tmp_order_detail on tmp_login.user_id=tmp_order_detail.
user_id;

with
tmp_order as
(
    select
        sku_id,
        count(*) order_count,
        sum(sku_num) order_num,
        sum(total_amount) order_amount
    from ${APP}.dwd_fact_order_detail
    where dt='$do_date'
```

```
    group by sku_id
),
tmp_payment as
(
    select
        sku_id,
        count(*) payment_count,
        sum(sku_num) payment_num,
        sum(total_amount) payment_amount
    from ${APP}.dwd_fact_order_detail
    where dt='$do_date'
    and order_id in
    (
        select
            id
        from ${APP}.dwd_fact_order_info
        where (dt='$do_date' or dt=date_add('$do_date',-1))
        and date_format(payment_time,'yyyy-MM-dd')='$do_date'
    )
    group by sku_id
),
tmp_refund as
(
    select
        sku_id,
        count(*) refund_count,
        sum(refund_num) refund_num,
        sum(refund_amount) refund_amount
    from ${APP}.dwd_fact_order_refund_info
    where dt='$do_date'
    group by sku_id
),
tmp_cart as
(
    select
        sku_id,
        count(*) cart_count,
        sum(sku_num) cart_num
    from ${APP}.dwd_fact_cart_info
```

```
    where dt='$do_date'
    and date_format(create_time,'yyyy-MM-dd')='$do_date'
    group by sku_id
),
tmp_favor as
(
    select
        sku_id,
        count(*) favor_count
    from ${APP}.dwd_fact_favor_info
    where dt='$do_date'
    and date_format(create_time,'yyyy-MM-dd')='$do_date'
    group by sku_id
),
tmp_appraise as
(
    select
        sku_id,
        sum(if(appraise='1201',1,0)) appraise_good_count,
        sum(if(appraise='1202',1,0)) appraise_mid_count,
        sum(if(appraise='1203',1,0)) appraise_bad_count,
        sum(if(appraise='1204',1,0)) appraise_default_count
    from ${APP}.dwd_fact_comment_info
    where dt='$do_date'
    group by sku_id
)

insert overwrite table ${APP}.dws_sku_action_daycount partition(dt=
'$do_date')
select
    sku_id,
    sum(order_count),
    sum(order_num),
    sum(order_amount),
    sum(payment_count),
    sum(payment_num),
    sum(payment_amount),
    sum(refund_count),
    sum(refund_num),
```

```
        sum(refund_amount),
        sum(cart_count),
        sum(cart_num),
        sum(favor_count),
        sum(appraise_good_count),
        sum(appraise_mid_count),
        sum(appraise_bad_count),
        sum(appraise_default_count)
from
(
    select
        sku_id,
        order_count,
        order_num,
        order_amount,
        0 payment_count,
        0 payment_num,
        0 payment_amount,
        0 refund_count,
        0 refund_num,
        0 refund_amount,
        0 cart_count,
        0 cart_num,
        0 favor_count,
        0 appraise_good_count,
        0 appraise_mid_count,
        0 appraise_bad_count,
        0 appraise_default_count
    from tmp_order
    union all
    select
        sku_id,
        0 order_count,
        0 order_num,
        0 order_amount,
        payment_count,
        payment_num,
        payment_amount,
        0 refund_count,
```

```
        0 refund_num,
        0 refund_amount,
        0 cart_count,
        0 cart_num,
        0 favor_count,
        0 appraise_good_count,
        0 appraise_mid_count,
        0 appraise_bad_count,
        0 appraise_default_count
from tmp_payment
union all
select
        sku_id,
        0 order_count,
        0 order_num,
        0 order_amount,
        0 payment_count,
        0 payment_num,
        0 payment_amount,
        refund_count,
        refund_num,
        refund_amount,
        0 cart_count,
        0 cart_num,
        0 favor_count,
        0 appraise_good_count,
        0 appraise_mid_count,
        0 appraise_bad_count,
        0 appraise_default_count
from tmp_refund
union all
select
        sku_id,
        0 order_count,
        0 order_num,
        0 order_amount,
        0 payment_count,
        0 payment_num,
        0 payment_amount,
```

```
            0 refund_count,
            0 refund_num,
            0 refund_amount,
            cart_count,
            cart_num,
            0 favor_count,
            0 appraise_good_count,
            0 appraise_mid_count,
            0 appraise_bad_count,
            0 appraise_default_count
    from tmp_cart
    union all
    select
            sku_id,
            0 order_count,
            0 order_num,
            0 order_amount,
            0 payment_count,
            0 payment_num,
            0 payment_amount,
            0 refund_count,
            0 refund_num,
            0 refund_amount,
            0 cart_count,
            0 cart_num,
            favor_count,
            0 appraise_good_count,
            0 appraise_mid_count,
            0 appraise_bad_count,
            0 appraise_default_count
    from tmp_favor
    union all
    select
            sku_id,
            0 order_count,
            0 order_num,
            0 order_amount,
            0 payment_count,
            0 payment_num,
```

```
        0 payment_amount,
        0 refund_count,
        0 refund_num,
        0 refund_amount,
        0 cart_count,
        0 cart_num,
        0 favor_count,
        appraise_good_count,
        appraise_mid_count,
        appraise_bad_count,
        appraise_default_count
    from tmp_appraise
) tmp
group by sku_id;

insert overwrite table ${APP}.dws_coupon_use_daycount partition(dt=
'$do_date')
select
    cu.coupon_id,
    ci.coupon_name,
    ci.coupon_type,
    ci.condition_amount,
    ci.condition_num,
    ci.activity_id,
    ci.benefit_amount,
    ci.benefit_discount,
    ci.create_time,
    ci.range_type,
    ci.spu_id,
    ci.tm_id,
    ci.category3_id,
    ci.limit_num,
    cu.get_count,
    cu.using_count,
    cu.used_count
from
(
    select
        coupon_id,
```

```
        sum(if(date_format(get_time,'yyyy-MM-dd')='$do_date',1,0)) get_count,
        sum(if(date_format(using_time,'yyyy-MM-dd')='$do_date',1,0))
using_count,
        sum(if(date_format(used_time,'yyyy-MM-dd')='$do_date',1,0))
used_count
    from ${APP}.dwd_fact_coupon_use
    where dt='$do_date'
    group by coupon_id
)cu
left join
(
    select
        *
    from ${APP}.dwd_dim_coupon_info
    where dt='$do_date'
)ci on cu.coupon_id=ci.id;

insert overwrite table ${APP}.dws_activity_info_daycount partition(dt=
'$do_date')
select
    oi.activity_id,
    ai.activity_name,
    ai.activity_type,
    ai.start_time,
    ai.end_time,
    ai.create_time,
    oi.order_count,
    oi.payment_count
from
(
    select
        activity_id,
        sum(if(date_format(create_time,'yyyy-MM-dd')='$do_date',1,0))
order_count,
        sum(if(date_format(payment_time,'yyyy-MM-dd')='$do_date',1,0))
payment_count
    from ${APP}.dwd_fact_order_info
    where (dt='$do_date' or dt=date_add('$do_date',-1))
    and activity_id is not null
```

```
    group by activity_id
)oi
join
(
    select
        *
    from ${APP}.dwd_dim_activity_info
    where dt='$do_date'
)ai
on oi.activity_id=ai.id;
"

$hive -e "$sql"
```

（2）增加指令稿執行許可權。

```
[atguigu@hadoop102 bin]$ chmod 777 dwd_to_dwo.sh
```

（3）執行指令稿，匯入資料。

```
[atguigu@hadoop102 bin]$ dwd_to_dws.sh 2020-03-11
```

（4）查詢結果資料。

```
hive (gmall)>
select * from dws_uv_detail_daycount where dt='2020-03-11';
select * from dws_user_action_daycount where dt='2020-03-11';
select * from dws_sku_action_daycount where dt='2020-03-11';
select * from dws_sale_detail_daycount where dt='2020-03-11';
select * from dws_coupon_use_daycount where dt='2020-03-11';
select * from dws_activity_info_daycount where dt='2020-03-11';
```

6.6 資料倉儲架設——DWT 層

在 DWS 層的架設中，我們把不同的主題按照天進行了聚合，獲得了每天每個主題的相關事實度量資料。在 DWT 層中，我們將把這些不同的主題進行進一步整理，獲得每個主題的全量資料表。

DWT 層主題寬表記錄的欄位包含每個維度連結的不同事實資料表度量值、累計某個時間段的度量值，以及第一次時間、末次時間、累計至今的度量值。

6.6.1 裝置主題寬表

DWT 層的裝置主題寬表將在每日裝置行為表的基礎上進行進一步整理，獲得每台裝置對應的詳細資訊，每天將新增加的裝置資訊增加到裝置主題寬表中，並增加第一次活躍時間、末次活躍時間、當日活躍次數和累計活躍天數資訊，方便後續實現與裝置相關的需求，如圖 6-33 所示。

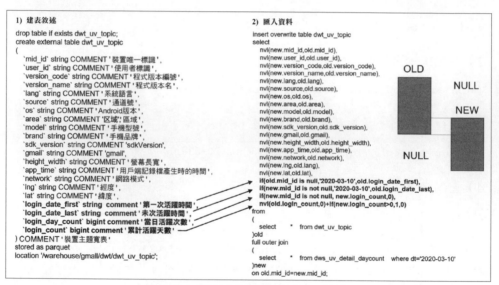

圖 6-33 裝置主題寬表資料匯入想法

1）建表敘述

```
hive (gmall)>
drop table if exists dwt_uv_topic;
create external table dwt_uv_topic
(
    `mid_id` string COMMENT '裝置唯一標識',
    `user_id` string COMMENT '使用者標識',
    `version_code` string COMMENT '程式版本編號',
```

```
    `version_name` string COMMENT '程式版本名',
    `lang` string COMMENT '系統語言',
    `source` string COMMENT '通路號',
    `os` string COMMENT 'Android版本',
    `area` string COMMENT '區域',
    `model` string COMMENT '手機型號',
    `brand` string COMMENT '手機品牌',
    `sdk_version` string COMMENT 'sdkVersion',
    `gmail` string COMMENT 'gmail',
    `height_width` string COMMENT '螢幕長寬',
    `app_time` string COMMENT '用戶端記錄檔產生時的時間',
    `network` string COMMENT '網路模式',
    `lng` string COMMENT '經度',
    `lat` string COMMENT '緯度',
    `login_date_first` string  comment '第一次活躍時間',
    `login_date_last` string  comment '末次活躍時間',
    `login_day_count` bigint comment '當日活躍次數',
    `login_count` bigint comment '累計活躍天數'
) COMMENT '裝置主題寬表'
stored as parquet
location '/warehouse/qmall/dwt/dwt_uv_topic';
```

2）匯入資料

```
hive (gmall)>
insert overwrite table dwt_uv_topic
select
    nvl(new.mid_id,old.mid_id),
    nvl(new.user_id,old.user_id),
    nvl(new.version_code,old.version_code),
    nvl(new.version_name,old.version_name),
    nvl(new.lang,old.lang),
    nvl(new.source,old.source),
    nvl(new.os,old.os),
    nvl(new.area,old.area),
    nvl(new.model,old.model),
    nvl(new.brand,old.brand),
```

```
    nvl(new.sdk_version,old.sdk_version),
    nvl(new.gmail,old.gmail),
    nvl(new.height_width,old.height_width),
    nvl(new.app_time,old.app_time),
    nvl(new.network,old.network),
    nvl(new.lng,old.lng),
    nvl(new.lat,old.lat),
    if(old.mid_id is null,'2020-03-10',old.login_date_first),
    if(new.mid_id is not null,'2020-03-10',old.login_date_last),
    if(new.mid_id is not null, new.login_count,0),
    nvl(old.login_count,0)+if(new.login_count>0,1,0)
from
(
    select
        *
    from dwt_uv_topic
)old
full outer join
(
    select
        *
    from dws_uv_detail_daycount
    where dt='2020-03-10'
)new
on old.mid_id=new.mid_id;
```

3）查詢結果資料

```
hive (gmall)> select * from dwt_uv_topic limit 5;
```

6.6.2 會員主題寬表

DWT 層會員主題寬表與多張事實資料表有連結，需要取得多個事實行為的第一次時間、末次時間和累計度量值。會員主題寬表資料匯入想法如圖 6-34 所示。

圖 6-34　會員主題寬表資料匯入想法

1）建表敘述

```
hive (gmall)>
drop table if exists dwt_user_topic;
create external table dwt_user_topic
(
    user_id string  comment '使用者id',
    login_date_first string  comment '第一次登入時間',
    login_date_last string  comment '末次登入時間',
    login_count bigint comment '累計登入天數',
    login_last_30d_count bigint comment '最近30日登入天數',
    order_date_first string  comment '第一次下單時間',
    order_date_last string  comment '末次下單時間',
    order_count bigint comment '累計下單次數',
    order_amount decimal(16,2) comment '累計下單金額',
    order_last_30d_count bigint comment '最近30日下單次數',
    order_last_30d_amount bigint comment '最近30日下單金額',
    payment_date_first string  comment '第一次支付時間',
    payment_date_last string  comment '末次支付時間',
    payment_count decimal(16,2) comment '累計支付次數',
    payment_amount decimal(16,2) comment '累計支付金額',
```

```
    payment_last_30d_count decimal(16,2) comment '最近30日支付次數',
    payment_last_30d_amount decimal(16,2) comment '最近30日支付金額'
)COMMENT '會員主題寬表'
stored as parquet
location '/warehouse/gmall/dwt/dwt_user_topic/'
```

2）匯入資料

```
hive (gmall)>
insert overwrite table dwt_user_topic
select
    nvl(new.user_id,old.user_id),
    if(old.login_date_first is null and new.login_count>0,'2020-03-10',
old.login_date_first),
    if(new.login_count>0,'2020-03-10',old.login_date_last),
    nvl(old.login_count,0)+if(new.login_count>0,1,0),
    nvl(new.login_last_30d_count,0),
    if(old.order_date_first is null and new.order_count>0,'2020-03-10',
old.order_date_first),
    if(new.order_count>0,'2020-03-10',old.order_date_last),
    nvl(old.order_count,0)+nvl(new.order_count,0),
    nvl(old.order_amount,0)+nvl(new.order_amount,0),
    nvl(new.order_last_30d_count,0),
    nvl(new.order_last_30d_amount,0),
    if(old.payment_date_first is null and new.payment_count>0,'2020-03-10',
old.payment_date_first),
    if(new.payment_count>0,'2020-03-10',old.payment_date_last),
    nvl(old.payment_count,0)+nvl(new.payment_count,0),
    nvl(old.payment_amount,0)+nvl(new.payment_amount,0),
    nvl(new.payment_last_30d_count,0),
    nvl(new.payment_last_30d_amount,0)
from
dwt_user_topic old
full outer join
(
    select
        user_id,
        sum(if(dt='2020-03-10',login_count,0)) login_count,
        sum(if(dt='2020-03-10',order_count,0)) order_count,
```

```
        sum(if(dt='2020-03-10',order_amount,0)) order_amount,
        sum(if(dt='2020-03-10',payment_count,0)) payment_count,
        sum(if(dt='2020-03-10',payment_amount,0)) payment_amount,
        sum(if(login_count>0,1,0)) login_last_30d_count,
        sum(order_count) order_last_30d_count,
        sum(order_amount) order_last_30d_amount,
        sum(payment_count) payment_last_30d_count,
        sum(payment_amount) payment_last_30d_amount
    from dws_user_action_daycount
    where dt>=date_add( '2020-03-10',-30)
    group by user_id
)new
on old.user_id=new.user_id;
```

3）查詢結果資料

```
hive (gmall)> select * from dwt_user_topic limit 5;
```

6.6.3 商品主題寬表

商品主題寬表與會員主題寬表稍有不同，商品的第一次被購買時間和末次被購買時間資料沒有太大的意義，重點需要取得多個事實行為的累計度量值和累計行為次數。

1）建表敘述

```
hive (gmall)>
drop table if exists dwt_sku_topic;
create external table dwt_sku_topic
(
    sku_id string comment '商品id',
    spu_id string comment '標準產品單位id',
    order_last_30d_count bigint comment '最近30日被下單次數',
    order_last_30d_num bigint comment '最近30日被下單例數',
    order_last_30d_amount decimal(16,2)  comment '最近30日被下單金額',
    order_count bigint comment '累計被下單次數',
    order_num bigint comment '累計被下單例數',
    order_amount decimal(16,2) comment '累計被下單金額',
```

```
    payment_last_30d_count    bigint   comment '最近30日被支付次數',
    payment_last_30d_num bigint comment '最近30日被支付件數',
    payment_last_30d_amount  decimal(16,2) comment '最近30日被支付金額',
    payment_count    bigint   comment '累計被支付次數',
    payment_num bigint comment '累計被支付件數',
    payment_amount   decimal(16,2) comment '累計被支付金額',
    refund_last_30d_count bigint comment '最近30日退款次數',
    refund_last_30d_num bigint comment '最近30日退款件數',
    refund_last_30d_amount decimal(10,2) comment '最近30日退款金額',
    refund_count bigint comment '累計退款次數',
    refund_num bigint comment '累計退款件數',
    refund_amount decimal(10,2) comment '累計退款金額',
    cart_last_30d_count bigint comment '最近30日被加入購物車次數',
    cart_last_30d_num bigint comment '最近30日被加入購物車件數',
    cart_count bigint comment '累計被加入購物車次數',
    cart_num bigint comment '累計被加入購物車件數',
    favor_last_30d_count bigint comment '最近30日被收藏次數',
    favor_count bigint comment '累計被收藏次數',
    appraise_last_30d_good_count bigint comment '最近30日好評數',
    appraise_last_30d_mid_count bigint comment '最近30日中評數',
    appraise_last_30d_bad_count bigint comment '最近30日差評數',
    appraise_last_30d_default_count bigint comment '最近30日預設評價數',
    appraise_good_count bigint comment '累計好評數',
    appraise_mid_count bigint comment '累計中評數',
    appraise_bad_count bigint comment '累計差評數',
    appraise_default_count bigint comment '累計預設評價數'
)COMMENT '商品主題寬表'
stored as parquet
location '/warehouse/gmall/dwt/dwt_sku_topic/'
```

2）匯入資料

```
hive (gmall)>
insert overwrite table dwt_sku_topic
select
    nvl(new.sku_id,old.sku_id),
    sku_info.spu_id,
```

```
    nvl(new.order_count30,0),
    nvl(new.order_num30,0),
    nvl(new.order_amount30,0),
    nvl(old.order_count,0) + nvl(new.order_count,0),
    nvl(old.order_num,0) + nvl(new.order_num,0),
    nvl(old.order_amount,0) + nvl(new.order_amount,0),
    nvl(new.payment_count30,0),
    nvl(new.payment_num30,0),
    nvl(new.payment_amount30,0),
    nvl(old.payment_count,0) + nvl(new.payment_count,0),
    nvl(old.payment_num,0) + nvl(new.payment_count,0),
    nvl(old.payment_amount,0) + nvl(new.payment_count,0),
    nvl(new.refund_count30,0),
    nvl(new.refund_num30,0),
    nvl(new.refund_amount30,0),
    nvl(old.refund_count,0) + nvl(new.refund_count,0),
    nvl(old.refund_num,0) + nvl(new.refund_num,0),
    nvl(old.refund_amount,0) + nvl(new.refund_amount,0),
    nvl(new.cart_count30,0),
    nvl(new.cart_num30,0),
    nvl(old.cart_count,0) + nvl(new.cart_count,0),
    nvl(old.cart_num,0) + nvl(new.cart_num,0),
    nvl(new.favor_count30,0),
    nvl(old.favor_count,0) + nvl(new.favor_count,0),
    nvl(new.appraise_good_count30,0),
    nvl(new.appraise_mid_count30,0),
    nvl(new.appraise_bad_count30,0),
    nvl(new.appraise_default_count30,0)   ,
    nvl(old.appraise_good_count,0) + nvl(new.appraise_good_count,0),
    nvl(old.appraise_mid_count,0) + nvl(new.appraise_mid_count,0),
    nvl(old.appraise_bad_count,0) + nvl(new.appraise_bad_count,0),
    nvl(old.appraise_default_count,0) + nvl(new.appraise_default_count,0)
from
(
    select
        sku_id,
        spu_id,
        order_last_30d_count,
        order_last_30d_num,
```

```
        order_last_30d_amount,
        order_count,
        order_num,
        order_amount   ,
        payment_last_30d_count,
        payment_last_30d_num,
        payment_last_30d_amount,
        payment_count,
        payment_num,
        payment_amount,
        refund_last_30d_count,
        refund_last_30d_num,
        refund_last_30d_amount,
        refund_count,
        refund_num,
        refund_amount,
        cart_last_30d_count,
        cart_last_30d_num,
        cart_count,
        cart_num,
        favor_last_30d_count,
        favor_count,
        appraise_last_30d_good_count,
        appraise_last_30d_mid_count,
        appraise_last_30d_bad_count,
        appraise_last_30d_default_count,
        appraise_good_count,
        appraise_mid_count,
        appraise_bad_count,
        appraise_default_count
    from dwt_sku_topic
)old
full outer join
(
    select
        sku_id,
        sum(if(dt='2020-03-10', order_count,0 )) order_count,
        sum(if(dt='2020-03-10',order_num ,0 ))  order_num,
        sum(if(dt='2020-03-10',order_amount,0 )) order_amount ,
```

```
        sum(if(dt='2020-03-10',payment_count,0 )) payment_count,
        sum(if(dt='2020-03-10',payment_num,0 )) payment_num,
        sum(if(dt='2020-03-10',payment_amount,0 )) payment_amount,
        sum(if(dt='2020-03-10',refund_count,0 )) refund_count,
        sum(if(dt='2020-03-10',refund_num,0 )) refund_num,
        sum(if(dt='2020-03-10',refund_amount,0 )) refund_amount,
        sum(if(dt='2020-03-10',cart_count,0 )) cart_count,
        sum(if(dt='2020-03-10',cart_num,0 )) cart_num,
        sum(if(dt='2020-03-10',favor_count,0 )) favor_count,
        sum(if(dt='2020-03-10',appraise_good_count,0 )) appraise_good_count,
        sum(if(dt='2020-03-10',appraise_mid_count,0 ) ) appraise_mid_count,
        sum(if(dt='2020-03-10',appraise_bad_count,0 )) appraise_bad_count,
        sum(if(dt='2020-03-10',appraise_default_count,0 )) appraise_default_
count,
        sum(order_count) order_count30 ,
        sum(order_num) order_num30,
        sum(order_amount) order_amount30,
        sum(payment_count) payment_count30,
        sum(payment_num) payment_num30,
        sum(payment_amount) payment_amount30,
        sum(refund_count) refund_count30,
        sum(refund_num) refund_num30,
        sum(refund_amount) refund_amount30,
        sum(cart_count) cart_count30,
        sum(cart_num) cart_num30,
        sum(favor_count) favor_count30,
        sum(appraise_good_count) appraise_good_count30,
        sum(appraise_mid_count) appraise_mid_count30,
        sum(appraise_bad_count) appraise_bad_count30,
        sum(appraise_default_count) appraise_default_count30
    from dws_sku_action_daycount
    where dt >= date_add ('2020-03-10', -30)
    group by sku_id
)new
on new.sku_id = old.sku_id
left join
(select * from dwd_dim_sku_info where dt='2020-03-10') sku_info
on nvl(new.sku_id,old.sku_id)= sku_info.id;
```

3）查詢結果資料

```
hive (gmall)> select * from dwt_sku_topic limit 5;
```

6.6.4 優惠券主題寬表

優惠券主題寬表主要取得優惠券的領用、下單使用、支付使用行為的累計發生次數和當日累計發生次數。優惠券主題寬表資料匯入想法如圖 6-35 所示。

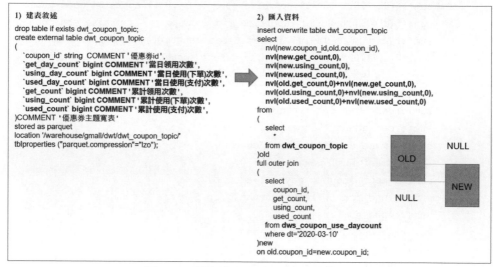

圖 6-35 優惠券主題寬表資料匯入想法

1）建表敘述

```
hive (gmall)>
drop table if exists dwt_coupon_topic;
create external table dwt_coupon_topic
(
    `coupon_id` string  COMMENT '優惠券id',
    `get_day_count` bigint COMMENT '當日領用次數',
    `using_day_count` bigint COMMENT '當日使用(下單)次數',
    `used_day_count` bigint COMMENT '當日使用(支付)次數',
    `get_count` bigint COMMENT '累計領用次數',
    `using_count` bigint COMMENT '累計使用(下單)次數',
```

```
     `used_count` bigint COMMENT '累計使用(支付)次數'
)COMMENT '優惠券主題寬表'
stored as parquet
location '/warehouse/gmall/dwt/dwt_coupon_topic/'
```

2）匯入資料

```
hive (gmall)>
insert overwrite table dwt_coupon_topic
select
    nvl(new.coupon_id,old.coupon_id),
    nvl(new.get_count,0),
    nvl(new.using_count,0),
    nvl(new.used_count,0),
    nvl(old.get_count,0)+nvl(new.get_count,0),
    nvl(old.using_count,0)+nvl(new.using_count,0),
    nvl(old.used_count,0)+nvl(new.used_count,0)
from
(
    select
        *
    from dwt_coupon_topic
)old
full outer join
(
    select
        coupon_id,
        get_count,
        using_count,
        used_count
    from dws_coupon_use_daycount
    where dt='2020-03-10'
)new
on old.coupon_id=new.coupon_id;
```

3）查詢結果資料

```
hive (gmall)> select * from dwt_coupon_topic limit 5;
```

6.6.5 活動主題寬表

活動主題寬表與優惠券主題寬表類似，主要取得下單、支付行為的當日行為
次數和累計行為次數。活動主題寬表資料匯入想法如圖 6-36 所示。

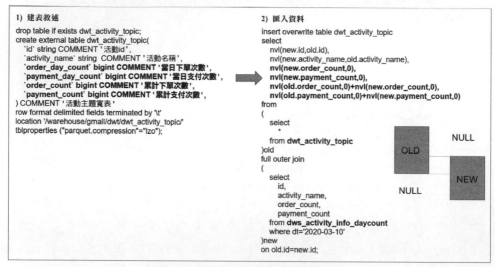

圖 6-36 活動主題寬表資料匯入想法

1）建表敘述

```
hive (gmall)>
drop table if exists dwt_activity_topic;
create external table dwt_activity_topic(
    `id` string COMMENT '活動id',
    `activity_name` string  COMMENT '活動名稱',
    `order_day_count` bigint COMMENT '當日下單次數',
    `payment_day_count` bigint COMMENT '當日支付次數',
    `order_count` bigint COMMENT '累計下單次數',
    `payment_count` bigint COMMENT '累計支付次數'
) COMMENT '活動主題寬表'
stored as parquet
location '/warehouse/gmall/dwt/dwt_activity_topic/'
```

2）匯入資料

```
hive (gmall)>
insert overwrite table dwt_activity_topic
select
    nvl(new.id,old.id),
    nvl(new.activity_name,old.activity_name),
    nvl(new.order_count,0),
    nvl(new.payment_count,0),
    nvl(old.order_count,0)+nvl(new.order_count,0),
    nvl(old.payment_count,0)+nvl(new.payment_count,0)
from
(
    select
        *
    from dwt_activity_topic
)old
full outer join
(
    select
        id,
        activity_name,
        order_count,
        payment_count
    from dws_activity_info_daycount
    where dt='2020-03-10'
)new
on old.id=new.id;
```

3）查詢結果資料

```
hive (gmall)> select * from dwt_activity_topic limit 5;
```

6.6.6 DWT 層資料匯入指令稿

將 DWT 層載入資料的過程撰寫成指令稿，方便每日呼叫執行。

（1）在 /home/atguigu/bin 目錄下建立指令稿 dws_to_dwt.sh。

```
[atguigu@hadoop102 bin]$ vim dws_to_dwt.sh
```

在指令稿中撰寫以下內容。

```bash
#!/bin/bash

APP=gmall
hive=/opt/module/hive/bin/hive

# 如果輸入了日期參數，則取輸入參數作為日期值；如果沒有輸入日期參數，則取目前
時間的前一天作為日期值
if [ -n "$1" ] ;then
    do_date=$1
else
    do_date=`date -d "-1 day" +%F`
fi

sql="
insert overwrite table ${APP}.dwt_uv_topic
select
    nvl(new.mid_id,old.mid_id),
    nvl(new.user_id,old.user_id),
    nvl(new.version_code,old.version_code),
    nvl(new.version_name,old.version_name),
    nvl(new.lang,old.lang),
    nvl(new.source,old.source),
    nvl(new.os,old.os),
    nvl(new.area,old.area),
    nvl(new.model,old.model),
    nvl(new.brand,old.brand),
    nvl(new.sdk_version,old.sdk_version),
    nvl(new.gmail,old.gmail),
    nvl(new.height_width,old.height_width),
    nvl(new.app_time,old.app_time),
    nvl(new.network,old.network),
    nvl(new.lng,old.lng),
    nvl(new.lat,old.lat),
    nvl(old.login_date_first,'$do_date'),
    if(new.login_count>0,'$do_date',old.login_date_last),
    nvl(new.login_count,0),
    nvl(new.login_count,0)+nvl(old.login_count,0)
```

```
from
(
    select
        *
    from ${APP}.dwt_uv_topic
)old
full outer join
(
    select
        *
    from ${APP}.dws_uv_detail_daycount
    where dt='$do_date'
)new
on old.mid_id=new.mid_id;

insert overwrite table ${APP}.dwt_user_topic
select
    nvl(new.user_id,old.user_id),
    if(old.login_date_first is null and new.login_count>0,'$do_date',
old.login_date_first),
    if(new.login_count>0,'$do_date',old.login_date_last),
    nvl(old.login_count,0)+if(new.login_count>0,1,0),
    nvl(new.login_last_30d_count,0),
    if(old.order_date_first is null and new.order_count>0,'$do_date',
old.order_date_first),
    if(new.order_count>0,'$do_date',old.order_date_last),
    nvl(old.order_count,0)+nvl(new.order_count,0),
    nvl(old.order_amount,0)+nvl(new.order_amount,0),
    nvl(new.order_last_30d_count,0),
    nvl(new.order_last_30d_amount,0),
    if(old.payment_date_first is null and new.payment_count>0,'$do_date',
old.payment_date_first),
    if(new.payment_count>0,'$do_date',old.payment_date_last),
    nvl(old.payment_count,0)+nvl(new.payment_count,0),
    nvl(old.payment_amount,0)+nvl(new.payment_amount,0),
    nvl(new.payment_last_30d_count,0),
    nvl(new.payment_last_30d_amount,0)
from
(
```

```
    select
        *
    from ${APP}.dwt_user_topic
)old
full outer join
(
    select
        user_id,
        sum(if(dt='$do_date',login_count,0)) login_count,
        sum(if(dt='$do_date',order_count,0)) order_count,
        sum(if(dt='$do_date',order_amount,0)) order_amount,
        sum(if(dt='$do_date',payment_count,0)) payment_count,
        sum(if(dt='$do_date',payment_amount,0)) payment_amount,
        sum(if(order_count>0,1,0)) login_last_30d_count,
        sum(order_count) order_last_30d_count,
        sum(order_amount) order_last_30d_amount,
        sum(payment_count) payment_last_30d_count,
        sum(payment_amount) payment_last_30d_amount
    from ${APP}.dws_user_action_daycount
    where dt>=date_add( '$do_date',-30)
    group by user_id
)new
on old.user_id=new.user_id;

with
sku_act as
(
select
    sku_id,
    sum(if(dt='$do_date', order_count,0 )) order_count,
    sum(if(dt='$do_date',order_num ,0 ))  order_num,
    sum(if(dt='$do_date',order_amount,0 )) order_amount ,
    sum(if(dt='$do_date',payment_count,0 )) payment_count,
    sum(if(dt='$do_date',payment_num,0 )) payment_num,
    sum(if(dt='$do_date',payment_amount,0 )) payment_amount,
    sum(if(dt='$do_date',refund_count,0 )) refund_count,
    sum(if(dt='$do_date',refund_num,0 )) refund_num,
    sum(if(dt='$do_date',refund_amount,0 )) refund_amount,
    sum(if(dt='$do_date',cart_count,0 )) cart_count,
```

```
    sum(if(dt='$do_date',cart_num,0 )) cart_num,
    sum(if(dt='$do_date',favor_count,0 )) favor_count,
    sum(if(dt='$do_date',appraise_good_count,0 )) appraise_good_count,
    sum(if(dt='$do_date',appraise_mid_count,0 ) ) appraise_mid_count ,
    sum(if(dt='$do_date',appraise_bad_count,0 )) appraise_bad_count,
    sum(if(dt='$do_date',appraise_default_count,0 )) appraise_default_count,
    sum( order_count  ) order_count30 ,
    sum( order_num  )  order_num30,
    sum(order_amount ) order_amount30,
    sum(payment_count ) payment_count30,
    sum(payment_num ) payment_num30,
    sum(payment_amount ) payment_amount30,
    sum(refund_count  ) refund_count30,
    sum(refund_num ) refund_num30,
    sum(refund_amount ) refund_amount30,
    sum(cart_count  ) cart_count30,
    sum(cart_num ) cart_num30,
    sum(favor_count ) favor_count30,
    sum(appraise_good_count ) appraise_good_count30,
    sum(appraise_mid_count  ) appraise_mid_count30,
    sum(appraise_bad_count ) appraise_bad_count30,
    sum(appraise_default_count )  appraise_default_count30
from ${APP}.dws_sku_action_daycount
where dt>=date_add ( '$do_date',-30)
group by sku_id
),
sku_topic
as
(
select
    sku_id,
    spu_id,
    order_last_30d_count,
    order_last_30d_num,
    order_last_30d_amount,
    order_count,
    order_num,
    order_amount  ,
    payment_last_30d_count,
```

```
    payment_last_30d_num,
    payment_last_30d_amount,
    payment_count,
    payment_num,
    payment_amount,
    refund_last_30d_count,
    refund_last_30d_num,
    refund_last_30d_amount ,
    refund_count ,
    refund_num ,
    refund_amount  ,
    cart_last_30d_count ,
    cart_last_30d_num ,
    cart_count ,
    cart_num ,
    favor_last_30d_count ,
    favor_count ,
    appraise_last_30d_good_count ,
    appraise_last_30d_mid_count ,
    appraise_last_30d_bad_count ,
    appraise_last_30d_default_count ,
    appraise_good_count ,
    appraise_mid_count ,
    appraise_bad_count ,
    appraise_default_count
from ${APP}.dwt_sku_topic
)
insert overwrite table ${APP}.dwt_sku_topic
select
    nvl(sku_act.sku_id,sku_topic.sku_id) ,
    sku_info.spu_id,
    nvl (sku_act.order_count30,0)        ,
    nvl (sku_act.order_num30,0)    ,
    nvl (sku_act.order_amount30,0)    ,
    nvl(sku_topic.order_count,0)+ nvl (sku_act.order_count,0) ,
    nvl(sku_topic.order_num,0)+ nvl (sku_act.order_num,0)    ,
    nvl(sku_topic.order_amount,0)+ nvl (sku_act.order_amount,0),
    nvl (sku_act.payment_count30,0),
    nvl (sku_act.payment_num30,0),
```

```
    nvl (sku_act.payment_amount30,0),
    nvl(sku_topic.payment_count,0)+ nvl (sku_act.payment_count,0) ,
    nvl(sku_topic.payment_num,0)+ nvl (sku_act.payment_count,0)  ,
    nvl(sku_topic.payment_amount,0)+ nvl (sku_act.payment_count,0)  ,
    nvl (refund_count30,0),
    nvl (sku_act.refund_num30,0),
    nvl (sku_act.refund_amount30,0),
    nvl(sku_topic.refund_count,0)+ nvl (sku_act.refund_count,0),
    nvl(sku_topic.refund_num,0)+ nvl (sku_act.refund_num,0),
    nvl(sku_topic.refund_amount,0)+ nvl (sku_act.refund_amount,0),
    nvl(sku_act.cart_count30,0)  ,
    nvl(sku_act.cart_num30,0)  ,
    nvl(sku_topic.cart_count  ,0)+ nvl (sku_act.cart_count,0),
    nvl( sku_topic.cart_num  ,0)+ nvl (sku_act.cart_num,0),
    nvl(sku_act.favor_count30 ,0)  ,
    nvl (sku_topic.favor_count  ,0)+ nvl (sku_act.favor_count,0),
    nvl (sku_act.appraise_good_count30 ,0)  ,
    nvl (sku_act.appraise_mid_count30 ,0)  ,
    nvl (sku_act.appraise_bad_count30 ,0)  ,
    nvl (sku_act.appraise_default_count30 ,0)  ,
    nvl (sku_topic.appraise_good_count  ,0)+ nvl (sku_act.appraise_good_
count,0) ,
    nvl (sku_topic.appraise_mid_count  ,0)+ nvl (sku_act.appraise_mid_
count,0) ,
    nvl (sku_topic.appraise_bad_count  ,0)+ nvl (sku_act.appraise_bad_
count,0) ,
    nvl (sku_topic.appraise_default_count  ,0)+ nvl (sku_act.appraise_
default_count,0)
from sku_act
full outer join sku_topic
on sku_act.sku_id =sku_topic.sku_id
left join
(select * from ${APP}.dwd_dim_sku_info where dt='$do_date') sku_info
on nvl(sku_topic.sku_id,sku_act.sku_id)= sku_info.id;

insert overwrite table ${APP}.dwt_coupon_topic
select
    nvl(new.coupon_id,old.coupon_id),
    nvl(new.get_count,0),
```

```
    nvl(new.using_count,0),
    nvl(new.used_count,0),
    nvl(old.get_count,0)+nvl(new.get_count,0),
    nvl(old.using_count,0)+nvl(new.using_count,0),
    nvl(old.used_count,0)+nvl(new.used_count,0)
from
(
    select
        *
    from ${APP}.dwt_coupon_topic
)old
full outer join
(
    select
        coupon_id,
        get_count,
        using_count,
        used_count
    from ${APP}.dws_coupon_use_daycount
    where dt='$do_date'
)new
on old.coupon_id=new.coupon_id;

insert overwrite table ${APP}.dwt_activity_topic
select
    nvl(new.id,old.id),
    nvl(new.activity_name,old.activity_name),
    nvl(new.order_count,0),
    nvl(new.payment_count,0),
    nvl(old.order_count,0)+nvl(new.order_count,0),
    nvl(old.payment_count,0)+nvl(new.payment_count,0)
from
(
    select
        *
    from ${APP}.dwt_activity_topic
)old
full outer join
(
```

```
    select
        id,
        activity_name,
        order_count,
        payment_count
    from ${APP}.dws_activity_info_daycount
    where dt='$do_date'
)new
on old.id=new.id;
"

$hive -e "$sql"
```

（2）增加指令稿執行許可權。

```
[atguigu@hadoop102 bin]$ chmod 777 dws_to_dwt.sh
```

（3）執行指令稿，匯入資料。

```
[atguigu@hadoop102 bin]$ dws_to_dwt.sh 2020 03 11
```

（4）查詢結果資料。

```
hive (gmall)>
select * from dwt_uv_topic limit 5;
select * from dwt_user_topic limit 5;
select * from dwt_sku_topic limit 5;
select * from dwt_coupon_topic limit 5;
select * from dwt_activity_topic limit 5;
```

6.7 資料倉儲架設──ADS 層

前面已完成 ODS、DWD、DWS、DWT 層資料倉儲的架設，本節主要實現實際需求。

6.7.1 裝置主題

本節主要實現與裝置主題相關的需求，可透過對裝置主題寬表進行適當維度的聚合獲得結果。

1. 活躍裝置數（日、周、月）

需求定義如下。

- 日活：當日活躍的裝置數。
- 周活：當周活躍的裝置數。
- 月活：當月活躍的裝置數。

1）建表敘述

```
hive (gmall)>
drop table if exists ads_uv_count;
create external table ads_uv_count(
    `dt` string COMMENT '統計日期',
    `day_count` bigint COMMENT '當日活躍裝置數量',
    `wk_count`  bigint COMMENT '當周活躍裝置數量',
    `mn_count`  bigint COMMENT '當月活躍裝置數量',
    `is_weekend` string COMMENT 'Y或N表示是否是週末,用於獲得本周最後結果',
    `is_monthend` string COMMENT 'Y或N表示是否是月末,用於獲得本月最後結果'
) COMMENT '活躍裝置數表'
row format delimited fields terminated by '\t'
location '/warehouse/gmall/ads/ads_uv_count/';
```

2）匯入資料

```
hive (gmall)>
insert into table ads_uv_count
select
    '2020-03-10' dt,
    daycount.ct,
    wkcount.ct,
    mncount.ct,
    if(date_add(next_day('2020-03-10','MO'),-1)='2020-03-10','Y','N') ,
```

```
    if(last_day('2020-03-10')='2020-03-10','Y','N')
from
(
    select
        '2020-03-10' dt,
        count(*) ct
    from dwt_uv_topic
    where login_date_last='2020-03-10'
)daycount join
(
    select
        '2020-03-10' dt,
        count (*) ct
    from dwt_uv_topic
    where login_date_last>=date_add(next_day('2020-03-10','MO'),-7)
    and login_date_last<= date_add(next_day('2020-03-10','MO'),-1)
) wkcount on daycount.dt=wkcount.dt
join
(
    select
        '2020-03-10' dt,
        count (*) ct
    from dwt_uv_topic
    where date_format(login_date_last,'yyyy-MM')=date_format('2020-03-10',
'yyyy-MM')
)mncount on daycount.dt=mncount.dt;
```

3）查詢結果資料

```
hive (gmall)> select * from ads_uv_count;
```

2. 每日新增裝置數

1）建表敘述

```
hive (gmall)>
drop table if exists ads_new_mid_count;
create external table ads_new_mid_count
(
    `create_date`      string comment '建立時間',
```

```
    `new_mid_count`    BIGINT comment '新增裝置數量'
)  COMMENT '每日新增裝置數表'
row format delimited fields terminated by '\t'
location '/warehouse/gmall/ads/ads_new_mid_count/';
```

2）匯入資料

```
hive (gmall)>
insert into table ads_new_mid_count
select
    login_date_first,
    count(*)
from dwt_uv_topic
where login_date_first='2020-03-10'
group by login_date_first;
```

3）查詢結果資料

```
hive (gmall)> select * from ads_new_mid_count;
```

3. 沉默裝置數

需求定義如下。

沉默裝置：只在安裝當天啟動過，且啟動時間在 7 天前。

1）建表敘述

```
hive (gmall)>
drop table if exists ads_silent_count;
create external table ads_silent_count(
    `dt` string COMMENT '統計日期',
    `silent_count` bigint COMMENT '沉默裝置數'
)COMMENT '沉默裝置數表'
row format delimited fields terminated by '\t'
location '/warehouse/gmall/ads/ads_silent_count';
```

2）匯入 2020-03-15 的資料

```
hive (gmall)>
insert into table ads_silent_count
```

```
select
    '2020-03-15',
    count(*)
from dwt_uv_topic
where login_date_first=login_date_last
and login_date_last<=date_add('2020-03-15',-7);
```

3）查詢結果資料

```
hive (gmall)> select * from ads_silent_count;
```

4. 本周回流裝置數

需求定義如下。

本周回流裝置：上周未活躍，而本周活躍的裝置，且不是本周新增的裝置。

1）建表敘述

```
hive (gmall)>
drop table if exists ads_back_count;
create external table ads_back_count(
    `dt` string COMMENT '統計日期',
    `wk_dt` string COMMENT '統計日期所在周',
    `wastage_count` bigint COMMENT '回流裝置數'
) COMMENT '本周回流裝置數表'
row format delimited fields terminated by '\t'
location '/warehouse/gmall/ads/ads_back_count';
```

2）匯入資料

```
hive (gmall)>
insert into table ads_back_count
select
    '2020-03-15',
    count(*)
from
(
    select
        mid_id
    from dwt_uv_topic
```

```
    where login_date_last>=date_add(next_day('2020-03-15','MO'),-7)
    and login_date_last<= date_add(next_day('2020-03-15','MO'),-1)
    and login_date_first<date_add(next_day('2020-03-15','MO'),-7)
)current_wk
left join
(
    select
        mid_id
    from dws_uv_detail_daycount
    where dt>=date_add(next_day('2020-03-15','MO'),-7*2)
    and dt<= date_add(next_day('2020-03-15','MO'),-7-1)
    group by mid_id
)last_wk
on current_wk.mid_id=last_wk.mid_id
where last_wk.mid_id is null;
```

3）查詢結果資料

```
hive (gmall)> select * from ads_back_count;
```

5. 流失裝置數

需求定義如下。

流失裝置：最近 7 天未活躍的裝置。

1）建表敘述

```
hive (gmall)>
drop table if exists ads_wastage_count;
create external table ads_wastage_count(
    `dt` string COMMENT '統計日期',
    `wastage_count` bigint COMMENT '流失裝置數'
) COMMENT '流失裝置數表'
row format delimited fields terminated by '\t'
location '/warehouse/gmall/ads/ads_wastage_count';
```

2）匯入 2020-03-20 的資料

```
hive (gmall)>
insert into table ads_wastage_count
```

```
select
    '2020-03-20',
    count(*)
from
(
    select
        mid_id
    from dwt_uv_topic
    where login_date_last<=date_add('2020-03-20',-7)
    group by mid_id
)t1;
```

3）查詢結果資料

```
hive (gmall)> select * from ads_wastage_count;
```

6. 留存率

1）概念

留存裝置是指某段時間內的新增裝置（活躍裝置），經過一段時間後，又被繼續使用的裝置；留存裝置佔當時新增裝置（活躍裝置）的比例是留存率。

舉例來說，2019-02-10 新增裝置 100 台，在這 100 台裝置上，2019-02-11 啟動過應用的有 30 台，2019-02-12 啟動過應用的有 25 台，2019-02-13 啟動過應用的有 32 台，則 2019-02-10 新增裝置的次日留存率是 30/100 = 30%，兩日留存率是 25/100=25%，三日留存率是 32/100=32%，如圖 6-37 所示。

時間	新增裝置	1天後	2天後	3天後
2019-02-10	100	30% (2019-02-11)	25% (2019-02-12)	32% (2019-02-13)
2019-02-11	200	20% (2019-02-12)	15% (2019-02-13)	
2019-02-12	100	25% (2019-02-13)		
2019-02-13				

圖 6-37 留存率計算範例

2）需求描述

需求：每天計算前 1、2、3、…n 天的留存率。

分析:假設今天是 2 月 11 日,統計前 1 天也就是 2 月 10 日新增裝置的留存率,公式如下:

2 月 10 日新增裝置的留存率 =2 月 10 日的新增裝置且 2 月 11 日活躍的裝置數 /2 月 10 日的新增裝置數

3)想法分析

在 ADS 層中,對裝置建立時間和留存天數進行整理,即可取得每天新增裝置的後期留存情況,再將 1、2、3 天留存率資料進行整理。

4)建表敘述

```
hive (gmall)>
drop table if exists ads_user_retention_day_rate;
create external table ads_user_retention_day_rate
(
    `stat_date`          string comment '統計日期',
    `create_date`        string  comment '裝置新增日期',
    `retention_day`      int comment '截至目前日期留存天數',
    `retention_count`    bigint comment  '留存數量',
    `new_mid_count`      bigint comment '裝置新增數量',
    `retention_ratio`    decimal(10,2) comment '留存率'
)  COMMENT '每日裝置留存率表'
row format delimited fields terminated by '\t'
location '/warehouse/gmall/ads/ads_user_retention_day_rate/';
```

5)匯入資料

```
hive (gmall)>
insert into table ads_user_retention_day_rate
select
    '2020-03-10',--統計日期
    date_add('2020-03-10',-1),--新增日期
    1,--留存天數
    sum(if(login_date_first=date_add('2020-03-10',-1) and login_date_last=
'2020-03-10',1,0)),--2020-03-09的1日留存數
    sum(if(login_date_first=date_add('2020-03-10',-1),1,0)),--2020-03-09新增
```

```
    sum(if(login_date_first=date_add('2020-03-10',-1) and login_date_last=
'2020-03-10',1,0))/sum(if(login_date_first=date_add('2020-03-10',-1),
1,0))*100
from dwt_uv_topic

union all

select
    '2020-03-10',--統計日期
    date_add('2020-03-10',-2),--新增日期
    2,--留存天數
    sum(if(login_date_first=date_add('2020-03-10',-2) and login_date_last=
'2020-03-10',1,0)),--2020-03-08的2日留存數
    sum(if(login_date_first=date_add('2020-03-10',-2),1,0)),--2020-03-08新增
    sum(if(login_date_first=date_add('2020-03-10',-2) and login_date_last=
'2020-03-10',1,0))/sum(if(login_date_first=date_add('2020-03-10',-2),
1,0))*100
from dwt_uv_topic

union all

select
    '2020-03-10',--統計日期
    date_add('2020-03-10',-3),--新增日期
    3,--留存天數
    sum(if(login_date_first=date_add('2020-03-10',-3) and login_date_last=
'2020-03-10',1,0)),--2020-03-07的3日留存數
    sum(if(login_date_first=date_add('2020-03-10',-3),1,0)),--2020-03-07新增
    sum(if(login_date_first=date_add('2020-03-10',-3) and login_date_last=
'2020-03-10',1,0))/sum(if(login_date_first=date_add('2020-03-10',-3),
1,0))*100
from dwt_uv_topic;
```

6）查詢結果資料

```
hive (gmall)>select * from ads_user_retention_day_rate;
```

7. 最近連續三周活躍裝置數

1）建表敘述

```
hive (gmall)>
drop table if exists ads_continuity_wk_count;
create external table ads_continuity_wk_count(
    `dt` string COMMENT '統計日期,一般用結束周周日日期,如果每天計算一次,
則可用當天日期',
    `wk_dt` string COMMENT '持續時間',
    `continuity_count` bigint COMMENT '活躍次數'
)COMMENT '最近連續三周活躍裝置數表'
row format delimited fields terminated by '\t'
location '/warehouse/gmall/ads/ads_continuity_wk_count';
```

2）匯入 2020-03-10 所在周的資料

```
hive (gmall)>
insert into table ads_continuity_wk_count
select
    '2020-03-10',
    concat(date_add(next_day('2020-03-10','MO'),-7*3),'_',date_add(next_
day('2020-03-10','MO'),-1)),
    count(*)
from
(
    select
        mid_id
    from
    (
        Select  --尋找本周活躍裝置
            mid_id
        from dws_uv_detail_daycount
        where dt>=date_add(next_day('2020-03-10','monday'),-7)
        and dt<=date_add(next_day('2020-03-10','monday'),-1)
        group by mid_id

        union all

        select  --尋找上周活躍裝置
```

```
            mid_id
    from dws_uv_detail_daycount
    where dt>=date_add(next_day('2020-03-10','monday'),-7*2)
    and dt<=date_add(next_day('2020-03-10','monday'),-7-1)
    group by mid_id


    union all


    select --尋找上上周活躍裝置
        mid_id
    from dws_uv_detail_daycount
    where dt>=date_add(next_day('2020-03-10','monday'),-7*3)
    and dt<=date_add(next_day('2020-03-10','monday'),-7*2-1)
    group by mid_id
    )t1
    group by mid_id  --對三周內的所有活躍裝置進行分組
    having count(*)=3 --分組後，mid_id個數為3的裝置為最近連續三周活躍裝置
)t2
```

3）查詢結果資料

```
hive (gmall)> select * from ads_continuity_wk_count;
```

8. 最近七天內連續三天活躍裝置數

1）建表敘述

```
hive (gmall)>
drop table if exists ads_continuity_uv_count;
create external table ads_continuity_uv_count(
    `dt` string COMMENT '統計日期',
    `wk_dt` string COMMENT '最近七天日期',
    `continuity_count` bigint
) COMMENT '最近七天內連續三天活躍裝置數表'
row format delimited fields terminated by '\t'
location '/warehouse/gmall/ads/ads_continuity_uv_count';
```

2）匯入資料

```
hive (gmall)>
```

```
insert into table ads_continuity_uv_count
select
    '2020-03-10',
    concat(date_add('2020-03-10',-6),'_','2020-03-10'),
    count(*)
from
(
    select mid_id
    from
    (
        select mid_id
        from
        (
            select
                mid_id,
                date_sub(dt,rank) date_dif --取排序值與日期值之差作為連續標示
            from
            (
                select
                    mid_id,
                    dt,
                    --對七天內登入過的裝置按照登入日期進行排序
                    rank() over(partition by mid_id order by dt) rank
                from dws_uv_detail_daycount
                where dt>=date_add('2020-03-10',-6) and dt<='2020-03-10'
            )t1
        )t2
        group by mid_id,date_dif --按照連續標示和裝置的mid_id進行分組
        --分組後，mid_id個數大於或等於3的裝置為最近七天內連續三天活躍的裝置
        having count(*)>=3
    )t3
    group by mid_id
)t4;
```

3）查詢結果資料

```
hive (gmall)> select * from ads_continuity_uv_count;
```

6.7.2 會員主題

本節主要實現與會員主題相關的需求，大部分需求可透過 DWT 層的會員主題寬表實現。

1. 會員主題資訊

在建立了 DWT 層的會員主題寬表之後，大部分需求的實現都比較簡單，我們將這些比較簡單的需求整理到一起實現。

1）建表敘述

```
hive (gmall)>
drop table if exists ads_user_topic;
create external table ads_user_topic(
    `dt` string COMMENT '統計日期',
    `day_users` string COMMENT '活躍會員數',
    `day_new_users` string COMMENT '新增會員數',
    `day_new_payment_users` string COMMENT '新增付費會員數',
    `payment_users` string COMMENT '總付費會員數',
    `users` string COMMENT '總會員數',
    `day_users2users` decimal(10,2) COMMENT '會員活躍率',
    `payment_users2users` decimal(10,2) COMMENT '會員付費率',
    `day_new_users2users` decimal(10,2) COMMENT '會員新鮮度'
) COMMENT '會員主題資訊表'
row format delimited fields terminated by '\t'
location '/warehouse/gmall/ads/ads_user_topic';
```

2）匯入資料

```
hive (gmall)>
insert into table ads_user_topic
select
    '2020-03-10',
    sum(if(login_date_last='2020-03-10',1,0)),
    sum(if(login_date_first='2020-03-10',1,0)),
    sum(if(payment_date_first='2020-03-10',1,0)),
```

```
    sum(if(payment_count>0,1,0)),
    count(*),
    sum(if(login_date_last='2020-03-10',1,0))/count(*),
    sum(if(payment_count>0,1,0))/count(*),
    sum(if(login_date_first='2020-03-10',1,0))/sum(if(login_date_last=
'2020-03-10',1,0))
from dwt_user_topic
```

3）查詢結果資料

```
hive (gmall)> select * from ads_user_topic;
```

2. 使用者行為漏斗分析

使用者行為漏斗分析也稱為轉換率，實際求何種轉換率視實際需求而定，舉例來說，消費使用者轉換率指的是單日日活中最後有多少使用者下單消費，即消費使用者轉換率＝單日消費使用者數／日活數。再如：

新存取使用者轉換率＝單日新存取使用者數／日活數
新註冊使用者轉換率＝單日新註冊使用者數／日活數
新付費使用者轉換率＝單日新付費使用者數／日活數

在實際業務中，我們通常關注一些特定頁面和行為的轉換率，圖 6-38 顯示了首頁存取到商品詳情頁瀏覽的轉換率，商品詳情頁瀏覽到加入購物車的轉換率，加入購物車到提交訂單的轉換率，提交訂單到最後支付成功的轉換率。透過這些數字，我們可以得知網頁設計是否存在缺陷，以及在哪一步損失了使用者，進一步對網頁設計進行改進。

在本需求中，主要計算三個轉換率：存取到加入購物車的轉換率、加入購物車到下單的轉換率和下單到支付的轉換率。存取人數可以從 ADS 層的活躍裝置數表 ads_uv_count 中取得，加入購物車人數、下單人數和支付人數從每日會員行為表中取得，再進一步計算轉換率，即使用者行為漏斗分析。

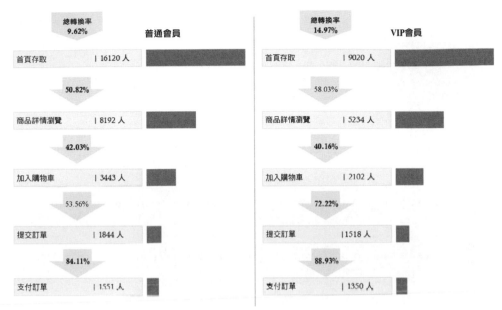

圖 6 38 轉換率示意

1）建表敘述

```
hive (gmall)>
drop table if exists ads_user_action_convert_day;
create external  table ads_user_action_convert_day(
    `dt` string COMMENT '統計日期',
    `total_visitor_m_count`  bigint COMMENT '總存取人數',
    `cart_u_count` bigint COMMENT '加入購物車的人數',
    `visitor2cart_convert_ratio` decimal(10,2) COMMENT '存取到加入購物車的
轉換率',
    `order_u_count` bigint      COMMENT '下單人數',
    `cart2order_convert_ratio`  decimal(10,2) COMMENT '加入購物車到下單的
轉換率',
    `payment_u_count` bigint       COMMENT '支付人數',
    `order2payment_convert_ratio` decimal(10,2) COMMENT '下單到支付的轉換率'
) COMMENT '使用者行為漏斗分析表'
row format delimited  fields terminated by '\t'
location '/warehouse/gmall/ads/ads_user_action_convert_day/';
```

2）匯入資料

```
hive (gmall)>
insert into table ads_user_action_convert_day
select
    '2020-03-10',
    uv.day_count,
    ua.cart_count,
    cast(ua.cart_count/uv.day_count as  decimal(10,2)) visitor2cart_
convert_ratio,
    ua.order_count,
    cast(ua.order_count/ua.cart_count as  decimal(10,2)) visitor2order_
convert_ratio,
    ua.payment_count,
    cast(ua.payment_count/ua.order_count as  decimal(10,2)) order2payment_
convert_ratio
from
(
    select
        dt,
        sum(if(cart_count>0,1,0)) cart_count,
        sum(if(order_count>0,1,0)) order_count,
        sum(if(payment_count>0,1,0)) payment_count
    from dws_user_action_daycount
where dt='2020-03-10'
group by dt
)ua join ads_uv_count uv on uv.dt=ua.dt;
```

3）查詢結果資料

```
hive (gmall)> select * from ads_user_action_convert_day;
```

6.7.3 商品主題

本節主要實現與商品主題相關的需求，大部分需求可透過 DWT 層的商品主題
寬表實現。

1. 商品個數資訊

1）建表敘述

```
hive (gmall)>
drop table if exists ads_product_info;
create external table ads_product_info(
    `dt` string COMMENT '統計日期',
    `sku_num` string COMMENT '商品個數',
    `spu_num` string COMMENT '標準產品單位個數'
) COMMENT '商品個數資訊表'
row format delimited fields terminated by '\t'
location '/warehouse/gmall/ads/ads_product_info';
```

2）匯入資料

```
hive (gmall)>
insert into table ads_product_info
select
    '2020-03-10' dt,
    sku_num,
    spu_num
from
(
    select
        '2020-03-10' dt,
        count(*) sku_num
    from
        dwt_sku_topic
) tmp_sku_num
join
(
    select
        '2020-03-10' dt,
        count(*) spu_num
    from
    (
        select
            spu_id
        from
```

```
          dwt_sku_topic
      group by
          spu_id
  ) tmp_spu_id
) tmp_spu_num
on
    tmp_sku_num.dt=tmp_spu_num.dt;
```

3）查詢結果資料

```
hive (gmall)> select * from ads_product_info;
```

2. 商品銷量排名

1）建表敘述

```
hive (gmall)>
drop table if exists ads_product_sale_topN;
create external table ads_product_sale_topN(
    `dt` string COMMENT '統計日期',
    `sku_id` string COMMENT '商品id',
    `payment_amount` bigint COMMENT '銷量'
) COMMENT '商品銷量排名表'
row format delimited fields terminated by '\t'
location '/warehouse/gmall/ads/ads_product_sale_topN';
```

2）匯入資料

```
hive (gmall)>
insert into table ads_product_sale_topN
select
    '2020-03-10' dt,
    sku_id,
    payment_amount
from
    dws_sku_action_daycount
where
    dt='2020-03-10'
order by payment_amount desc
limit 10;
```

3）查詢結果資料

```
hive (gmall)> select * from ads_product_sale_topN;
```

3. 商品收藏排名

1）建表敘述

```
hive (gmall)>
drop table if exists ads_product_favor_topN;
create external table ads_product_favor_topN(
    `dt` string COMMENT '統計日期',
    `sku_id` string COMMENT '商品id',
    `favor_count` bigint COMMENT '收藏量'
) COMMENT '商品收藏排名表'
row format delimited fields terminated by '\t'
location '/warehouse/gmall/ads/ads_product_favor_topN';
```

2）匯入資料

```
hive (gmall)>
insert into table ads_product_favor_topN
select
    '2020-03-10' dt,
    sku_id,
    favor_count
from
    dws_sku_action_daycount
where
    dt='2020-03-10'
order by favor_count desc
limit 10;
```

3）查詢結果資料

```
hive (gmall)> select * from ads_product_favor_topN;
```

4. 商品加入購物車排名

1）建表敘述

```
hive (gmall)>
drop table if exists ads_product_cart_topN;
create external table ads_product_cart_topN(
    `dt` string COMMENT '統計日期',
    `sku_id` string COMMENT '商品id',
    `cart_num` bigint COMMENT '加入購物車數量'
) COMMENT '商品加入購物車排名表'
row format delimited fields terminated by '\t'
location '/warehouse/gmall/ads/ads_product_cart_topN';
```

2）匯入資料

```
hive (gmall)>
insert into table ads_product_cart_topN
select
    '2020-03-10' dt,
    sku_id,
    cart_num
from
    dws_sku_action_daycount
where
    dt='2020-03-10'
order by cart_num desc
limit 10;
```

3）查詢結果資料

```
hive (gmall)> select * from ads_product_cart_topN;
```

5. 商品退款率排名（最近 30 天）

1）建表敘述

```
hive (gmall)>
drop table if exists ads_product_refund_topN;
create external table ads_product_refund_topN(
    `dt` string COMMENT '統計日期',
```

```
    `sku_id` string COMMENT '商品id',
    `refund_ratio` decimal(10,2) COMMENT '退款率'
) COMMENT '商品退款率排名表'
row format delimited fields terminated by '\t'
location '/warehouse/gmall/ads/ads_product_refund_topN';
```

2）匯入資料

```
hive (gmall)>
insert into table ads_product_refund_topN
select
    '2020-03-10',
    sku_id,
    refund_last_30d_count/payment_last_30d_count*100 refund_ratio
from dwt_sku_topic
order by refund_ratio desc
limit 10;
```

3）查詢結果資料

```
hive (gmall)> select * from ads_product_refund_topN;
```

6. 商品差評率

1）建表敘述

```
hive (gmall)>
drop table if exists ads_appraise_bad_topN;
create external table ads_appraise_bad_topN(
    `dt` string COMMENT '統計日期',
    `sku_id` string COMMENT '商品id',
    `appraise_bad_ratio` decimal(10,2) COMMENT '差評率'
) COMMENT '商品差評率排名表'
row format delimited fields terminated by '\t'
location '/warehouse/gmall/ads/ads_appraise_bad_topN';
```

2）匯入資料

```
hive (gmall)>
insert into table ads_appraise_bad_topN
```

```
select
    '2020-03-10' dt,
    sku_id,
appraise_bad_count/(appraise_good_count+appraise_mid_count+appraise_bad_
count+appraise_default_count) appraise_bad_ratio
from
    dws_sku_action_daycount
where
    dt='2020-03-10'
order by appraise_bad_ratio desc
limit 10;
```

3）查詢結果資料

```
hive (gmall)> select * from ads_appraise_bad_topN;
```

6.7.4 行銷主題

本節主要實現同時有關使用者和商品 2 個主題的需求，在實際工作中，這種需求很常見，有時我們需要整合多張主題寬表來實現。

1. 下單數目統計

需求分析：統計每日下單筆數、下單金額及下單使用者數。

1）建表敘述

```
hive (gmall)>
drop table if exists ads_order_daycount;
create external table ads_order_daycount(
    dt string comment '統計日期',
    order_count bigint comment '每日下單筆數',
    order_amount bigint comment '每日下單金額',
    order_users bigint comment '每日下單使用者數'
) comment '每日訂單總計表'
row format delimited fields terminated by '\t'
location '/warehouse/gmall/ads/ads_order_daycount';
```

2）匯入資料

```
hive (gmall)>
insert into table ads_order_daycount
select
    '2020-03-10',
    sum(order_count),
    sum(order_amount),
    sum(if(order_count>0,1,0))
from dws_user_action_daycount
where dt='2020-03-10';
```

3）查詢結果資料

```
hive (gmall)> select * from ads_order_daycount;
```

2. 支付資訊統計

需求分析：統計每日支付金額、支付人數、支付商品數、支付筆數，以及下單到支付的平均時長（取 DWD 層的資料）。

1）建表敘述

```
hive (gmall)>
drop table if exists ads_payment_daycount;
create external table ads_payment_daycount(
    dt string comment '統計日期',
    payment_count bigint comment '每日支付筆數',
    payment_amount bigint comment '每日支付金額',
    payment_user_count bigint comment '每日支付人數',
    payment_sku_count bigint comment '每日支付商品數',
    payment_avg_time double comment '下單到支付的平均時長,取分鐘數'
) comment '每日支付總計表'
row format delimited fields terminated by '\t'
location '/warehouse/gmall/ads/ads_payment_daycount';
```

2）匯入資料

```
hive (gmall)>
insert into table ads_payment_daycount
```

```
select
    tmp_payment.dt,
    tmp_payment.payment_count,
    tmp_payment.payment_amount,
    tmp_payment.payment_user_count,
    tmp_skucount.payment_sku_count,
    tmp_time.payment_avg_time
from
(
    select
        '2020-03-10' dt,
        sum(payment_count) payment_count,
        sum(payment_amount) payment_amount,
        sum(if(payment_count>0,1,0)) payment_user_count
    from dws_user_action_daycount
    where dt='2020-03-10'
)tmp_payment
join
(
    select
        '2020-03-10' dt,
        sum(if(payment_count>0,1,0)) payment_sku_count
    from dws_sku_action_daycount
    where dt='2020-03-10'
)tmp_skucount on tmp_payment.dt=tmp_skucount.dt
join
(
    select
        '2020-03-10' dt,
        sum(unix_timestamp(payment_time)-unix_timestamp(create_time))/
count(*)/60 payment_avg_time
    from dwd_fact_order_info
    where dt='2020-03-10'
    and payment_time is not null
)tmp_time on tmp_payment.dt=tmp_time.dt
```

3）查詢結果資料

```
hive (gmall)> select * from ads_payment_daycount;
```

3. 複購率

1）建表敘述

```
hive (gmall)>
drop table ads_sale_tm_category1_stat_mn;
create external table ads_sale_tm_category1_stat_mn
(
    tm_id string comment '品牌id',
    category1_id string comment '一級品項id ',
    category1_name string comment '一級品項名稱 ',
    buycount    bigint comment  '購買人數',
    buy_twice_last bigint  comment '兩次以上購買人數',
    buy_twice_last_ratio decimal(10,2)  comment  '單次複購率',
    buy_3times_last   bigint comment   '三次以上購買人數',
    buy_3times_last_ratio decimal(10,2)  comment  '多次複購率',
    stat_mn string comment '統計月份',
    stat_date string comment '統計日期'
) COMMENT '複購率統計表'
row format delimited fields terminated by '\t'
location '/warehouse/gmall/ads/ads_sale_tm_category1_stat_mn/';
```

2）匯入資料

```
hive (gmall)>
with
tmp_order as
(
    select
        user_id,
        order_stats_struct.sku_id sku_id,
        order_stats_struct.order_count order_count
    from dws_user_action_daycount lateral view explode(order_stats) tmp
as order_stats_struct
    where date_format(dt,'yyyy-MM')=date_format('2020-03-10','yyyy-MM')
),
tmp_sku as
(
    select
```

```
        id,
        tm_id,
        category1_id,
        category1_name
    from dwd_dim_sku_info
    where dt-'2020-03-10'
)
insert into table ads_sale_tm_category1_stat_mn
select
    tm_id,
    category1_id,
    category1_name,
    sum(if(order_count>=1,1,0)) buycount,
    sum(if(order_count>=2,1,0)) buyTwiceLast,
    sum(if(order_count>=2,1,0))/sum( if(order_count>=1,1,0))
buyTwiceLastRatio,
    sum(if(order_count>=3,1,0))  buy3timeLast  ,
    sum(if(order_count>=3,1,0))/sum( if(order_count>=1,1,0))
buy3timeLastRatio,
    date_format('2020-03-10' ,'yyyy-MM') stat_mn,
    '2020-03-10' stat_date
from
(
    select
        tmp_order.user_id,
        tmp_sku.category1_id,
        tmp_sku.category1_name,
        tmp_sku.tm_id,
        sum(order_count) order_count
    from tmp_order
    join tmp_sku
    on tmp_order.sku_id=tmp_sku.id
    group by tmp_order.user_id,tmp_sku.category1_id,tmp_sku.category1_name,
tmp_sku.tm_id
)tmp
group by tm_id, category1_id, category1_name
```

6.7.5 ADS 層資料匯入指令稿

（1）在 /home/atguigu/bin 目錄下建立指令稿 dwt_to_ads.sh。

```
[atguigu@hadoop102 bin]$ vim dwt_to_ads.sh
```

在指令稿中撰寫以下內容。

```bash
#!/bin/bash

hive=/opt/module/hive/bin/hive

# 如果輸入了日期參數，則取輸入參數作為日期值；如果沒有輸入日期參數，則取目前
時間的前一天作為日期值
if [ -n "$1" ] ;then
    do_date=$1
else
    do_date=`date -d "-1 day" +%F`
fi

sql="use gmall;
insert into table ads_uv_count
select
    '$do_date',
    sum(if(login_date_last='$do_date',1,0)),
    sum(if(login_date_last>=date_add(next_day('$do_date','monday'),-7)
and login_date_last<=date_add(next_day('$do_date','monday'),-1) ,1,0)),
    sum(if(date_format(login_date_last,'yyyy-MM')=date_format('$do_date',
'yyyy-MM'),1,0)),
    if('$do_date'=date_add(next_day('$do_date','monday'),-1),'Y','N'),
    if('$do_date'=last_day('$do_date'),'Y','N')
from dwt_uv_topic;

insert into table ads_new_mid_count
select
    '$do_date',
    count(*)
from dwt_uv_topic
where login_date_first='$do_date';
```

```
insert into table ads_silent_count
select
    '$do_date',
    count(*)
from dwt_uv_topic
where login_date_first=login_date_last
and login_date_last<=date_add('$do_date',-7);

insert into table ads_back_count
select
    '$do_date',
    concat(date_add(next_day('2020-03-10','MO'),-7),'_',date_add(next_day
('2020-03-10','MO'),-1)),
    count(*)
from
(
    select
        mid_id
    from dwt_uv_topic
    where login_date_last>=date_add(next_day('$do_date','MO'),-7)
    and login_date_last<= date_add(next_day('$do_date','MO'),-1)
    and login_date_first<date_add(next_day('$do_date','MO'),-7)
)current_wk
left join
(
    select
        mid_id
    from dws_uv_detail_daycount
    where dt>=date_add(next_day('$do_date','MO'),-7*2)
    and dt<= date_add(next_day('$do_date','MO'),-7-1)
    group by mid_id
)last_wk
on current_wk.mid_id=last_wk.mid_id
where last_wk.mid_id is null;

insert into table ads_wastage_count
select
    '$do_date',
```

```
    count(*)
from dwt_uv_topic
where login_date_last<=date_add('$do_date',-7);


insert into table ads_user_retention_day_rate
select
    '$do_date',
    date_add('$do_date',-3),
    3,
    sum(if(login_date_first=date_add('$do_date',-3) and login_date_last=
'$do_date',1,0)),
    sum(if(login_date_first=date_add('$do_date',-3),1,0)),
    sum(if(login_date_first=date_add('$do_date',-3) and login_date_last=
'$do_date',1,0))/sum(if(login_date_first=date_add('$do_date',-3),1,0))*100
from dwt_uv_topic
union all
select
    '$do_date',
    date_add('$do_date',-2),
    2,
    sum(if(login_date_first=date_add('$do_date',-2) and login_date_last=
'$do_date',1,0)),
    sum(if(login_date_first=date_add('$do_date',-2),1,0)),
    sum(if(login_date_first=date_add('$do_date',-2) and login_date_last=
'$do_date',1,0))/sum(if(login_date_first=date_add('$do_date',-2),1,0))*100
from dwt_uv_topic
union all
select
    '$do_date',
    date_add('$do_date',-1),
    1,
    sum(if(login_date_first=date_add('$do_date',-1) and login_date_last=
'$do_date',1,0)),
    sum(if(login_date_first=date_add('$do_date',-1),1,0)),
    sum(if(login_date_first=date_add('$do_date',-1) and login_date_last=
'$do_date',1,0))/sum(if(login_date_first=date_add('$do_date',-1),1,0))*100
from dwt_uv_topic;


insert into table ads_continuity_wk_count
```

```
select
    '$do_date',
    concat(date_add(next_day('$do_date','MO'),-7*3),'_',date_add(next_day
('$do_date','MO'),-1)),
    count(*)
from
(
    select
        mid_id
    from
    (
        select
            mid_id
        from dws_uv_detail_daycount
        where dt>=date_add(next_day('$do_date','monday'),-7)
        and dt<=date_add(next_day('$do_date','monday'),-1)
        group by mid_id

        union all

        select
            mid_id
        from dws_uv_detail_daycount
        where dt>=date_add(next_day('$do_date','monday'),-7*2)
        and dt<=date_add(next_day('$do_date','monday'),-7-1)
        group by mid_id

        union all

        select
            mid_id
        from dws_uv_detail_daycount
        where dt>=date_add(next_day('$do_date','monday'),-7*3)
        and dt<=date_add(next_day('$do_date','monday'),-7*2-1)
        group by mid_id
    )t1
    group by mid_id
    having count(*)=3
)t2;
```

```
insert into table ads_continuity_uv_count
select
    '$do_date',
    concat(date_add('$do_date',-6),'_','$do_date'),
    count(*)
from
(
    select mid_id
    from
    (
        select mid_id
        from
        (
            select
                mid_id,
                date_sub(dt,rank) date_dif
            from
            (
                select
                    mid_id,
                    dt,
                    rank() over(partition by mid_id order by dt) rank
                from dws_uv_detail_daycount
                where dt>=date_add('$do_date',-6) and dt<='$do_date'
            )t1
        )t2
        group by mid_id,date_dif
        having count(*)>=3
    )t3
    group by mid_id
)t4;

insert into table ads_user_topic
select
    '$do_date',
    sum(if(login_date_last='$do_date',1,0)),
    sum(if(login_date_first='$do_date',1,0)),
    sum(if(payment_date_first='$do_date',1,0)),
```

```
    sum(if(payment_count>0,1,0)),
    count(*),
    sum(if(login_date_last='$do_date',1,0))/count(*),
    sum(if(payment_count>0,1,0))/count(*),
    sum(if(login_date_first='$do_date',1,0))/sum(if(login_date_last=
'$do_date',1,0))
from dwt_user_topic;

insert into table ads_user_action_convert_day
select
    '$do_date',
    uv.day_count,
    ua.cart_count,
    ua.cart_count/uv.day_count*100 visitor2cart_convert_ratio,
    ua.order_count,
    ua.order_count/ua.cart_count*100  visitor2order_convert_ratio,
    ua.payment_count,
    ua.payment_count/ua.order_count*100 order2payment_convert_ratio
from
(
    select
        '$do_date' dt,
        sum(if(cart_count>0,1,0)) cart_count,
        sum(if(order_count>0,1,0)) order_count,
        sum(if(payment_count>0,1,0)) payment_count
    from dws_user_action_daycount
    where dt='$do_date'
)ua join ads_uv_count uv on uv.dt=ua.dt;

insert into table ads_product_info
select
    '$do_date' dt,
    sku_num,
    spu_num
from
(
    select
        '$do_date' dt,
        count(*) sku_num
```

```
        from
            dwt_sku_topic
    ) tmp_sku_num
    join
    (
        select
            '$do_date' dt,
            count(*) spu_num
        from
        (
            select
                spu_id
            from
                dwt_sku_topic
            group by
                spu_id
        ) tmp_spu_id
    ) tmp_spu_num
    on tmp_sku_num.dt=tmp_spu_num.dt;

    insert into table ads_product_sale_topN
    select
        '$do_date',
        sku_id,
        payment_amount
    from dws_sku_action_daycount
    where dt='$do_date'
    order by payment_amount desc
    limit 10;

    insert into table ads_product_favor_topN
    select
        '$do_date',
        sku_id,
        favor_count
    from dws_sku_action_daycount
    where dt='$do_date'
    order by favor_count
    limit 10;
```

```sql
insert into table ads_product_cart_topN
select
    '$do_date' dt,
    sku_id,
    cart_num
from dws_sku_action_daycount
where dt='$do_date'
order by cart_num
limit 10;

insert into table ads_product_refund_topN
select
    '$do_date',
    sku_id,
    refund_last_30d_count/payment_last_30d_count*100 refund_ratio
from dwt_sku_topic
order by refund_ratio desc
limit 10;

insert into table ads_appraise_bad_topN
select
    '$do_date' dt,
    sku_id,
    appraise_bad_count/(appraise_bad_count+appraise_good_count+appraise_
mid_count+appraise_default_count)*100 appraise_bad_ratio
from dws_sku_action_daycount
where dt='$do_date'
order by appraise_bad_ratio desc
limit 10;

insert into table ads_order_daycount
select
    '$do_date',
    sum(order_count),
    sum(order_amount),
    sum(if(order_count>0,1,0))
from dws_user_action_daycount
where dt='$do_date';
```

```
insert into table ads_payment_daycount
select
    tmp_payment.dt,
    tmp_payment.payment_count,
    tmp_payment.payment_amount,
    tmp_payment.payment_user_count,
    tmp_skucount.payment_sku_count,
    tmp_time.payment_avg_time
from
(
    select
        '$do_date' dt,
        sum(payment_count) payment_count,
        sum(payment_amount) payment_amount,
        sum(if(payment_count>0,1,0)) payment_user_count
    from dws_user_action_daycount
    where dt='$do_date'
)tmp_payment
join
(
    select
        '$do_date' dt,
        sum(if(payment_count>0,1,0)) payment_sku_count
    from dws_sku_action_daycount
    where dt='$do_date'
)tmp_skucount on tmp_payment.dt=tmp_skucount.dt
join
(
    select
        '$do_date' dt,
        sum(unix_timestamp(payment_time)-unix_timestamp(create_time))/
count(*)/60 payment_avg_time
    from dwd_fact_order_info
    where dt='$do_date'
    and payment_time is not null
)tmp_time on tmp_payment.dt=tmp_time.dt;

with
```

```
tmp_order as
(
    select
        user_id,
        order_stats_struct.sku_id sku_id,
        order_stats_struct.order_count order_count
    from dws_user_action_daycount lateral view explode(order_stats) tmp
as order_stats_struct
    where date_format(dt,'yyyy-MM')=date_format('$do_date','yyyy-MM')
),
tmp_sku as
(
    select
        id,
        tm_id,
        category1_id,
        category1_name
    from dwd_dim_sku_info
    where dt='$do_date'
)
insert into table ads_sale_tm_category1_stat_mn
select
    tm_id,
    category1_id,
    category1_name,
    sum(if(order_count>=1,1,0)) buycount,
    sum(if(order_count>=2,1,0)) buyTwiceLast,
    sum(if(order_count>=2,1,0))/sum( if(order_count>=1,1,0))
buyTwiceLastRatio,
    sum(if(order_count>=3,1,0))  buy3timeLast  ,
    sum(if(order_count>=3,1,0))/sum( if(order_count>=1,1,0))
buy3timeLastRatio,
    date_format('$do_date' ,'yyyy-MM') stat_mn,
    '$do_date' stat_date
from
(
    select
        tmp_order.user_id,
        tmp_sku.category1_id,
```

```
        tmp_sku.category1_name,
        tmp_sku.tm_id,
        sum(order_count) order_count
    from tmp_order
    join tmp_sku
    on tmp_order.sku_id=tmp_sku.id
    group by tmp_order.user_id,tmp_sku.category1_id,tmp_sku.category1_
name,tmp_sku.tm_id
)tmp
group by tm_id, category1_id, category1_name;
"

$hive -e "$sql"
```

6.8 結果資料匯出指令稿

想要對資料倉儲中的結果資料進行視覺化，需要使用 Sqoop 將結果資料匯出
到關聯式資料庫（MySQL）中。使用 Sqoop 可以方便地在關聯式資料庫和巨
量資料儲存系統之間實現資料傳輸。透過撰寫資料匯出指令稿即可實現。指
令稿撰寫步驟如下所示。

1. 撰寫結果資料的 Sqoop 匯出指令稿

在 /home/atguigu/bin 目錄下建立指令稿 hdfs_to_mysql.sh。

```
[atguigu@hadoop102 bin]$ vim hdfs_to_mysql.sh
```

在指令稿中撰寫以下內容。

```
#!/bin/bash

hive_db_name=gmall
mysql_db_name=gmall_report

export_data() {
/opt/module/sqoop/bin/sqoop export \
```

```
--connect "jdbc:mysql://hadoop102:3306/${mysql_db_name}?useUnicode=
true&characterEncoding=utf-8"  \
--username root \
--password 000000 \
--table $1 \
--num-mappers 1 \
--export-dir /warehouse/$hive_db_name/ads/$1 \
--input-fields-terminated-by "\t" \
--update-mode allowinsert \
--update-key $2 \
--input-null-string '\\N'    \
--input-null-non-string '\\N'
}

case $1 in
  "ads_uv_count")
    export_data "ads_uv_count" "dt"
;;
  "ads_user_action_convert_day")
    export_data "ads_user_action_convert_day" "dt"
;;
  "ads_user_topic")
    export_data "ads_user_topic" "dt"
;;
  "all")
    export_data "ads_uv_count" "dt"
    export_data "ads_user_action_convert_day" "dt"
    export_data "ads_user_topic" "dt"
;;
esac
```

匯出指令稿中的指令參數說明如下。

- --connect、--username、--password 參數與匯入指令中的參數含義相同，參見 5.2.5 節相關內容。
- --table：匯出到 MySQL 中的表名。
- --num-mappers：匯出資料的作業個數。
- --export-dir：匯出資料所在目錄。

- --input-fields-terminated-by：指定分隔符號，需要與建立 Hive 表時使用的分隔符號一致。
- --update-mode：若傳入 updateonly，則匯出資料後只根據 update-key 進行更新，不允許插入新資料；若傳入 allowinsert，則可以根據 update-key 進行更新，並允許插入新資料。
- --update-key：在允許更新的情況下，指定一個或多個欄位，在進行資料匯出時，若這些欄位全部相同，則說明是同一筆資料，該筆資料將進行更新操作，而非插入一筆新資料。多個欄位用逗點分隔。舉例來說，一個表中存在 10 個欄位，我們可以指定其中 2 個欄位為 update-key，在進行資料匯出時，新舊資料間這 2 個欄位完全相同，則視新資料與舊資料為同一筆資料，就會執行更新操作，否則執行插入操作。
- --input-null-string 和 --input-null-non-string：將字串列與非字串列的空字串和 "NULL" 轉換成 '\\N'。

Hive 中的 NULL 在底層是以 "\N" 儲存的，而 MySQL 中的 NULL 在底層就是 NULL，為了確保資料兩端的一致性，在匯出資料時使用 --input-null-string 和 --input-null-non-string 兩個參數，匯入資料時則使用 --null-string 和 --null-non-string 兩個參數，這樣可避免匯出資料時出錯。

2. Sqoop 匯出指令稿用法

```
[atguigu@hadoop102 bin]$ chmod 777 hdfs_to_mysql.sh
[atguigu@hadoop102 bin]$ hdfs_to_mysql.sh all
```

6.9 會員主題指標取得的全排程流程

6.9.1 Azkaban 安裝

Azkaban 是由 Linkedin 公司推出的批次工作流任務排程器，主要用於在一個工作流內以一個特定的順序執行一組工作和流程。它透過簡單的 key-value 對的方式進行設定，透過設定中的 dependencies 來設定相依關係。Azkaban 使用

job 描述檔案建立任務之間的相依關係,並提供一個易用的 Web 介面維護和追蹤使用者的工作流。

1. 安裝前準備

(1)將 Azkaban Web 伺服器(azkaban-web-server-2.5.0.tar.gz)、Azkaban 執行伺服器(azkaban-executor-server-2.5.0.tar.gz)、Azkaban 的 SQL 執行指令稿(azkaban-sql-script-2.5.0.tar.gz)及 MySQL 安裝套件(mysql-libs.zip)複製到 hadoop102 虛擬機器的 /opt/software 目錄下。

(2)Azkaban 建立了一些 MySQL 連接增強功能,所以選擇 MySQL 作為 Azkaban 資料庫,以方便 Azkaban 的設定,並可增強服務可用性。

2. 安裝 Azkaban

(1)在 /opt/module 目錄下建立 azkaban 目錄。

```
[atguigu@hadoop102 module]$ mkdir azkaban
```

(2)解壓 azkaban-web-server-2.5.0.tar.gz、azkaban-executor-server-2.5.0.tar.gz、azkaban-sql-script-2.5.0.tar.gz 到 /opt/module/azkaban 目錄下。

```
[atguigu@hadoop102 software]$ tar -zxvf azkaban-web-server-2.5.0.tar.gz
-C /opt/module/azkaban
[atguigu@hadoop102 software]$ tar -zxvf azkaban-executor-server-
2.5.0.tar.gz -C /opt/module/azkaban
[atguigu@hadoop102 software]$ tar -zxvf azkaban-sql-script-2.5.0.tar.gz
-C /opt/module/azkaban
```

(3)對解壓後的 azkaban-web-server-2.5.0 和 azkaban-executor-server-2.5.0 檔案重新命名。

```
[atguigu@hadoop102 azkaban]$ mv azkaban-web-server-2.5.0 server
[atguigu@hadoop102 azkaban]$ mv azkaban-executor-server-2.5.0 executor
```

(4)Azkaban 指令稿匯入。
進入 MySQL,建立 azkaban 資料庫,並將解壓的指令稿匯入 azkaban 資料庫。

```
[atguigu@hadoop102 azkaban]$ mysql -uroot -p000000
mysql> create database azkaban;
mysql> use azkaban;
mysql> source /opt/module/azkaban/azkaban-2.5.0/create-all-sql-2.5.0.sql
```

> ★ **注意**：source 後跟 .sql 檔案，用於批次處理 .sql 檔案中的 SQL 敘述。

3. 產生金鑰庫

- keytool：Java 資料憑證的管理工具，讓使用者能夠管理自己的公開金鑰 / 私密金鑰對及相關憑證。
- -keystore：指定金鑰庫的名稱及位置。
- -genkey：在用戶家目錄中建立一個預設檔案 ".keystore"。
- -alias：對產生的 ".keystore" 檔案指定別名；如果沒有，則預設是 mykcy。
- -keyalg：指定金鑰的演算法 RSA/DSA，預設是 DSA。

（1）產生金鑰庫的密碼及對應資訊。

```
[atguigu@hadoop102 azkaban]$ keytool -keystore keystore -alias jetty
-genkey -keyalg RSA
輸入金鑰庫密碼：
再次輸入新密碼：
您的名字與姓氏是什麼?
  [Unknown]:
您的組織單位名稱是什麼?
  [Unknown]:
您的組織名稱是什麼?
  [Unknown]:
您所在的城市或區域名稱是什麼?
  [Unknown]:
您所在的省/市/自治區名稱是什麼?
  [Unknown]:
該單位的雙字母國家/地區程式是什麼?
  [Unknown]:
CN=Unknown, OU=Unknown, O=Unknown, L=Unknown, ST=Unknown, C=Unknown是否正確?
```

```
   [否]: y

輸入 <jetty> 的金鑰密碼
         (如果和金鑰庫密碼相同，則按Enter鍵):
再次輸入新密碼:
```

> ★ **注意**：金鑰庫的密碼至少是 6 個字元，可以是純數字、字母或數字和字母
> 的組合等。
> 金鑰庫的密碼最好與 <jetty> 的金鑰密碼相同，方便記憶。

（2）將金鑰庫所在資料夾 keystore 複製到 Azkaban Web 伺服器根目錄下。

```
[atguigu@hadoop102 azkaban]$ mv keystore /opt/module/azkaban/server/
```

4. 時間同步設定

設定節點伺服器上的時區。

（1）如果在 /usr/share/zoneinfo 目錄下不存在時區設定檔 Asia/Shanghai，則透過執行 tzselect 指令來產生。

```
[atguigu@hadoop102 azkaban]$ tzselect
Please identify a location so that time zone rules can be set correctly.
Please select a continent or ocean.
 1) Africa
 2) Americas
 3) Antarctica
 4) Arctic Ocean
 5) Asia
 6) Atlantic Ocean
 7) Australia
 8) Europe
 9) Indian Ocean
10) Pacific Ocean
11) none - I want to specify the time zone using the Posix TZ format.
#? 5
Please select a country.
```

```
 1) Afghanistan      18) Israel         35) Palestine
 2) Armenia          19) Japan          36) Philippines
 3) Azerbaijan       20) Jordan         37) Qatar
 4) Bahrain          21) Kazakhstan     38) Russia
 5) Bangladesh       22) Korea (North)  39) Saudi Arabia
 6) Bhutan           23) Korea (South)  40) Singapore
 7) Brunei           24) Kuwait         41) Sri Lanka
 8) Cambodia         25) Kyrgyzstan     42) Syria
 9) China            26) Laos           43) Taiwan
10) Cyprus           27) Lebanon        44) Tajikistan
11) East Timor       28) Macau          45) Thailand
12) Georgia          29) Malaysia       46) Turkmenistan
13) Hong Kong        30) Mongolia       47) United Arab Emirates
14) India            31) Myanmar (Burma) 48) Uzbekistan
15) Indonesia        32) Nepal          49) Vietnam
16) Iran             33) Oman           50) Yemen
17) Iraq             34) Pakistan
#? 9
Please select one of the following time zone regions.
1) Beijing Time
2) Xinjiang Time
#? 1


The following information has been given:

        China
        Beijing Time


Therefore TZ='Asia/Shanghai' will be used.
Local time is now:      Thu Oct 18 16:24:23 CST 2018.
Universal Time is now:  Thu Oct 18 08:24:23 UTC 2018.
Is the above information OK?
1) Yes
2) No
#? 1


You can make this change permanent for yourself by appending the line
        TZ='Asia/Shanghai'; export TZ
to the file '.profile' in your home directory; then log out and log in again.
```

```
Here is that TZ value again, this time on standard output so that you
can use the /usr/bin/tzselect command in shell scripts:
Asia/Shanghai
```

（2）複製該時區檔案，覆蓋系統本機時區的設定。

```
[atguigu@hadoop102 azkaban]$ cp /usr/share/zoneinfo/Asia/Shanghai /etc/
localtime
```

（3）叢集時間同步設定。

叢集時間同步設定可參考 3.3.1 節中的內容。

5. Web 伺服器設定

（1）進入 Azkaban Web 伺服器安裝目錄 conf，開啟 azkaban.properties 檔案。

```
[atguigu@hadoop102 conf]$ pwd
/opt/module/azkaban/server/conf
[atguigu@hadoop102 conf]$ vim azkaban.properties
```

（2）按照以下設定，修改 azkaban.properties 檔案。

```
#Azkaban 個性化設定
#伺服器UI名稱，在伺服器上方顯示
azkaban.name=Test
#描述
azkaban.label=My Local Azkaban
#UI顏色
azkaban.color=#FF3601
azkaban.default.servlet.path=/index
#預設Web 伺服器儲存Web檔案的目錄
web.resource.dir=/opt/module/azkaban/server/web/
#預設時區為美國，已改為亞洲/上海
default.timezone.id=Asia/Shanghai

#Azkaban 使用者管理訂製
user.manager.class=azkaban.user.XmlUserManager
#使用者許可權管理預設類別（絕對路徑）
user.manager.xml.file=/opt/module/azkaban/server/conf/azkaban-users.xml
```

```
#專案載入設定
#global設定檔所在位置（絕對路徑）
executor.global.properties=/opt/module/azkaban/executor/conf/global.properties
azkaban.project.dir=projects

#資料庫類型
database.type=mysql
#通訊埠
mysql.port=3306
#資料庫連接IP位址
mysql.host=hadoop102
#資料庫實例名稱
mysql.database=azkaban
#資料庫使用者名稱
mysql.user=root
#資料庫密碼
mysql.password=000000
#最大連接數
mysql.numconnections=100


velocity.dev.mode=false


#Jetty伺服器屬性
#最大執行緒數
jetty.maxThreads=25
#Jetty SSL通訊埠
jetty.ssl.port=8443
#Jetty通訊埠
jetty.port=8081


#金鑰庫設定
#SSL檔案名稱（絕對路徑）
jetty.keystore=/opt/module/azkaban/server/keystore
#SSL檔案密碼
jetty.password=000000
#Jetty主密碼與.keystore檔案相同
jetty.keypassword=000000
```

```
#可信金鑰庫設定
#SSL檔案名稱（絕對路徑）
jetty.truststore=/opt/module/azkaban/server/keystore
#SSL檔案密碼
jetty.trustpassword=000000

#Azkaban Executor 設定
executor.port=12321

#郵件發送設定
mail.sender=
mail.host=
job.failure.email=
job.success.email=

lockdown.create.projects=false

cache.directory=cache
```

（3）Web 伺服器使用者設定。在 Azkaban Web 伺服器安裝目錄 conf 下，按照以下設定修改 azkaban-users.xml 檔案，增加管理員使用者。

```
[atguigu@hadoop102 conf]$ vim azkaban-users.xml
<azkaban-users>
 <user username="azkaban" password="azkaban" roles="admin" groups="azkaban" />
 <user username="metrics" password="metrics" roles="metrics"/>
 <user username="admin" password="admin" roles="admin" />
 <role name="admin" permissions="ADMIN" />
 <role name="metrics" permissions="METRICS"/>
</azkaban-users>
```

6. 執行伺服器設定

（1）進入伺服器安裝目錄 conf，開啟 azkaban.properties 檔案。

```
[atguigu@hadoop102 conf]$ pwd
/opt/module/azkaban/executor/conf
[atguigu@hadoop102 conf]$ vim azkaban.properties
```

（2）按照以下設定，修改 azkaban.properties 檔案。

```
#Azkaban
#時區
default.timezone.id=Asia/Shanghai

#Azkaban 作業類型外掛程式設定
#JobTypes 外掛程式所在位置
azkaban.jobtype.plugin.dir=plugins/jobtypes

#Loader for projects
executor.global.properties=/opt/module/azkaban/executor/conf/global.properties
azkaban.project.dir=projects

database.type=mysql
mysql.port=3306
mysql.host=hadoop102
mysql.database=azkaban
mysql.user=root
mysql.password=000000
mysql.numconnections=100

# Azkaban Executor 設定
#最大執行緒數
executor.maxThreads=50
#通訊埠（如需修改，要與Web伺服器中的通訊埠保持一致）
executor.port=12321
#執行緒數
executor.flow.threads=30
```

7. 啟動 Executor 伺服器

在 Executor 伺服器目錄下執行啟動指令。

```
[atguigu@hadoop102 executor]$ pwd
/opt/module/azkaban/executor
[atguigu@hadoop102 executor]$ bin/azkaban-executor-start.sh
```

8. 啟動 Web 伺服器

在 Azkaban Web 伺服器目錄下執行啟動指令。

```
[atguigu@hadoop102 server]$ pwd
/opt/module/azkaban/server
[atguigu@hadoop102 server]$ bin/azkaban-web-start.sh
```

> ✦ **注意：**先啟動 Executor 伺服器，再啟動 Web 伺服器，避免 Web 伺服器因找不到執行器而啟動失敗。

執行 jps 指令，檢視處理程序。

```
[atguigu@hadoop102 server]$ jps
3601 AzkabanExecutorServer
5880 Jps
3661 AzkabanWebServer
```

啟動完成後，在瀏覽器（建議使用 Google 瀏覽器）網址列中輸入 https:// 伺服器 IP 位址 :8443，即可存取 Azkaban 服務。

在登入頁面中，輸入剛才在 azkaban-users.xml 檔案中新增加的使用者名稱及密碼，點擊 "Login" 按鈕，如圖 6-39 所示。

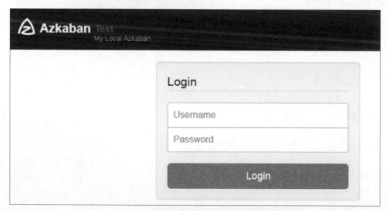

圖 6-39　Azkaban 登入

Azkaban 登入成功的頁面如圖 6-40 所示。

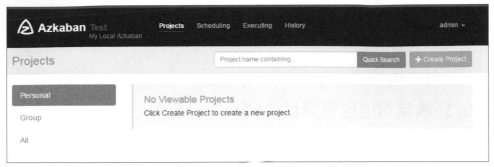

圖 6-40　Azkaban 登入成功的頁面

6.9.2　建立視覺化的 **MySQL** 資料庫和表

（1）建立 gmall_report 資料庫，如圖 6-41 所示。

資料庫名稱:	gmall_report	
字元集:	utf8	∨
排序規則:	utf8_general_ci	∨

圖 6-41　建立 gmall_report 資料庫

建立 gmall_report 資料庫的 SQL 敘述如下。

```
CREATE DATABASE `gmall_report` CHARACTER SET 'utf8' COLLATE
'utf8_general_ci';
```

（2）建立表，敘述如下。

```
DROP TABLE IF EXISTS `ads_user_topic`;
CREATE TABLE `ads_user_topic` (
  `dt` date NOT NULL,
  `day_users` bigint(255) NULL DEFAULT NULL,
  `day_new_users` bigint(255) NULL DEFAULT NULL,
  `day_new_payment_users` bigint(255) NULL DEFAULT NULL,
  `payment_users` bigint(255) NULL DEFAULT NULL,
  `users` bigint(255) NULL DEFAULT NULL,
  `day_users2users` double(255, 2) NULL DEFAULT NULL,
  `payment_users2users` double(255, 2) NULL DEFAULT NULL,
```

```
   `day_new_users2users` double(255, 2) NULL DEFAULT NULL,
   PRIMARY KEY (`dt`) USING BTREE
) ENGINE = InnoDB CHARACTER SET = utf8 COLLATE = utf8_general_ci
ROW_FORMAT = Compact;
```

6.9.3 撰寫指標取得排程流程

Azkaban 安裝完成之後，即可撰寫 job 描述檔案，然後將 job 描述檔案包裝成壓縮檔提交到 Azkaban 並執行。

Azkaban 內建的任務類型支援 command、Java，任務類型可以透過 type 指定，若該任務即時執行需要傳入參數，則可以透過 &{PARAMETER} 的形式傳入，若任務之間存在相依關係，則可以透過 dependencies 指定。

（1）產生 2020-03-13 的資料。

（2）撰寫 Azkaban 程式，執行 job 描述檔案。

① mysql_to_hdfs.job 檔案。

```
type=command
command=/home/atguigu/bin/mysql_to_hdfs.sh all ${dt}
```

② hdfs_to_ods_log.job 檔案。

```
type=command
command=/home/atguigu/bin/hdfs_to_ods_log.sh ${dt}
```

③ hdfs_to_ods_db.job 檔案。

```
type=command
command=/home/atguigu/bin/hdfs_to_ods_db.sh all ${dt}
dependencies=mysql_to_hdfs
```

④ ods_to_dwd_start_log.job 檔案。

```
type=command
command=/home/atguigu/bin/ods_to_dwd_start_log.sh ${dt}
dependencies=hdfs_to_ods_log
```

⑤ ods_to_dwd_db.job 檔案。

```
type=command
command=/home/atguigu/bin/ods_to_dwd_db.sh ${dt}
dependencies=hdfs_to_ods_db
```

⑥ dwd_to_dws.job 檔案。

```
type=command
command=/home/atguigu/bin/dwd_to_dws.sh ${dt}
dependencies=ods_to_dwd_db,ods_to_dwd_start_log
```

⑦ dws_to_dwt.job 檔案。

```
type=command
command=/home/atguigu/bin/dws_to_dwt.sh ${dt}
dependencies=dwd_to_dws
```

⑧ dwt_to_ads.job 檔案。

```
type=command
command=/home/atguigu/bin/dwt_to_ads.sh ${dt}
dependencies=dws_to_dwt
```

⑨ hdfs_to_mysql.job 檔案。

```
type=command
command=/home/atguigu/bin/hdfs_to_mysql.sh ads_user_topic
dependencies=dwt_to_ads
```

⑩ 將以上 9 個 job 描述檔案壓縮成 gmall.zip 檔案，如圖 6-42 所示。

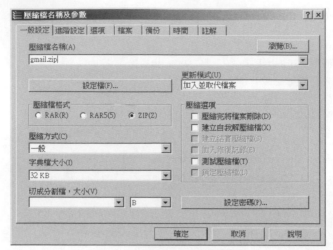

圖 6-42 壓縮 job 描述檔案

（3）任務排程執行。

① 在瀏覽器網址列中輸入 https://hadoop102:8443，登入 Azkaban，登入成功後，在頁面上點擊 "Create Project" 按鈕，建立專案，如圖 6-43 所示。

圖 6-43 建立專案入口

② 對該排程專案進行命名並增加描述資訊，然後點擊 "Create Project" 按鈕，
如圖 6-44 所示。

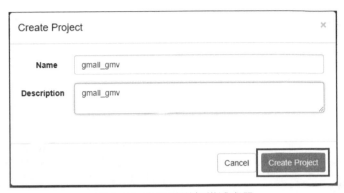

圖 6-44 命名並增加描述資訊

③ 在開啟的頁面上點擊 "Upload" 按鈕，上傳 job 描述 zip 壓縮檔，如圖 6-45
所示。

圖 6-45 上傳 job 描述 zip 壓縮檔

④ 上傳 job 描述 zip 壓縮檔之後,專案即建立成功,使用者可以點擊 "Execute Flow" 按鈕對排程流程進行設定並執行,如圖 6-46 所示。

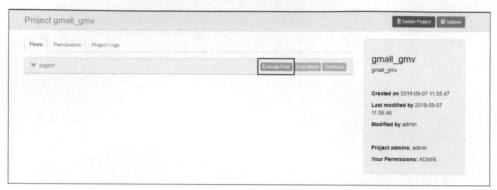

圖 6-46 專案建立成功

⑤ 點擊 "Flow Parameters" 按鈕,傳入參數,如圖 6-47 和圖 6-48 所示。

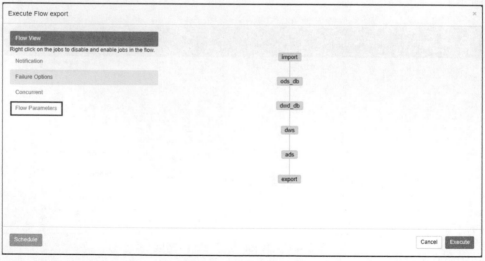

圖 6-47 傳入參數入口

圖 6-48 傳入參數成功

⑥ 參數設定完畢後，可以點擊 "Schedule" 按鈕，進行定時任務排程（企業中通常採用此方案），如圖 6-49 所示，也可以點擊 "Execute" 按鈕，使任務立即執行，如圖 6-50 所示。

圖 6-49 提交定時任務入口

圖 6-50 任務立即執行入口

⑦ 任務執行過程中的頁面顯示如圖 6-51 所示，任務執行成功的頁面顯示如圖
6-52 所示。

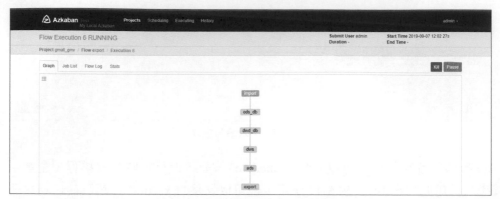

圖 6-51 任務執行過程中的頁面顯示

圖 6-52 任務執行成功的頁面顯示

⑧ 任務執行成功後，在 MySQL 中檢視結果。

```
select * from ads_gmv_sum_day;
```

6.10 本章歸納

本章主要對資料倉儲架設模組進行了說明，從資料倉儲架設所需的理論基礎開始，說明了資料倉儲的建模分層理念，以及本資料倉儲專案對理論的實際應用，然後說明了資料倉儲架設所需的基本環境，接著重點說明了資料倉儲如何按照分層的理念一層層地架設並實現需求，最後為讀者簡單介紹了資料倉儲的任務排程流程。本章重點說明了資料倉儲的分層理論及分層架設過程，這是無論架設多大規模的資料倉儲都需要首先考慮的問題。

資料視覺化模組

將需求實現，取得到最後的結果資料之後，僅讓結果資料儲存於資料倉儲中是遠遠不夠的，還需要將資料進行視覺化。通常視覺化的想法是從巨量資料的儲存系統中將資料匯出到關聯式資料庫中，再使用 Java 程式進行展示。

7.1 模擬視覺化資料

在 MySQL 中，根據視覺化的需求進行建表，要求與 ADS 層的表結構完全一致。本章只對會員主題中的 ads_user_topic 表和地區主題中的 ads_area_topic 表進行視覺化，為了能儘快看到展示效果，我們在建立表的同時在表中插入許多筆資料，建表及資料插入敘述如下。

7.1.1 會員主題

1. 建立會員主題結果表 ads_user_topic

```
DROP TABLE IF EXISTS `ads_user_topic`;
CREATE TABLE `ads_user_topic` (
  `dt` date NOT NULL,
  `day_users` bigint(255) NULL DEFAULT NULL,
  `day_new_users` bigint(255) NULL DEFAULT NULL,
```

```
`day_new_payment_users` bigint(255) NULL DEFAULT NULL,
`payment_users` bigint(255) NULL DEFAULT NULL,
`users` bigint(255) NULL DEFAULT NULL,
`day_users2users` double(255, 2) NULL DEFAULT NULL,
`payment_users2users` double(255, 2) NULL DEFAULT NULL,
`day_new_users2users` double(255, 2) NULL DEFAULT NULL,
PRIMARY KEY (`dt`) USING BTREE
) ENGINE = InnoDB CHARACTER SET = utf8 COLLATE = utf8_general_ci
ROW_FORMAT = Compact;
```

2. 匯入資料

```
INSERT INTO `ads_user_topic` VALUES ('2020-03-12', 761, 52, 0, 8, 989,
0.77, 0.01, 0.07);
INSERT INTO `ads_user_topic` VALUES ('2020-03-13', 840, 69, 10, 5, 1000,
0.59, 0.01, 0.05);
INSERT INTO `ads_user_topic` VALUES ('2020-03-14', 900, 69, 10, 5, 1000,
0.59, 0.01, 0.05);
INSERT INTO `ads_user_topic` VALUES ('2020-03-15', 890, 69, 10, 5, 1000,
0.59, 0.01, 0.05);
INSERT INTO `ads_user_topic` VALUES ('2020-03-16', 607, 69, 10, 5, 1000,
0.59, 0.01, 0.05);
INSERT INTO `ads_user_topic` VALUES ('2020-03-17', 812, 69, 10, 5, 1000,
0.59, 0.01, 0.05);
INSERT INTO `ads_user_topic` VALUES ('2020-03-18', 640, 69, 10, 5, 1000,
0.59, 0.01, 0.05);
INSERT INTO `ads_user_topic` VALUES ('2020-03-19', 740, 69, 10, 5, 1000,
0.59, 0.01, 0.05);
INSERT INTO `ads_user_topic` VALUES ('2020-03-20', 540, 69, 10, 5, 1000,
0.59, 0.01, 0.05);
INSERT INTO `ads_user_topic` VALUES ('2020-03-21', 940, 69, 10, 5, 1000,
0.59, 0.01, 0.05);
INSERT INTO `ads_user_topic` VALUES ('2020-03-22', 840, 69, 10, 5, 1000,
0.59, 0.01, 0.05);
INSERT INTO `ads_user_topic` VALUES ('2020-03-23', 1000, 32, 32, 32,
23432, 22.00, 0.11, 0.55);
```

7.1.2 地區主題

1. 建立地區主題結果表 ads_area_topic

```
DROP TABLE IF EXISTS `ads_area_topic`;
CREATE TABLE `ads_area_topic` (
  `dt` date NOT NULL,
  `iso_code` varchar(255) CHARACTER SET utf8 COLLATE utf8_general_ci NOT
NULL,
  `province_name` varchar(255) CHARACTER SET utf8 COLLATE utf8_general_ci
NULL DEFAULT NULL,
  `area_name` varchar(255) CHARACTER SET utf8 COLLATE utf8_general_ci
NULL DEFAULT NULL,
  `order_count` bigint(255) NULL DEFAULT NULL,
  `order_amount` double(255, 2) NULL DEFAULT NULL,
  `payment_count` bigint(255) NULL DEFAULT NULL,
  `payment_amount` double(255, 2) NULL DEFAULT NULL,
  PRIMARY KEY (`dt`, `iso_code`) USING BTREE
) ENGINE = InnoDB CHARACTER SET = utf8 COLLATE = utf8_general_ci ROW_
FORMAT = Compact;
```

2. 匯入資料

```
INSERT INTO `ads_area_topic` VALUES ('2020-03-10', 'CN-11', '北京市',
'華北', 652, 652.00, 652, 652.00);
INSERT INTO `ads_area_topic` VALUES ('2020-03-10', 'CN-12', '天津市',
'華北', 42, 42.00, 42, 42.00);
INSERT INTO `ads_area_topic` VALUES ('2020-03-10', 'CN-13', '河北',
'華北', 3435, 3435.00, 3435, 3435.00);
INSERT INTO `ads_area_topic` VALUES ('2020-03-10', 'CN-14', '山西',
'華北', 4, 4.00, 4, 4.00);
INSERT INTO `ads_area_topic` VALUES ('2020-03-10', 'CN-15', '內蒙古',
'華北', 44, 44.00, 44, 44.00);
INSERT INTO `ads_area_topic` VALUES ('2020-03-10', 'CN-21', '遼寧',
'東北', 335, 335.00, 335, 335.00);
INSERT INTO `ads_area_topic` VALUES ('2020-03-10', 'CN-22', '吉林',
'東北', 44, 44.00, 44, 44.00);
INSERT INTO `ads_area_topic` VALUES ('2020-03-10', 'CN-23', '黑龍江',
'東北', 337, 337.00, 337, 337.00);
```

```
INSERT INTO `ads_area_topic` VALUES ('2020-03-10', 'CN-31', '上海',
'華東', 4, 4.00, 4, 4.00);
INSERT INTO `ads_area_topic` VALUES ('2020-03-10', 'CN-32', '江蘇',
'華東', 4545, 4545.00, 4545, 4545.00);
INSERT INTO `ads_area_topic` VALUES ('2020-03-10', 'CN-33', '浙江',
'華東', 43, 43.00, 43, 43.00);
INSERT INTO `ads_area_topic` VALUES ('2020-03-10', 'CN-34', '安徽',
'華東', 12345, 2134.00, 324, 252.00);
INSERT INTO `ads_area_topic` VALUES ('2020-03-10', 'CN-35', '福建',
'華東', 435, 435.00, 435, 435.00);
INSERT INTO `ads_area_topic` VALUES ('2020-03-10', 'CN-36', '江西',
'華東', 4453, 4453.00, 4453, 4453.00);
INSERT INTO `ads_area_topic` VALUES ('2020-03-10', 'CN-37', '山東',
'華東', 34, 34.00, 34, 34.00);
INSERT INTO `ads_area_topic` VALUES ('2020-03-10', 'CN-41', '河南',
'華中', 34, 34.00, 34, 34.00);
INSERT INTO `ads_area_topic` VALUES ('2020-03-10', 'CN-42', '湖北',
'華中', 4, 4.00, 4, 4.00);
INSERT INTO `ads_area_topic` VALUES ('2020-03-10', 'CN-43', '湖南',
'華中', 54, 54.00, 54, 54.00);
INSERT INTO `ads_area_topic` VALUES ('2020-03-10', 'CN-44', '廣東',
'華南', 24, 24.00, 24, 24.00);
INSERT INTO `ads_area_topic` VALUES ('2020-03-10', 'CN-45', '廣西',
'華南', 4, 4.00, 4, 4.00);
INSERT INTO `ads_area_topic` VALUES ('2020-03-10', 'CN-46', '海南',
'華南', 42, 42.00, 42, 42.00);
INSERT INTO `ads_area_topic` VALUES ('2020-03-10', 'CN-50', '重慶',
'西南', 4532, 4532.00, 4532, 4532.00);
INSERT INTO `ads_area_topic` VALUES ('2020-03-10', 'CN-51', '四川',
'西南', 3435, 3435.00, 3435, 3435.00);
INSERT INTO `ads_area_topic` VALUES ('2020-03-10', 'CN-52', '貴州',
'西南', 5725, 5725.00, 5725, 5725.00);
INSERT INTO `ads_area_topic` VALUES ('2020-03-10', 'CN-53', '雲南',
'西南', 4357, 4357.00, 4357, 4357.00);
INSERT INTO `ads_area_topic` VALUES ('2020-03-10', 'CN-54', '西藏',
'西南', 54, 54.00, 54, 54.00);
INSERT INTO `ads_area_topic` VALUES ('2020-03-10', 'CN-61', '陝西',
'西北', 44, 44.00, 44, 44.00);
```

```
INSERT INTO `ads_area_topic` VALUES ('2020-03-10', 'CN-62', '甘肅',
'西北', 78, 78.00, 78, 78.00);
INSERT INTO `ads_area_topic` VALUES ('2020-03-10', 'CN-63', '青海',
'西北', 3444, 3444.00, 3444, 3444.00);
INSERT INTO `ads_area_topic` VALUES ('2020-03-10', 'CN-64', '寧夏',
'西北', 445, 445.00, 445, 445.00);
INSERT INTO `ads_area_topic` VALUES ('2020-03-10', 'CN-65', '新疆',
'西北', 4442, 4442.00, 4442, 4442.00);
INSERT INTO `ads_area_topic` VALUES ('2020-03-10', 'CN-91', '香港',
'華南', 44, 44.00, 44, 44.00);
INSERT INTO `ads_area_topic` VALUES ('2020-03-10', 'CN-92', '澳門',
'華南', 34, 34.00, 34, 34.00);
```

7.2 Superset 部署

Apache Superset 是一個開放原始碼的、現代的、輕量級的 BI 分析工具，能夠對接多種資料來源，擁有豐富的圖示展示形式，支援自訂儀表板，且擁有人性化的使用者介面，十分好用。

7.2.1 環境準備

Superset 是由 Python 撰寫的 Web 應用，要求使用 Python3.6 的環境，但是 CentOS 系統附帶的 Python 環境是 2.x 版本的，所以我們需要先安裝 Python3 環境。

1. 安裝 Miniconda

Conda 是一個開放原始碼的套件和環境管理員，可以用於在同一台機器上安裝不同版本的 Python 軟體套件及依賴，並能在不同的 Python 環境之間切換。Anaconda 和 Miniconda 都整合了 Conda，而 Anaconda 包含更多的工具套件，如 numpy、pandas，Miniconda 則只包含 Conda 和 Python。

此處，我們不需要太多的工具套件，故選擇 Miniconda。

1) 下載 Miniconda（Python3 版本）

2) 安裝 Miniconda

（1）執行以下指令，安裝 Miniconda，並按照提示操作，直到安裝完成。

```
[atguigu@hadoop102 lib]$ bash Miniconda3-latest-Linux-x86_64.sh
```

（2）在安裝過程中，出現如圖 7-1 所示的提示時，需指定安裝路徑。

```
Miniconda3 will now be installed into this location:
/home/atguigu/miniconda3

  - Press ENTER to confirm the location
  - Press CTRL-C to abort the installation
  - Or specify a different location below

[/home/atguigu/miniconda3] >>> /opt/module/miniconda3
```

圖 7-1　Miniconda 安裝提示

（3）出現如圖 7-2 所示的提示時，表示安裝完成。

```
Thank you for installing Miniconda3!
```

圖 7-2　Miniconda 安裝完成提示

3) 載入環境變數檔案

```
[atguigu@hadoop102 ~]$ source ~/.bashrc
```

4) 禁止啟動預設的 base 環境

Miniconda 安裝完成後，每次開啟終端都會啟動其預設的 base 環境，我們可透過以下指令，禁止啟動預設的 base 環境。

```
[atguigu@hadoop102 ~]$ conda config --set auto_activate_base false
```

2. 使用 Python 3.6 環境

（1）建立 Python 3.6 環境。

```
[atguigu@hadoop102 ~]$ conda create --name superset python=3.6
```

Conda 環境管理員常用指令如下。

- 建立環境：conda create -n env_name。
- 檢視所有環境：conda info --envs。
- 刪除一個環境：conda remove -n env_name --all。

（2）啟動 Superset 環境。

```
[atguigu@hadoop102 ~]$ conda activate superset
```

啟動後的效果如圖 7-3 所示。

```
(superset) [atguigu@hadoop102 ~]$
```
圖 7-3　Superset 環境啟動後的效果

（3）退出目前環境。

```
[atguigu@hadoop102 ~]$ conda deactivate
```

（4）執行 python 指令，檢視 Python 版本，如圖 7-4 所示。

```
(superset) [atguigu@hadoop102 ~]$ python
Python 3.6.10 |Anaconda, Inc.| (default, Jan  7 2020, 21:14:29)
[GCC 7.3.0] on linux
Type "help", "copyright", "credits" or "license" for more information.
>>> quit();
```
圖 7-4　檢視 Python 版本

7.2.2　Superset 安裝

1. 安裝依賴

在安裝 Superset 之前，需要先安裝以下所需的依賴。

```
(superset) [atguigu@hadoop102 ~]$ sudo yum install -y python-setuptools
(superset) [atguigu@hadoop102 ~]$ sudo yum install -y gcc gcc-c++ libffi-
devel python-devel python-pip python-wheel openssl-devel cyrus-sasl-devel
openldap-devel
```

2. 安裝 Superset

1）安裝（更新）setuptools 和 pip

```
(superset) [atguigu@hadoop102 ~]$ pip install --upgrade setuptools pip -i
```

```
https://pypi.douban.com/simple/
```

說明：pip 是 Python 的套件管理工具，與 CentOS 中的 yum 類似。

2）安裝 Superset

```
(superset) [atguigu@hadoop102 ~]$ pip install apache-superset -i https://
pypi.douban.com/simple/
```

說明：-i 的作用是指定映像檔。

3）初始化 superset 資料庫

```
(superset) [atguigu@hadoop102 ~]$ superset db upgrade
```

4）建立管理員使用者

```
(superset) [atguigu@hadoop102 ~]$ export FLASK_APP=superset
(superset) [atguigu@hadoop102 ~]$ flask fab create-admin
```

說明：flask 是一個 Python Web 架構，Superset 使用的就是 flask。

5）初始化 Superset

```
(superset) [atguigu@hadoop102 ~]$ superset init
```

3. 操作 Superset

1）安裝 gunicorn

```
(superset) [atguigu@hadoop102 ~]$ pip install gunicorn -i https://
pypi.douban.com/simple/
```

說明：gunicorn 是一個 Python Web Server，與 Java 中的 TomCat 類似。

2）啟動 Superset
（1）確保目前 Conda 的環境為 Superset。
（2）啟動。

```
(superset) [atguigu@hadoop102 ~]$ gunicorn --workers 5 --timeout 120
--bind hadoop102:8787  --daemon "superset.app:create_app()"
```

參數說明如下。

- --workers：指定處理程序個數。
- --timeout：Worker 處理程序逾時，逾時會自動重新啟動。
- --bind：綁定本機位址，即 Superset 的造訪網址。
- --daemon：後台執行。

3）停止執行 Superset

（1）停掉 gunicorn 處理程序。

```
(superset) [atguigu@hadoop102 ~]$ ps -ef | awk '/gunicorn/ && !/awk/
{print $2}' | xargs kill -9
```

（2）退出 Superset 環境。

```
(superset) [atguigu@hadoop102 ~]$ conda deactivate
```

4）啟動後登入 Superset

存取 http://hadoop102:8787，進入 Superset 登入頁面，如圖 7-5 所示，並使用前面建立的管理員帳號進行登入。

圖 7-5 Superset 登入頁面

7.3 Superset 使用

7.3.1 對接 MySQL 資料來源

1. 安裝依賴

```
(superset) [atguigu@hadoop102 ~]$ conda install mysqlclient
```

說明：對接不同的資料來源，需要安裝不同的依賴。

2. 重新啟動 Superset

1）停掉 gunicorn 處理程序

```
(superset) [atguigu@hadoop102 ~]$ ps -ef | awk '/gunicorn/ && !/awk/
{print $2}' | xargs kill -9
```

2）啟動 Superset

```
(superset) [atguigu@hadoop102 ~]$ gunicorn --workers 5 --timeout 120
--bind hadoop102:8787  --daemon "superset.app:create_app()"
```

3. 資料來源設定

1）Database 設定

步驟 1　選擇 "Sources" → "Databases" 選項，如圖 7-6 所示。

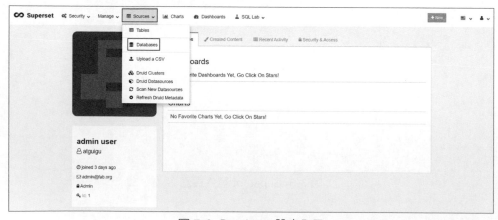

圖 7-6 Database 設定入口

步驟 2　點擊 ⊕ 圖示增加資料庫，如圖 7-7 所示。

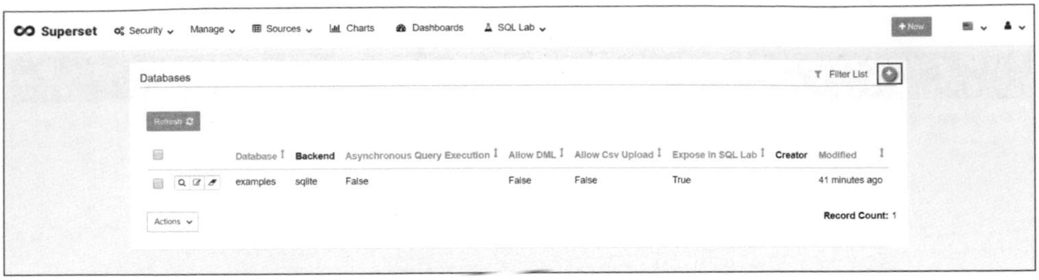

圖 7-7　增加資料庫操作

步驟 3　填寫 "Database" 及 "SQLAlchemy URI"，如圖 7-8 所示。

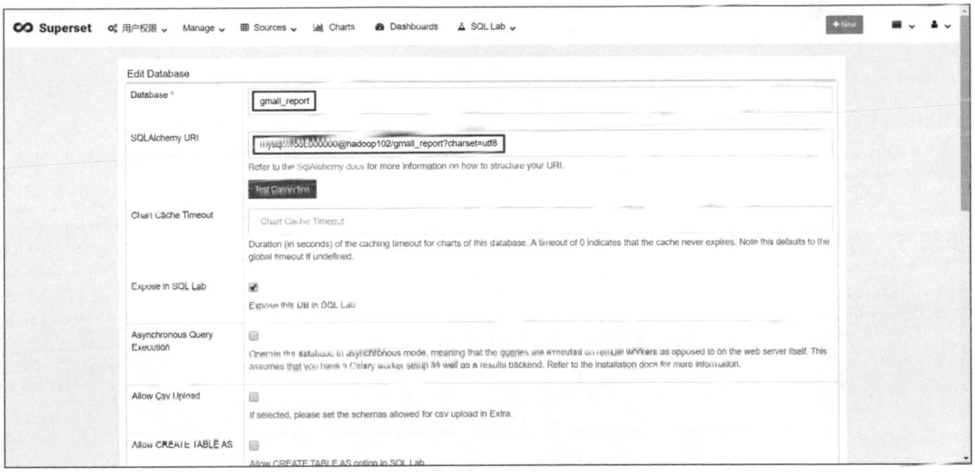

圖 7-8　編輯 Database 相關設定

"SQLAlchemy URI" 的撰寫標準為 mysql:// 帳號 : 密碼 @ 伺服器 IP 位址 / 資料庫名稱 ?charset=utf8。

步驟4 點擊 "Test Connection" 按鈕,出現 "Seems Ok !" 提示,表示連接成功,如圖 7-9 所示。

圖 7-9 測試是否連接成功

步驟5 儲存設定,如圖 7-10 所示。

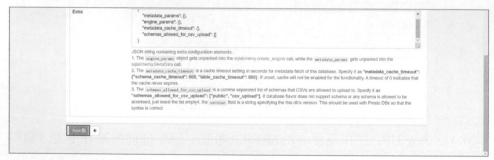

圖 7-10 儲存設定

2)Table 設定

步驟1 選擇 "Sources" → "Tables" 選項,如圖 7-11 所示。

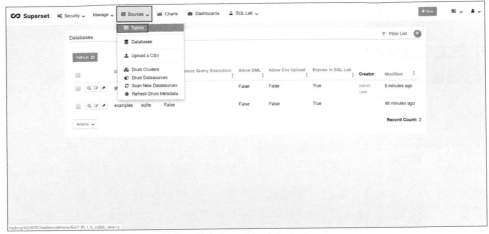

圖 7-11 Table 設定入口

步驟2 點擊 ⊕ 圖示增加 Table，如圖 7-12 所示。

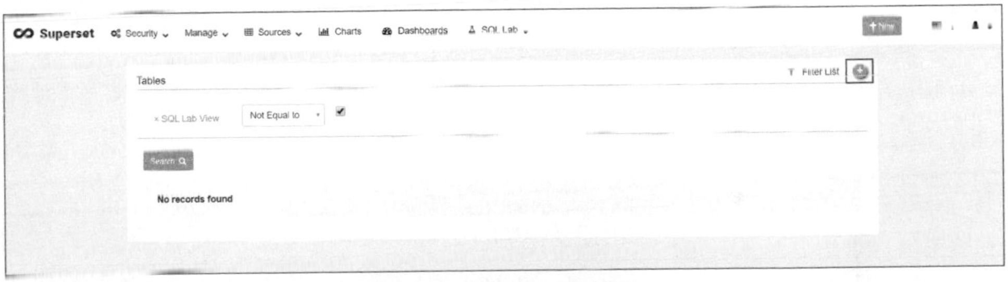

圖 7-12 增加 Table 操作

步驟3 設定 Table，如圖 7-13 所示。

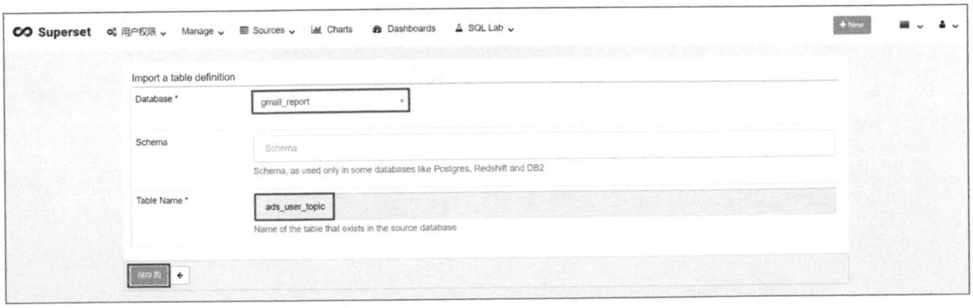

圖 7-13 設定 Table

7.3.2 製作儀表板

1. 建立空白儀表板

（1）選擇 "Dashboards" 選項並點擊 ⊕ 圖示，如圖 7-14 所示。

圖 7-14 建立空白儀表板入口

（2）設定儀表板，如圖 7-15 所示。

圖 7-15 設定儀表板

（3）儲存儀表板，如圖 7-16 所示。

圖 7-16 儲存儀表板

2. 建立圖表

（1）選擇 "Charts" 選項並點擊 ⊕ 圖示，如圖 7-17 所示。

圖 7-17 建立圖表入口

（2）選擇資料來源及圖表類型，如圖 7-18 所示。

圖 7-18 選擇資料來源及圖表類型

（3）選擇合適的圖表樣式，如圖 7-19 所示。

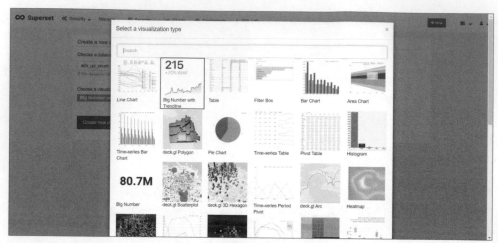

圖 7-19 選擇合適的圖表樣式

（4）建立圖表，如圖 7-20 所示。

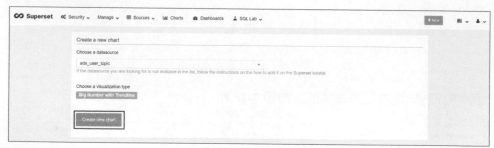

圖 7-20 建立圖表

7.4 本章歸納

本章對會員主題和地區主題 2 個主題指標進行了視覺化，讓讀者對資料倉儲的實際作用有了更具體的體會。資料倉儲在實現資料需求，取得最後的結果資料之餘，有的時候，也需要承擔資料視覺化的責任。業務實現程式不屬於巨量資料範圍，此處不再贅述。資料倉儲關鍵在於為視覺化提供正確的結果資料。

即席查詢模組

除了前幾章說明的需求實現，資料倉儲系統還需要滿足簡單的即席查詢需求。本章為讀者說明 3 個即席查詢的架構，分別是 Presto、Druid 和 Kylin，這 3 個架構在巨量資料領域應用十分廣泛，效能上各有千秋，下面分別說明。

8.1 Presto

Presto 是 Facebook 推出的開放原始碼的分散式 SQL 查詢引擎，資料規模可以支援 GB 到 PB 級，主要應用於處理秒級查詢的場景。

> ★ 注意：雖然 Presto 可以解析 SQL，但它不是一個標準的資料庫，不是 MySQL、Oracle 的代替品，也不能用來處理線上交易（OLTP）。

8.1.1 Presto 特點

Presto 是一個 master-slave 架構，由一個 Coordinator 節點、一個 Discovery Server 節點、多個 Worker 節點組成。Discovery Server 通常內嵌於 Coordinator 節點中；Coordinator 節點負責解析 SQL 敘述，產生執行計畫，分發執行任務給 Worker 節點；Worker 節點負責執行查詢任務。Worker 節點啟動反向 Discovery Server 服務註冊，Coordinator 節點從 Discovery Server 節點獲得可

以正常執行的 Worker 節點。如果設定了 Hive Connector，則需要設定一個 Hive Metastore 服務，為 Presto 提供 Hive 詮譯資訊，Worker 節點與 HDFS 互動讀取資料，Presto 架構如圖 8-1 所示。

Presto 基於記憶體運算，減少了磁碟 I/O 操作，計算速度更快。它能夠連接多個資料來源，並跨資料來源進行連表查詢，如從 Hive 中查詢大量網站存取記錄，然後從 MySQL 中比對出裝置資訊。

雖然 Presto 能夠處理 PB 級的巨量資料，但 Presto 並不是把 PB 級的資料都放在記憶體中進行計算，而是根據場景，如 count、AVG 等聚合運算，邊讀取資料邊計算，再清記憶體，再讀取資料再計算的，這種方式消耗的記憶體並不高。如果多表連接，就可能產生大量的臨時資料，因此，計算速度會變慢，此時使用 Hive 反而更好。

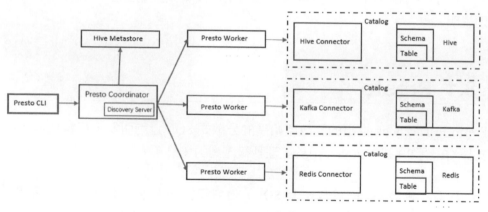

圖 8-1　Presto 架構

8.1.2　Presto 安裝

1. Presto Server 的安裝

（1）下載 Presto Server 安裝套件。

（2）將下載的 presto-server-0.196.tar.gz 匯入 hadoop102 的 /opt/software 目錄下，並解壓到 /opt/module 目錄下。

```
[atguigu@hadoop102 software]$ tar -zxvf presto-server-0.196.tar.gz -C
/opt/module
```

（3）修改 presto-server-0.196 的名稱為 presto。

```
[atguigu@hadoop102 module]$ mv presto-server-0.196/ presto
```

（4）進入 /opt/module/presto 目錄，建立儲存資料的資料夾。

```
[atguigu@hadoop102 presto]$ mkdir data
```

（5）進入 /opt/module/presto 目錄，建立儲存設定檔的資料夾。

```
[atguigu@hadoop102 presto]$ mkdir etc
```

（6）在 /opt/module/presto/etc 目錄下建立 jvm.config 設定檔。

```
[atguigu@hadoop102 etc]$ vim jvm.config
```

在設定檔中增加以下內容。

```
-server
-Xmx16G
-XX:+UseG1GC
-XX:G1HeapRegionSize=32M
-XX:+UseGCOverheadLimit
-XX:+ExplicitGCInvokesConcurrent
-XX:+HeapDumpOnOutOfMemoryError
-XX:+ExitOnOutOfMemoryError
```

（7）Presto 可以支援多個資料來源，在 Presto 中稱為 Catalog，這裡我們設定支援 Hive 的資料來源，設定一個 Hive 的 Catalog，需要新增一個目錄 catalog，在 catalog 目錄下新增檔案 hive.properties。

```
[atguigu@hadoop102 etc]$ mkdir catalog
[atguigu@hadoop102 catalog]$ vim hive.properties
```

在檔案中增加以下內容。

```
connector.name=hive-hadoop2
hive.metastore.uri=thrift://hadoop102:9083
```

（8）將 hadoop102 上的 presto 分發到 hadoop103、hadoop104。

```
[atguigu@hadoop102 module]$ xsync presto
```

（9）分發之後，分別進入 hadoop102、hadoop103、hadoop104 三台節點伺服器
的 /opt/module/presto/etc 目錄，設定 node 屬性，每台節點伺服器中 node 屬性
的 id 都不一樣。

```
[atguigu@hadoop102 etc]$vim node.properties
node.environment=production
node.id=ffffffff-ffff-ffff-ffff-ffffffffffff
node.data-dir=/opt/module/presto/data

[atguigu@hadoop103 etc]$vim node.properties
node.environment=production
node.id=ffffffff-ffff-ffff-ffff-fffffffffffe
node.data-dir=/opt/module/presto/data

[atguigu@hadoop104 etc]$vim node.properties
node.environment=production
node.id=ffffffff-ffff-ffff-ffff-fffffffffffd
node.data-dir=/opt/module/presto/data
```

（10）Presto 是 由 一 個 Coordinator 節 點 和 多 個 Worker 節 點 組 成 的。 在
hadoop102 上設定為 Coordinator，在 hadoop103、hadoop104 上設定為 Worker。

① 在 hadoop102 上設定 Coordinator 節點。

```
[atguigu@hadoop102 etc]$ vim config.properties
```

增加內容如下。

```
coordinator=true
node-scheduler.include-coordinator=false
http-server.http.port=8881
query.max-memory=50GB
discovery-server.enabled=true
discovery.uri=http://hadoop102:8881
```

② 在 hadoop103 上設定 Worker 節點。

```
[atguigu@hadoop103 etc]$ vim config.properties
```

增加內容如下。

```
coordinator=false
http-server.http.port=8881
query.max-memory=50GB
discovery.uri=http://hadoop102:8881
```

在 hadoop104 上設定 Worker 節點。

```
[atguigu@hadoop104 etc]$ vim config.properties
```

增加內容如下。

```
coordinator=false
http-server.http.port=8881
query.max-memory=50GB
discovery.uri=http://hadoop102:8881
```

（11）在 hadoop102 的 /opt/module/hive 目錄下，以 atguigu 角色啟動 Hive Metastore。

```
[atguigu@hadoop102 hive]$
nohup bin/hive --service metastore >/dev/null 2>&1 &
```

（12）分別在 hadoop102、hadoop103、hadoop104 上啟動 Presto Server。

① 在前台啟動 Presto，主控台顯示的記錄檔如下。

```
[atguigu@hadoop102 presto]$ bin/launcher run
[atguigu@hadoop103 presto]$ bin/launcher run
[atguigu@hadoop104 presto]$ bin/launcher run
```

② 在後台啟動 Presto。

```
[atguigu@hadoop102 presto]$ bin/launcher start
[atguigu@hadoop103 presto]$ bin/launcher start
[atguigu@hadoop104 presto]$ bin/launcher start
```

（13）檢視記錄檔的路徑為 /opt/module/presto/data/var/log。

2. Presto 命令列用戶端的安裝

（1）下載 Presto 的用戶端。

（2）將 presto-cli-0.196-executable.jar 上傳到 hadoop102 的 /opt/module/presto 目錄下。

（3）修改檔案名稱。

```
[atguigu@hadoop102 presto]$ mv presto-cli-0.196-executable.jar  prestocli
```

（4）增加執行許可權。

```
[atguigu@hadoop102 presto]$ chmod +x prestocli
```

（5）執行 prestocli 指令，啟動 Presto 用戶端。

```
[atguigu@hadoop102 presto]$ ./prestocli --server hadoop102:8881 --catalog
hive --schema default
```

（6）Presto 命令列操作。
Presto 的命令列操作相當於 Hive 的命令列操作。每張表必須加上 Schema。
例如：

```
select * from schema.table limit 100
```

3. Presto 視覺化用戶端安裝

（1）將 yanagishima-18.0.zip 上傳到 hadoop102 的 /opt/module 目錄下。

（2）解壓 yanagishima-18.0.zip。

```
[atguigu@hadoop102 module]$ unzip yanagishima-18.0.zip
cd yanagishima-18.0
```

（3）進入 /opt/module/yanagishima-18.0/conf 目錄，建立 yanagishima.properties 設定檔。

```
[atguigu@hadoop102 conf]$ vim yanagishima.properties
```

在設定檔中增加以下內容。

```
jetty.port=7080
presto.datasources=atiguigu-presto
presto.coordinator.server.atiguigu-presto=http://hadoop102:8881
catalog.atiguigu-presto=hive
schema.atiguigu-presto=default
sql.query.engines=presto
```

（4）在 /opt/module/yanagishima-18.0 目錄下執行以下指令啟動 Presto。

```
[atguigu@hadoop102 yanagishima-18.0]$
nohup bin/yanagishima-start.sh >y.log 2>&1 &
```

（5）啟動 Web 頁面（http://hadoop102:7080），即可查詢相關資訊。

（6）檢視 Presto 表結構，如圖 8-2 所示。

圖 8-2 檢視 Presto 表結構

在 "Treeview" 頁面下可以檢視所有表的結構，包含 Schema、Table、Column 等。

舉例來說，執行 select * from hive.dw_weather.mid_daily_news_dt limit 10。

每張表後面都有一個複製圖示，點擊此圖示會複製完整的表名，然後在上面的文字標籤中輸入 SQL 敘述即可，如圖 8-3 所示。

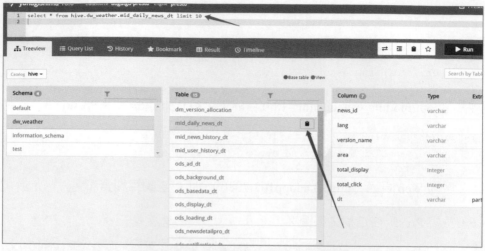

圖 8-3　Presto 複製錶名輸入 SQL 敘述

還可以查詢清單中其他的表格，舉例來說，執行 select * from hive.dw_weather.tmp_news_click limit 10，按 Ctrl+Enter 組合鍵顯示查詢結果，如圖 8-4 所示。

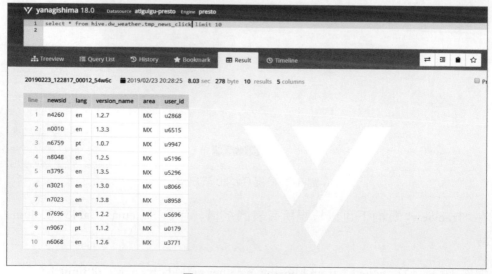

圖 8-4　Presto 查詢結果

8.1.3 Presto 最佳化之資料儲存

想要使用 Presto 更高效率地查詢資料，需要在資料儲存方面利用一些最佳化方法。

1. 合理設定分區

與 Hive 類似，Presto 會根據中繼資料資訊讀取分區資料，合理地設定分區能減少 Presto 資料讀取量，提升查詢效能。

2. 使用 ORC 格式儲存

Presto 對 ORC 檔案讀取進行了特定最佳化，因此，在 Hive 中建立 Presto 使用的表時，建議採用 ORC 格式儲存。相對於 Parquet 格式，Presto 對 ORC 格式支援得更好。

3. 使用壓縮

對資料進行壓縮可以減少節點伺服器間的資料傳輸對 I/O 頻寬產生的壓力，即席查詢需要快速解壓，建議採用 Snappy 格式壓縮。

8.1.4 Presto 最佳化之查詢 SQL

想要使用 Presto 更高效率地查詢資料，需要在撰寫查詢 SQL 敘述方面利用一些最佳化方法。

1. 只選擇需要的欄位

由於採用列式儲存，所以只選擇需要的欄位可加快欄位的讀取速度，減少資料量。避免採用 * 讀取所有欄位。

```
[GOOD]: select time, user, host from tbl


[BAD]:  select * from tbl
```

2. 過濾條件必須加上分區欄位

對於有分區的表，where 敘述中優先使用分區欄位進行過濾。acct_day 是分區欄位，visit_time 是實際存取時間。

```
[GOOD]: select time, user, host from tbl where acct_day=20171101

[BAD]:  select * from tbl where visit_time=20171101
```

3. group by 敘述最佳化

合理安排 group by 敘述中欄位的執行順序，對效能有一定的提升。將 group by 敘述中的欄位按照每個欄位去重後的資料量進行降冪排列。

```
[GOOD]: select group by uid, gender

[BAD]:  select group by gender, uid
```

4. order by 時使用 limit

order by 時需要掃描資料到單一 Worker 節點進行排序，導致單一 Worker 節點需要佔用大量記憶體。如果是查詢 Top N 或 Bottom N，則使用 limit 可減少排序計算時間和記憶體壓力。

```
[GOOD]: select * from tbl order by time limit 100

[BAD]:  select * from tbl order by time
```

5. 使用 join 敘述時將大表放在左邊

Presto 中 join 的預設演算法是 broadcast join，即將 join 左邊表的資料分割到多個 Worker 節點，然後將 join 右邊表的資料整個複製一份發送到每個 Worker 節點進行計算。如果右邊的表資料量太大，則可能會報記憶體溢位錯誤。

```
[GOOD] select ... from large_table l join small_table s on l.id = s.id
[BAD] select ... from small_table s join large_table l on l.id = s.id
```

8.1.5 Presto 注意事項

使用 Presto 需要注意以下幾項。

1. 欄位名稱參考

避免欄位名稱與關鍵字衝突：MySQL 對與關鍵字衝突的欄位名稱加反引號，Presto 對與關鍵字衝突的欄位名稱加雙引號。當然，如果欄位名稱不與關鍵字衝突，則可以不加雙引號。

2. 時間函數

在 Presto 中比較時間的時候，需要增加 timestamp 關鍵字，而在 MySQL 中可以直接進行比較。

```
/*MySQL中的寫法*/
select t from a where t > '2017 01-01 00:00:00';

/*Presto中的寫法*/
select t from a where t > timestamp '2017-01-01 00:00:00';
```

3. 不支援 insert overwrite 語法

Presto 中不支援 insert overwrite 語法，只能先刪除資料，然後插入資料。

4. Parquet 格式

Presto 目前支援 Parquet 格式，支援查詢，但不支援 insert。

8.2 Druid

Druid 是一個高性能的即時分析資料庫。它在 PB 級資料處理、毫秒級查詢、資料即時處理方面,比傳統的 OLAP 系統有顯著的效能提升。

8.2.1 Druid 簡介

Druid 具有以下技術特點。

- 列式儲存格式。Druid 使用列式儲存格式,它只需要載入特定查詢所需要的列。查詢速度快。
- 可擴充的分散式系統。Druid 通常部署在數十到數百台伺服器的叢集中,並且提供數百萬筆 / 秒的攝取率,保留數百萬筆記錄,以及具有微秒到幾秒的查詢延遲。
- 大規模的平行查詢。Druid 可以在整個叢集中進行大規模的平行查詢。
- 即時攝取或批次處理。Druid 可以即時攝取資料(即時取得的資料可立即用於查詢)或批次處理資料。
- 自愈、自平衡、易操作。叢集的擴充和縮小,只需增加或刪除伺服器,叢集將在後台自動重新平衡,無須停機。
- 資料有效地進行了預聚合或預計算,查詢速度快。
- 結果資料應用了 Bitmap 壓縮技術。

Druid 適用於以下幾種場景。

- 適用於將清洗好的記錄即時輸入,但不需要更新操作的場景。
- 適用於支援寬表,不用 join 的場景(換句話說就是一張單表)。
- 適用於即時性要求高的場景。
- 適用於對資料品質的敏感度不高的場景。

8.2.2 Druid 架構原理

Druid 整體架構如圖 8-5 所示。

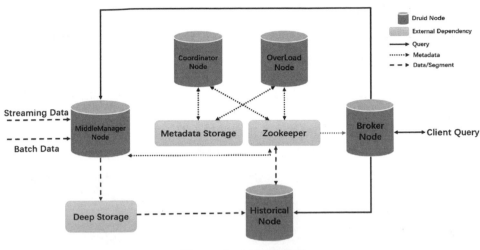

圖 8-5 Druid 整體架構

Druid 的整體架構包含以下四種節點。

- 中間管埋節點（MiddleManager Node）：及時攝入即時資料，並產生 Segment 資料檔案。
- 歷史節點（Historical Node）：載入已產生的資料檔案，以供使用者查詢資料。
- 查詢節點（Broker Node）：對外提供資料查詢服務，並同時從中間管理節點與歷史節點中查詢資料，合併後傳回呼叫方。
- 協調節點（Coordinator Node）：負責歷史節點的資料負載平衡，以及透過規則（Rule）管理資料的生命週期。

叢集還包含以下三種外部依賴。

- 中繼資料庫（Metadata Storage）：儲存 Druid 叢集的中繼資料資訊，舉例來說，Segment 的相關資訊，一般使用 MySQL 或 PostgreSQL 儲存。
- 分散式協調服務（Zookeeper）：為 Druid 叢集提供一致性協調服務的元件，通常為 Zookeeper。
- 資料檔案儲存函數庫（Deep Storage）：儲存產生的 Segment 資料檔案，並提供歷史伺服器下載功能，對於單節點叢集，可以是本機磁碟，而對於分散式叢集，一般是 HDFS 或 NFS。

8.2.3 Druid 資料結構

以 DataSource 和 Segment 為基礎的 Druid 資料結構與 Druid 架構相輔相成，它們共同成就了 Druid 的高性能優勢。

DataSource 相當於關聯式資料庫中的表（Table）。DataSource 的結構如下。

- 時間列：表明每行資料的時間值，預設使用 UTC 時間格式且精確到毫秒級。
- 維度列：維度來自 OLAP 的概念，用來標識資料行的各個類別資訊。
- 指標列：用於聚合和計算的列。通常是一些數字，計算操作包含 Count、Sum 等。

DataSource 結構如表 8-1 所示。

表 8-1　DataSource 結構

時間列（Timestamp）	維 度 列		指標列（click）
	publisher	country	
2019-01-01T01:01:35Z	www.atguigu.com	China	0
2019-01-01T01:03:35Z	www.atguigu.com	China	1
2019-01-01T02:05:35Z	www.baidu.com	China	1
2019-01-01T02:07:35Z	www.baidu.com	China	0
2019-01-01T03:09:35Z	www.google.com	USA	1
2019-01-01T03:10:35Z	www.google.com	USA	0

無論是即時攝取資料還是批次處理資料，Druid 在基於 DataSource 結構儲存資料時，可選擇對任意的指標列進行聚合操作。該聚合操作主要以維度列與時間列為基礎。表 8-2 顯示的是 DataSource 聚合後的資料結構。

表 8-2　DataSource 聚合後的資料結構

時間列（Timestamp）	維 度 列		指標列（click）
	publisher	country	
2019-01-01T01:03:35Z	www.atguigu.com	China	1
2019-01-01T02:05:35Z	www.baidu.com	China	1
2019-01-01T03:09:35Z	www.google.com	USA	1

在資料儲存時便可對資料進行聚合操作是 Druid 的特點,該特點使得 Druid 不僅能夠節省儲存空間,而且能夠加強匯總查詢的效率。

DataSource 是一個邏輯概念,Segment 是資料的實際物理儲存格式。Druid 將不同時間範圍內的資料儲存在不同的 Segment 資料區塊中,這便是所謂的資料水平切割。按照時間水平切割資料,避免了全表查詢,相當大地加強了效率。

在 Segment 中,採用列式儲存格式對資料進行壓縮儲存(Bitmap 壓縮技術),這便是所謂的資料垂直切割。

8.2.4 Druid 安裝(單機版)

1. 安裝部署

從 imply 貞面下載 Druid 最新版本的安裝套件,imply 整合了 Druid,提供了 Druid 從部署到設定再到各種視覺化工具的完整解決方案。

(1)將 imply-2.7.10.tar.gz 上傳到 hadoop102 的 /opt/software 目錄下,並解壓到 /opt/module 目錄下。

```
[atguigu@hadoop102 software]$ tar -zxvf imply-2.7.10.tar.gz -C /opt/module
```

(2)修改 imply-2.7.10 的名稱為 imply。

```
[atguigu@hadoop102 module]$ mv imply-2.7.10 imply
```

(3)修改設定檔。
① 修改 Druid 的 ZK 設定。

```
[atguigu@hadoop102 _common]$ vim /opt/module/imply/conf/druid/_common/
common.runtime.properties
```

需要修改的內容如下。

```
druid.zk.service.host=hadoop102:2181,hadoop103:2181,hadoop104:2181
```

② 修改啟動指令參數，使其不驗證、不啟動內建 ZK。

```
[atguigu@hadoop102 supervise]$
vim /opt/module/imply/conf/supervise/quickstart.conf
```

需要修改的內容如下。

```
:verify bin/verify-java
#:verify bin/verify-default-ports
#:verify bin/verify-version-check
:kill-timeout 10

#!p10 zk bin/run-zk conf-quickstart
```

（4）啟動。

① 啟動 Zookeeper。

```
[atguigu@hadoop102 imply]$ zk.sh start
```

② 透過 bin 目錄下的 supervise 指令啟動 imply。

```
[atguigu@hadoop102 imply]$ bin/supervise  -c conf/supervise/quickstart.conf
```

說明：每啟動一個服務均會列印出一筆記錄檔。讀者可以在 /opt/module/imply/var/sv/ 目錄下檢視服務啟動時的記錄檔資訊。

③ 啟動擷取 Flume 和 Kafka（主要是為了節省記憶體負擔，同時將 hadoop102 的記憶體調整為 8GB）。

```
[atguigu@hadoop102 imply]$ f1.sh start
[atguigu@hadoop102 imply]$ kf.sh start
```

按 Ctrl + C 組合鍵中斷監督處理程序，如果想在中斷服務後重新啟動，則需要刪除 /opt/module/imply/var 目錄。

2. Web 頁面使用

（1）啟動記錄產生程式（延遲時間 1s 發送一筆記錄檔）。

```
[atguigu@hadoop102 server]$ lg.sh 1000 5000
```

（2）在瀏覽器網址列中輸入 hadoop102:9095/datasets/，開啟 imply Web UI 頁面，如圖 8-6 所示。

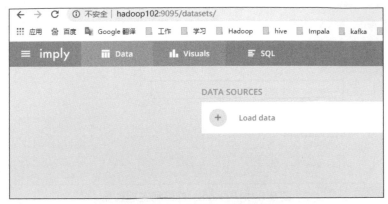

圖 8-6　imply Web UI 頁面

（3）點擊 "Load data" 按鈕，然後點擊 "Apache Kafka" 按鈕，如圖 8-7 所示。

圖 8-7　載入 Kafka 來源資料

（4）增加 Kafka 叢集和主題資訊，如圖 8-8 所示。

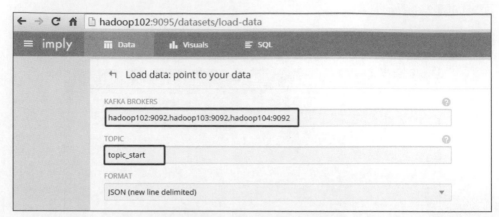

圖 8-8 增加 Kafka 叢集和主題資訊

（5）確認資料樣本格式，如圖 8-9 所示。

圖 8-9 確認資料樣本格式

（6）載入資料，必須有時間欄位，如圖 8-10 所示。

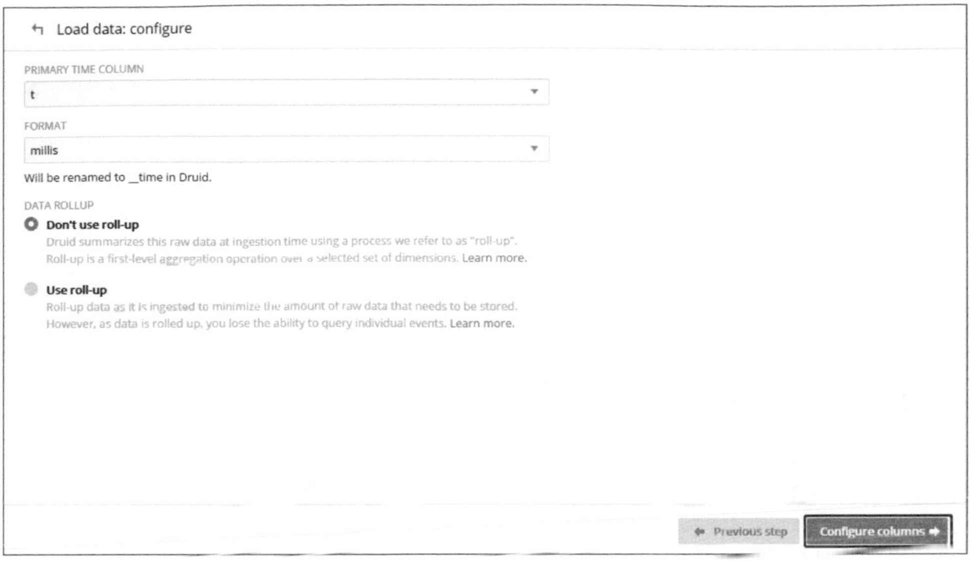

圖 8-10 載入資料設定

（7）設定要載入的列，如圖 8-11 所示。

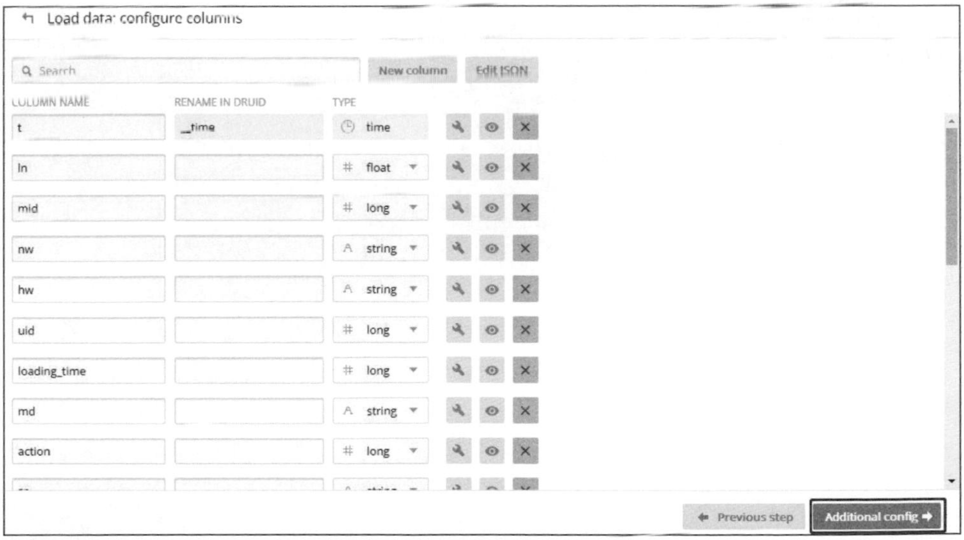

圖 8-11 設定要載入的列

（8）設定 Kafka 資料的主題名稱，如圖 8-12 所示。

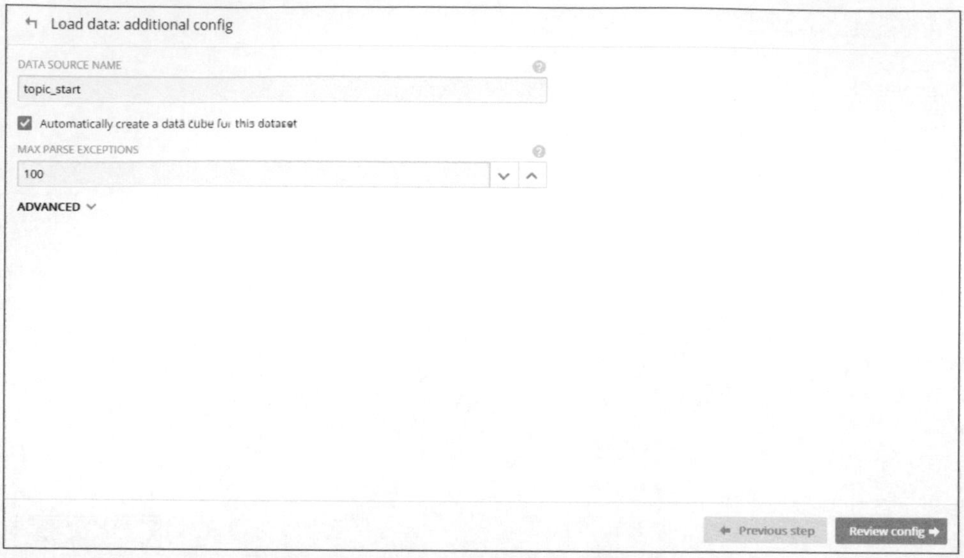

圖 8-12　設定 Kafka 資料的主題名稱

（9）確認載入資料的設定，如圖 8-13 所示。

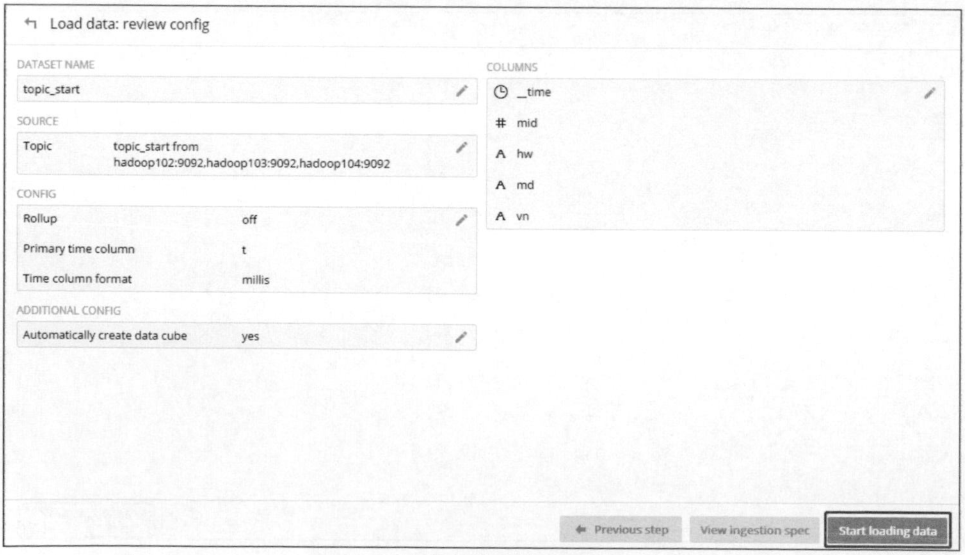

圖 8-13　確認載入資料的設定

（10）連接 Kafka 的 topic_start，如圖 8-14 和圖 8-15 所示。

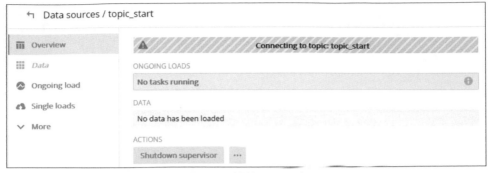

圖 8-14　連接 Kafka 中

圖 8-15　連接 Kafka 成功

（11）選擇 "SQL"，如圖 8-16 所示。

圖 8-16　選擇 "SQL" 選項

（12）查詢指標，如圖 8-17 所示。

圖 8-17 查詢指標

8.3 Kylin

Apache Kylin 是一個開放原始碼的分散式分析引擎，提供 Hadoop/Spark 之上的 SQL 查詢介面及多維分析功能，支援超大規模資料，最初由 eBay 開發並貢獻至開放原始碼社區。它能在微秒內查詢極大的 Hive 表。

8.3.1 Kylin 簡介

Kylin 架構如圖 8-18 所示。

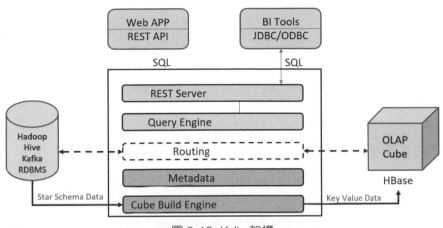

圖 8-18　Kylin 架構

Kylin 具有以下幾個關鍵元件。

1. REST Server

REST Server 是針對應用程式開發的進入點，旨在完成針對 Kylin 平台的應用程式開發工作，可以提供查詢、取得結果、觸發 Cube 建置任務、取得中繼資料及取得使用者許可權等功能，另外可以透過 REST API 介面實現 SQL 查詢。

2. Query Engine

當 Cube 準備就緒後，查詢引擎即可取得並解析使用者所查詢的問題。它隨後會與系統中的其他元件進行互動，進一步向使用者傳回對應的結果。

3. Routing

在最初設計時，設計者曾考慮將 Kylin 不能執行的查詢啟動到 Hive 中繼續執行，但在實作後發現，Hive 與 Kylin 的查詢速度差異過大，導致使用者無法對查詢的速度有一致的期望，大多數查詢幾秒內就傳回結果了，而有些查詢則要等幾分鐘到幾十分鐘，因此，使用者體驗非常糟糕。最後這個路由功能在發行版本中被預設關閉。

4. Metadata

Kylin 是一款中繼資料驅動型應用程式。中繼資料管理工具是一大關鍵性元件，用於對保存在 Kylin 中的所有中繼資料進行管理，其中包含最為重要的 Cube 中繼資料。其他元件的正常運作都需要以中繼資料管理工具為基礎。Kylin 的中繼資料儲存在 HBase 中。

5. Cube Build Engine

Cube 建置引擎的設計目的在於處理所有離線任務，其中包含 Shell 指令稿、Java API、Map Reduce 任務等。Cube Build Engine 對 Kylin 中的全部任務加以管理與協調，進一步確保每項任務都能獲得切實執行並解決期間出現的故障。

Kylin 的主要特點及說明如下。

- 支援標準 SQL 介面：Kylin 以標準的 SQL 作為對外服務的介面。
- 支援超大規模資料集：Kylin 對巨量資料的支撐能力是目前所有技術中較為領先的。早在 2015 年 eBay 的生產環境中就能支援百億筆記錄的秒級查詢，之後在行動的應用場景中又有了支援千億筆記錄的秒級查詢的案例。
- 微秒級回應：Kylin 擁有優異的查詢回應速度，這點得益於預計算，很多複雜的計算，舉例來說，連接、聚合，在離線的預計算過程中就已經完成，這大幅降低了查詢時刻所需的計算量，加強了回應速度。
- 高伸縮性和高吞吐量：單節點 Kylin 可實現每秒 70 個查詢，還可以架設 Kylin 的叢集。
- BI 工具整合。Kylin 可以與現有的 BI 工具整合。

Kylin 開發團隊還貢獻了 Zeppelin 的外掛程式，使用者可以使用 Zeppelin 存取 Kylin 服務。

8.3.2 HBase 安裝

在安裝 Kylin 前需要先安裝部署好 Hadoop、Hive、Zookeeper 和 HBase，並且需要在 /etc/profile 目錄下設定 HADOOP_HOME、HIVE_HOME、HBASE_HOME 環境變數，注意執行 source/etc/profile 指令使其生效。

（1）保障 Zookeeper 叢集的正常部署，並啟動它。

```
[atguigu@hadoop102 zookeeper-3.4.10]$ bin/zkServer.sh start
[atguigu@hadoop103 zookeeper-3.4.10]$ bin/zkServer.sh start
[atguigu@hadoop104 zookeeper-3.4.10]$ bin/zkServer.sh start
```

（2）保障 Hadoop 叢集的正常部署，並啟動它。

```
[atguigu@hadoop102 hadoop-2.7.2]$ sbin/start-dfs.sh
[atguigu@hadoop103 hadoop-2.7.2]$ sbin/start-yarn.sh
```

（3）解壓 HBase 安裝套件到指定目錄。

```
[atguigu@hadoop102 software]$ tar -zxvf hbase-1.3.1-bin.tar.gz -C /opt/
module
```

（4）修改 HBase 對應的設定檔。

①hbase-env.sh 檔案的修改內容如下。

```
export JAVA_HOME=/opt/module/jdk1.8.0_144
export HBASE_MANAGES_ZK=false
```

②hbase-site.xml 檔案的修改內容如下。

```
<configuration>
 <property>
  <name>hbase.rootdir</name>
  <value>hdfs://hadoop102:9000/hbase</value>
 </property>

 <property>
  <name>hbase.cluster.distributed</name>
  <value>true</value>
 </property>

 <!-- 0.98版本後新變動的內容，之前版本沒有.port，預設通訊埠為60000 -->
 <property>
  <name>hbase.master.port</name>
  <value>16000</value>
 </property>
```

```
<property>
  <name>hbase.zookeeper.quorum</name>
      <value>hadoop102,hadoop103,hadoop104</value>
</property>

<property>
  <name>hbase.zookeeper.property.dataDir</name>
      <value>/opt/module/zookeeper-3.4.10/zkData</value>
</property>
</configuration>
```

③ 在 regionservers 檔案中增加以下內容。

```
hadoop102
hadoop103
hadoop104
```

④ 軟連接 Hadoop 設定檔到 HBase。

```
[atguigu@hadoop102 module]$ ln -s /opt/module/hadoop-2.7.2/etc/hadoop/
core-site.xml
/opt/module/hbase/conf/core-site.xml
[atguigu@hadoop102 module]$ ln -s /opt/module/hadoop-2.7.2/etc/hadoop/
hdfs-site.xml
/opt/module/hbase/conf/hdfs-site.xml
```

⑤ 將 HBase 遠端發送到其他叢集。

```
[atguigu@hadoop102 module]$ xsync hbase/
```

（5）啟動 HBase 服務。

① 啟動方式 1，敘述如下。

```
[atguigu@hadoop102 hbase]$ bin/hbase-daemon.sh start master
[atguigu@hadoop102 hbase]$ bin/hbase-daemon.sh start regionserver
```

> ✦ 提示：如果叢集之間的節點時間不同步，會導致 regionserver 無法啟動，並
> 拋出 ClockOutOfSyncException 異常。讀者可參考 3.3.1 節中的相關內容。

② 啟動方式 2，敘述如下。

```
[atguigu@hadoop102 hbase]$ bin/start-hbase.sh
```

（6）停止 HBase 服務，敘述如下。

```
[atguigu@hadoop102 hbase]$ bin/stop-hbase.sh
```

（7）檢視 HBase 頁面。

HBase 服務啟動成功後，可以透過 "host:port" 的方式來存取 HBase 頁面：
http://hadoop102:16010。

8.3.3 Kylin 安裝

（1）下載 Kylin 安裝套件。

（2）解壓 apache-kylin-2.5.1-bin-hbase1x.tar.gz 到 /opt/module/ 目錄下。

```
[atguigu@hadoop102 software]$ tar -zxvf apache-kylin-2.5.1-bin-hbase1x.
tar.gz -C /opt/module/
```

> ✦ **注意**：啟動前需檢查 /etc/profile 目錄中的 HADOOP_HOME、HIVE_HOME
> 和 HBASE_HOME 是否設定完畢。

（3）啟動。

① 在啟動 Kylin 之前，需要先啟動 Hadoop（HDFS、YARN、JobHistoryServer）、
Zookeeper 和 HBase。需要注意的是，要同時啟動 Hadoop 的歷史伺服器，對
Hadoop 叢集設定進行以下修改。

■ 設定 mapred-site.xml 檔案。

```
[atguigu@hadoop102 hadoop]$ vim mapred-site.xml
<property>
    <name>mapreduce.jobhistory.address</name>
    <value>hadoop102:10020</value>
</property>
<property>
```

```
    <name>mapreduce.jobhistory.webapp.address</name>
    <value>hadoop102:19888</value>
</property>
```

■ 設定 yarn-site.xml 檔案。

```
[atguigu@hadoop102 hadoop]$ vim yarn-site.xml
<!-- 記錄檔聚集功能開啟 -->
<property>
    <name>yarn.log-aggregation-enable</name>
    <value>true</value>
</property>
<!-- 記錄檔保留時間設定為7天 -->
<property>
    <name>yarn.log-aggregation.retain-seconds</name>
    <value>604800</value>
</property>
```

■ 修改設定後，分發設定檔，重新啟動 Hadoop 的 HDFS 和 YARN 的所有處理程序。
■ 啟動 Hadoop 的歷史伺服器。

```
[atguigu@hadoop102 hadoop-2.7.2]$ sbin/mr-jobhistory-daemon.sh start
historyserver
```

② 啟動 Kylin。

```
[atguigu@hadoop102 kylin]$ bin/kylin.sh start
```

啟動之後檢視各台節點伺服器的處理程序。

```
-------------------- hadoop102 ----------------
3360 JobHistoryServer
31425 HMaster
3282 NodeManager
3026 DataNode
53283 Jps
2886 NameNode
44007 RunJar
2728 QuorumPeerMain
```

```
31566 HRegionServer
-------------------- hadoop103 ----------------
5040 HMaster
2864 ResourceManager
9729 Jps
2657 QuorumPeerMain
4946 HRegionServer
2979 NodeManager
2727 DataNode
-------------------- hadoop104 ----------------
4688 HRegionServer
2900 NodeManager
9848 Jps
2636 QuorumPeerMain
2700 DataNode
2815 SecondaryNameNode
```

在瀏覽器網址列中輸入 hadoop102:7070/kylin/login，檢視 Web 頁面，如圖 8-19 所示。

圖 8-19 Kylin 的 Web 頁面

使用者名稱為 ADMIN，密碼為 KYLIN（系統已填）。

（4）關閉 Kylin。

```
[atguigu@hadoop102 kylin]$ bin/kylin.sh stop
```

8.3.4 Kylin 使用

以 gmall 資料庫中的 dwd_fact_payment_info 表作為事實資料表，以 dwd_dim_base_province、dwd_dim_user_info_his 表作為維度資料表，建置星形模型，並示範如何使用 Kylin 進行 OLAP 分析。

1. 建立專案

（1）點擊 ➕ 圖示建立專案，如圖 8-20 所示。

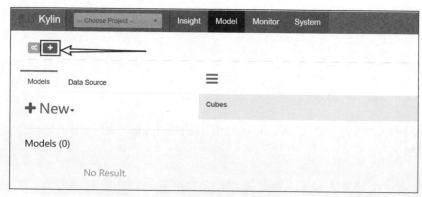

圖 8-20　建立專案入口

（2）填寫專案名稱和描述資訊，並點擊 "Submit" 按鈕提交，如圖 8-21 所示。

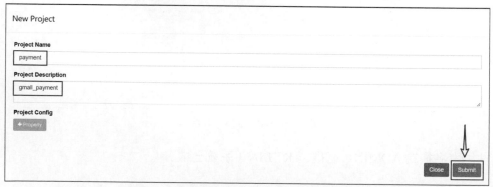

圖 8-21　填寫專案名稱和描述資訊並提交

2. 取得資料來源

（1）選擇 "Data Source" 標籤，如圖 8-22 所示。

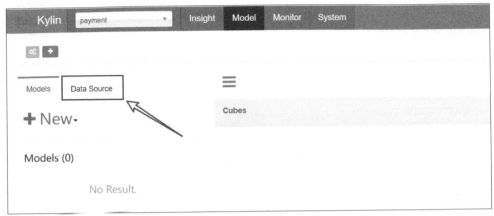

圖 8-22　選擇 "Data Source" 標籤

（2）點擊圖 8-23 中箭頭所指的圖示，進入 Hive 表。

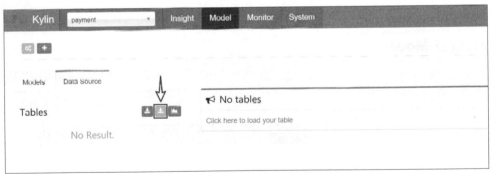

圖 8-23　Hive 表匯入圖示

（3）選擇所需資料表,並點擊 "Sync" 按鈕,如圖 8-24 所示。

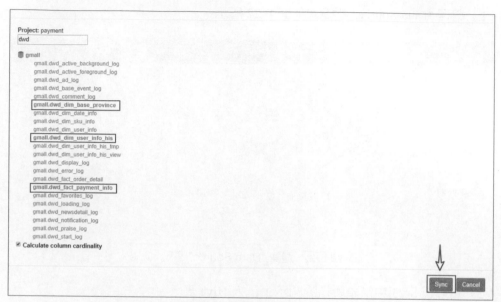

圖 8-24 選擇表格並同步

3. 建立 Model

（1）選擇 "Models" 標籤,然後點擊 "New" 按鈕,接著點擊 "New Model" 按鈕,如圖 8-25 所示。

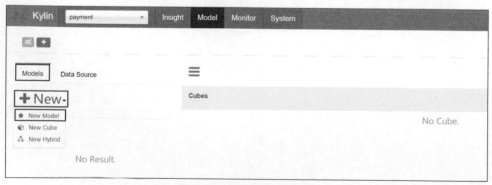

圖 8-25 建立 Model 示意

（2）填寫 Model 資訊，然後點擊 "Next" 按鈕，如圖 8-26 所示。

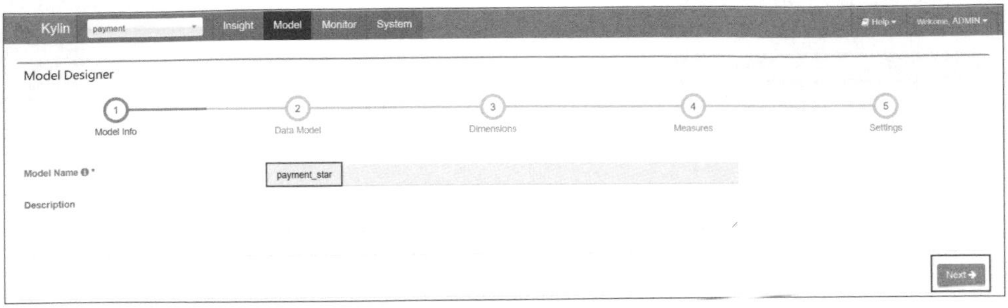

圖 8-26　填寫 Model 資訊

（3）選擇事實資料表，如圖 8-27 所示。

圖 8-27　選擇事實資料表

（4）選擇維度資料表，並指定事實資料表和維度資料表的連接條件，然後點擊 "OK" 按鈕，如圖 8-28 所示。

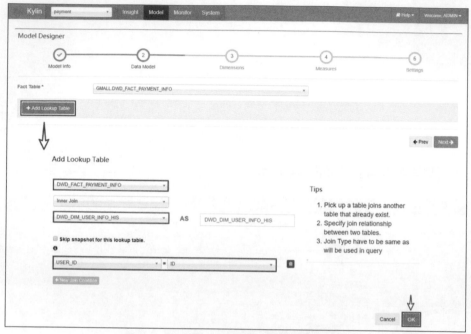

圖 8-28 選擇維度資料表並指定連接條件

維度資料表增加完畢之後，點擊 "Next" 按鈕，如圖 8-29 所示。

圖 8-29 維度資料表增加完畢

（5）指定維度欄位，並點擊 "Next" 按鈕，如圖 8-30 所示。

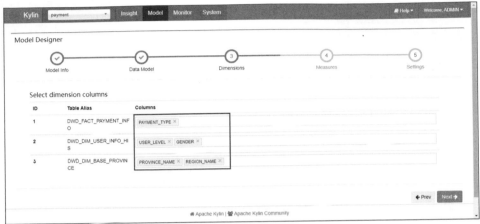

圖 8-30 指定維度欄位

（6）指定度量欄位，並點擊 "Next" 按鈕，如圖 8-31 所示。

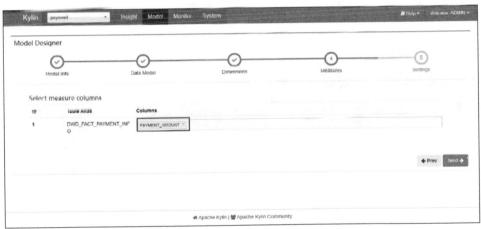

圖 8-31 指定度量欄位

（7）指定事實資料表分區欄位（僅支援時間分區），並點擊 "Save" 按鈕，如圖 8-32 所示，Model 建立完畢。

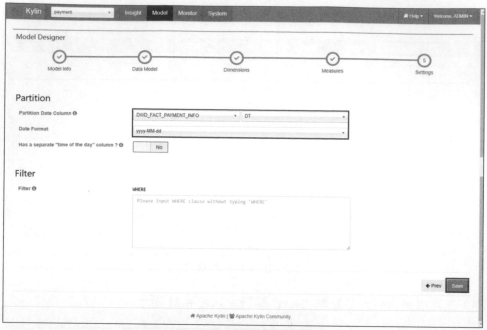

圖 8-32 指定事實資料表分區欄位

4. 建置 Cube

（1）點擊 "New Cube" 按鈕，如圖 8-33 所示。

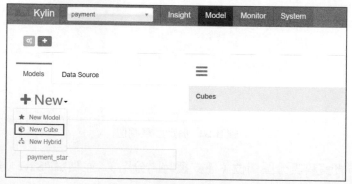

圖 8-33 建置 Cube 入口

（2）填寫 Cube 資訊，選擇 Cube 所依賴的 Model，並點擊 "Next" 按鈕，如圖 8-34 所示。

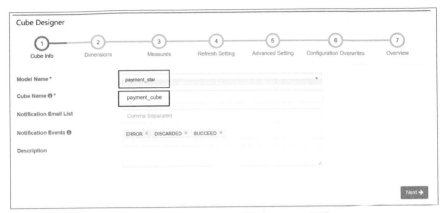

圖 8-34 選擇 Model 並設定 Cube 名稱

（3）選擇 Cube 所需的維度，如圖 8-35 所示。

圖 8-35 選擇 Cube 所需的維度

（4）選擇 Cube 所需的度量值，如圖 8-36 所示。

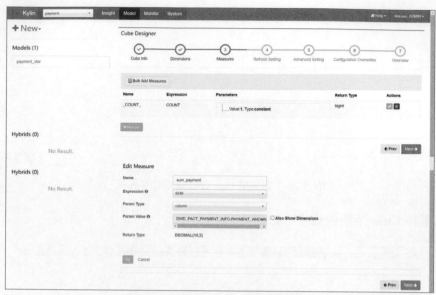

圖 8-36　選擇 Cube 所需的度量值

（5）Cube 自動合併設定。每天 Cube 需按照日期分區欄位進行建置，每次建置的結果會儲存到 HBase 的一張表中，為加強查詢效率，需將每日的 Cube 進行合併，此處可設定合併週期，如圖 8-37 所示。

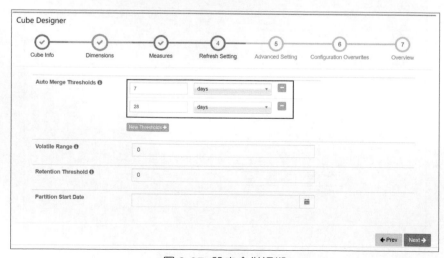

圖 8-37　設定合併週期

（6）Kylin 進階設定（最佳化相關設定，暫時跳過），如圖 8-38 所示。

圖 8-38 Kylin 進階設定

（7）Kylin 屬性值覆蓋相關設定，如圖 8-39 所示。

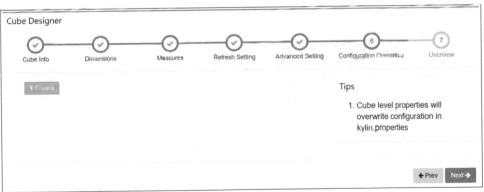

圖 8-39 Kylin 屬性值覆蓋相關設定

（8）Cube 設計資訊總覽，如圖 8-40 所示，點擊 "Save" 按鈕，Cube 建立完成。

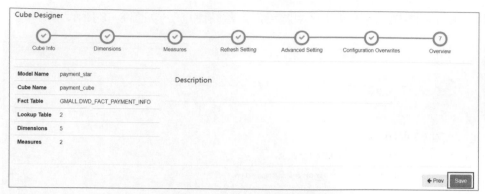

圖 8-40　Cube 設計資訊總覽

（9）建置 Cube（計算），點擊對應 Cube 的 "Action" 下拉按鈕，選擇 "Build" 選項，如圖 8-41 所示。

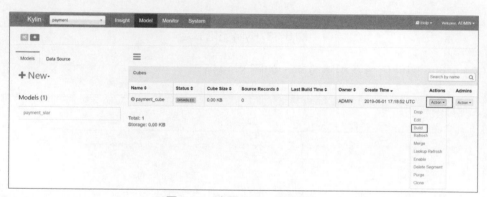

圖 8-41　建置 Cube（計算）

（10）選擇要建置的時間區間，並點擊 "Submit" 按鈕，如圖 8-42 所示。

圖 8-42　選擇要建置的時間區間

（11）選擇 "Monitor"，檢視建置進度，如圖 8-43 所示。

圖 8-43　檢視建置進度

5. 使用進階

1）如何處理每日全量維度資料表

如果按照上述流程建置 Cube，則會出現如圖 8-44 所示的錯誤。

```
Output

org.apache.kylin.engine.mr.exrenties !!.JoupJHU11Exception: java.lang.RuntimeException: Checking snapshot of TableRef[DWD_ORDER_INFO] failed.
    at org.apache.kylin.cube.cli.DictionaryGeneratorCLI.processSegment(DictionaryGeneratorCLI.java:103)
    at org.apache.kylin.cube.cli.DictionaryGeneratorCLI.processSegment(DictionaryGeneratorCLI.java:50)
    at org.apache.kylin.engine.mr.steps.CreateDictionaryJob.run(CreateDictionaryJob.java:73)
    at org.apache.kylin.engine.mr.MRUtil.runMRJob(MRUtil.java:92)
    at org.apache.kylin.engine.mr.common.HadoopShellExecutable.doWork(HadoopShellExecutable.java:63)
    at org.apache.kylin.job.execution.AbstractExecutable.execute(AbstractExecutable.java:164)
    at org.apache.kylin.job.execution.DefaultChainedExecutable.doWork(DefaultChainedExecutable.java:70)
    at org.apache.kylin.job.execution.AbstractExecutable.execute(AbstractExecutable.java:164)
    at org.apache.kylin.job.impl.threadpool.DefaultScheduler$JobRunner.run(DefaultScheduler.java:113)
    at java.util.concurrent.ThreadPoolExecutor.runWorker(ThreadPoolExecutor.java:1149)
    at java.util.concurrent.ThreadPoolExecutor$Worker.run(ThreadPoolExecutor.java:624)
    at java.lang.Thread.run(Thread.java:748)
Caused by: java.lang.IllegalStateException: The table: DWD_DIM_USER_INFO DUP key found, key=[1], value1=[1,651,3,1,1,2828202075,2019-02-10 08:10:13,0,2019-02-1
1 12:47:17.0,2019-02-10], value2=[1,651,3,1,1,2828202075,2019-02-10 08:10:13.0,2019-02-11 12:47:17.0,2019-02-11]
    at org.apache.kylin.dict.lookup.LookupTable.initRow(LookupTable.java:86)
    at org.apache.kylin.dict.lookup.LookupTable.init(LookupTable.java:69)
```

圖 8-44　建置流程顯示出錯

錯誤原因分析如下。

出現上述錯誤的原因是 Model 中的維度資料表 dwd_dim_user_info_his 為拉鏈表，所以使用整張表作為維度資料表，必然會出現同一個 user_id 對應多筆資料的問題。針對上述問題，有以下兩種解決方案。

方案一：在 Hive 中建立維度資料表的臨時表，並在該臨時表中儲存前一天的分區資料，在 Kylin 中建立模型時選擇該臨時表作為維度資料表。

方案二：與方案一想法相同，但不使用物理臨時表，而使用視圖（view）實現相同的功能。

此處採用方案二。

（1）建立維度資料表格視圖（使用視圖取得前一天的分區資料）。

```
-- 拉鍊維度資料表格視圖
create view dwd_dim_user_info_his_view as select * from dwd_dim_user_
info_his where end_date='9999-99-99';

-- 全量維度資料表格視圖
create view dwd_dim_sku_info_view as select * from dwd_dim_sku_info where
dt=date_add(current_date,-1);
```

（2）在 DataSource 中匯入新建立的視圖，如圖 8-45 所示，可選擇將之前的維度資料表刪除。

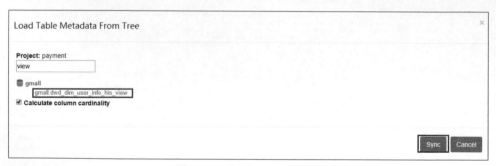

圖 8-45　匯入新建立的視圖

（3）重新建立 Model、Cube。

（4）重新查詢結果，如圖 8-46 所示。

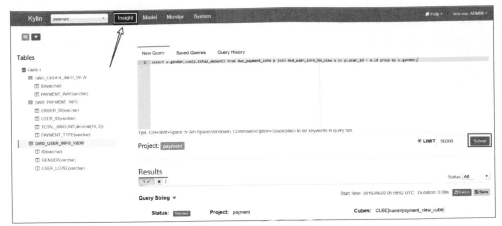

圖 8-46 重新查詢結果

2）如何實現每日自動建置 Cube

Kylin 提供了 REST API，因此，我們可以將建置 Cube 的指令寫到指令稿中，然後將指令稿交給 Azkaban 或 Oozie 排程工具，以實現定時排程的功能。

指令稿如下。

```bash
#! /bin/bash
cube_name=payment_view_cube
do_date=`date -d '-1 day' +%F`

# 取得00:00的時間戳記
start_date_unix=`date -d "$do_date 08:00:00" +%s`
start_date=$(($start_date_unix*1000))

# 取得24:00的時間戳記
stop_date=$(($start_date+86400000))

curl -X PUT -H "Authorization: Basic QURNSU46S1lMSU4=" -H 'Content-Type:
application/json' -d '{"startTime":'$start_date', "endTime":'$stop_date',
"buildType":"BUILD"}' http://hadoop102:7070/kylin/api/cubes/$cube_name/
build
```

8.3.5 Kylin Cube 建置原理

Apache Kylin 的工作原理本質上是 MOLAP（Multidimension On-Line Analysis Processing）Cube，也就是多維立方體分析，是資料分析中非常經典的理論，下面簡介。

維度：觀察資料的角度。舉例來說，關於員工資料，可以從性別角度來觀察，也可以更加細化，從入職時間或地區的角度來觀察。維度是一組離散的值，舉例來說，性別中的男和女，或時間維度上的每個獨立的日期。因此，在統計時可以將維度值相同的記錄聚合在一起，然後應用匯總函數進行累加，以及求平均值、最大值和最小值等聚合計算。

度量：被聚合（觀察）的統計值，也就是聚合運算的結果。舉例來說，員工資料中不同性別員工的人數，又如，在同一年入職的員工數。

有了維度和度量，就可以對一張資料表或一個資料模型上的所有欄位進行分類了，它們不是維度，就是度量（可以被聚合）。於是就有了根據維度和度量進行預計算的 Cube 理論，如圖 8-47 所示。

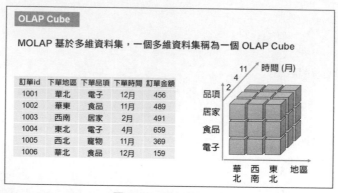

圖 8-47 OLAP Cube

指定一個資料模型，我們可以對其上的所有維度進行聚合，對 N 個維度來說，組合的所有可能性共有 2^N 種。對每種維度組合的度量值進行聚合計算，然後將結果儲存為一個物化視圖，稱為 Cuboid。所有維度組合的 Cuboid 作為一個整體，稱為 Cube，如圖 8-48 所示。

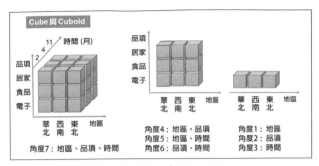

圖 8-48 Cube 與 Cuboid

接下來介紹 Kylin Cube 的建置演算法，主要分為兩種：逐層建置演算法和快速建置演算法。

1. 逐層建置演算法

我們知道，一個 N 維的 Cube 是由 1 個 N 維子立方體、N 個 N-1 維子立方體、$N \times (N-1)/2$ 個 N-2 維子立方體、…、N 個 1 維子立方體和 1 個 0 維子立方體組成的，一共由 2^N 個子立方體組成，在逐層建置演算法中，按維度數逐層減少來計算，每個層級的結果（除了第一層，第一層是從原始資料聚合而來的）是以它上一層級的結果為基礎聚合得出的。舉例來說，[Group by A, B] 的結果，可以以 [Group by A, B, C] 的結果為基礎，透過去掉 C 後聚合得出，這樣可以減少重複計算。當 0 維度 Cuboid 計算出來的時候，整個 Cube 的計算也就完成了，計算過程如圖 8-49 所示。

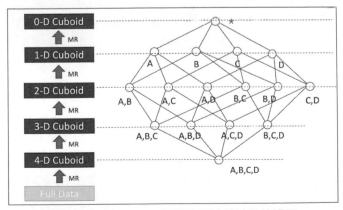

圖 8-49 Kylin Cube 逐層建置演算法計算過程

每輪的計算都是一個 MapReduce 任務，且串列執行；一個 N 維的 Cube，至少需要執行 N 次 MapReduce 任務，如圖 8-50 所示。

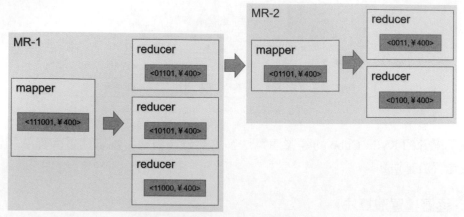

圖 8-50 MapReduce 任務建置 Cube

逐層建置演算法的優點如下。

（1）此演算法充分利用 MapReduce 的優點，處理了中間複雜的排序和 Shuffle 工作，所以建置演算法的程式清晰簡單、易於維護。

（2）受益於 Hadoop 的日趨成熟，此演算法非常穩定，即使在叢集資源緊張時，也能保障工作最後完成。

逐層建置演算法的缺點如下。

（1）當 Cube 有較多維度的時候，所需要的 MapReduce 任務也會對應增加；由於 Hadoop 的任務排程需要耗費額外資源，特別是叢集較龐大的時候，反覆遞交任務造成的額外負擔會相當大。

（2）由於 mapper 邏輯中並未進行聚合操作，所以每輪 MR 的 Shuffle 工作量都很大，進一步導致效率不佳。

（3）對 HDFS 的讀 / 寫入操作較多：由於每層計算的輸出會被用作下一層計算的輸入，所以這些 key-value 需要寫到 HDFS 上；當所有計算都完成後，Kylin 還需要執行額外的一輪任務將這些檔案轉成 HBase 的 HFile 格式，以匯入 HBase 中。

整體而言，該演算法的效率較低，尤其是當 Cube 維度數較多的時候。

2. 快速建置演算法

快速建置演算法也被稱作「逐段」（By Segment）或「逐塊」（By Split）演算法，Kylin Cube 從 1.5.x 版本開始引用該演算法，該演算法的主要思想是每個 mapper 將其所分配到的資料區塊計算成一個完整的小 Cube 段（包含所有的 Cuboid）。每個 mapper 將計算完成的 Cube 段輸出給 reducer 並進行合併，產生大 Cube，也就是最後結果。Kylin 快速建置演算法的計算過程如圖 8-51 所示。

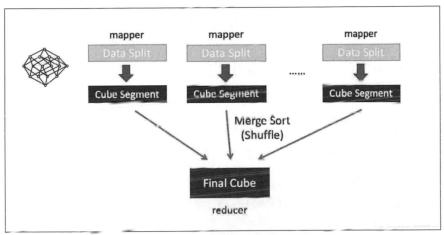

圖 8-51 Kylin 快速建置演算法的計算過程

與逐層建置演算法相比，快速建置演算法主要有兩點不同。

（1）mapper 會利用記憶體進行預聚合，算出所有組合；mapper 輸出的每個 key 都是不同的，這樣會減少輸出到 Hadoop MapReduce 的資料量，也不再需要 Combiner。

（2）執行一輪 MapReduce 任務便會完成所有層次的計算，減少了 Hadoop 任務的轉換，如圖 8-52 所示。

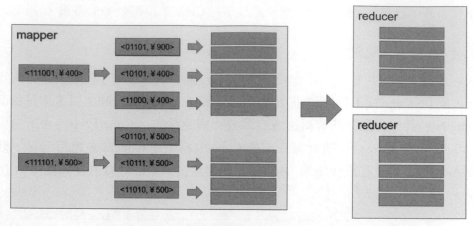

圖 8-52 快速建置演算法 MapReduce 示意

8.3.6 Kylin Cube 建置最佳化

1. 衍生維度

衍生維度用於在有效維度內將維度資料表中的非主鍵維度排除，並使用維度資料表中的主鍵（其實是事實資料表中對應的外鍵）來代替它們。Kylin 會在底層記錄維度資料表主鍵與維度資料表其他維度之間的對映關係，以便在查詢時能夠動態地將維度資料表中的主鍵「翻譯」成非主鍵維度，並進行即時聚合，如圖 8-53 所示。

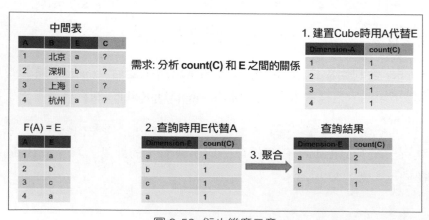

圖 8-53 衍生維度示意

雖然衍生維度具有非常大的優勢，但並不是說所有維度資料表中的維度都需要變成衍生維度，如果從維度資料表主鍵到維度資料表某個維度所需要的聚合工作量非常大，則不建議使用衍生維度。

2. 聚合組

聚合組（Aggregation Group）是一個強大的剪枝工具。聚合組假設一個 Cube 的所有維度均可以根據業務需求劃分成許多組（當然也可以是一個組），由於同一個分組內的維度更可能同時被同一個查詢用到，因此會表現出更加緊密的內在連結。每個分組的維度集合均是 Cube 所有維度的子集，不同的分組各自擁有一個維度集合，它們可能與其他分組有相同的維度，也可能沒有相同的維度。每個分組各自獨立地根據本身的規則貢獻出一批需要被物化的 Cuboid，所有分組貢獻的 Cuboid 的聯集就成了目前 Cube 中所有需要物化的 Cuboid 的集合。不同的分組有可能會貢獻出相同的 Cuboid，建置引擎會察覺到這一點，並保障每個 Cuboid 無論在多少個分組中出現，它們都只會被物化一次。

對於每個分組內部的維度，使用者可以使用以下三種可選的方式定義它們之間的關係。

（1）強制維度：如果一個維度被定義為強制維度，那麼在這個分組產生的所有 Cuboid 中每個 Cuboid 都會包含該維度。每個分組中都可以有 0 個、1 個或多個強制維度。如果根據這個分組的業務邏輯，該維度一定會在過濾條件或分組條件中，則可以在該分組中把該維度設定為強制維度，如圖 8-54 所示。

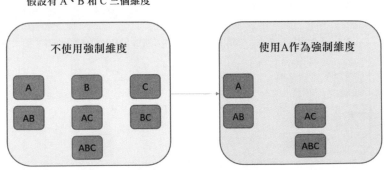

圖 8-54　強制維度示意

（2）層級維度：每個層級包含兩個或多個維度。假設一個層級中包含 $D_1,D_2,\cdots,D_n$ n 個維度，那麼在該分組產生的任何 Cuboid 中，這 n 個維度只會以（ ）,（D_1）,（D_1,D_2）,$\cdots$,（$D_1,D_2,\cdots,D_n$）$n+1$ 種形式中的一種出現。每個分組中可以有 0 個、1 個或多個層級，不同的層級之間不應有共用的維度。如果根據這個分組的業務邏輯，多個維度之間存在層級關係，則可以在該分組中把這些維度設定為層級維度，如圖 8-55 所示。

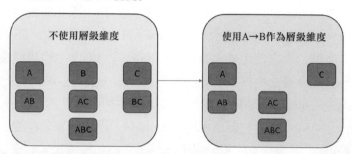

圖 8-55 層級維度示意

（3）聯合維度，每個聯合中包含兩個或多個維度，如果某些列形成一個聯合，那麼在該分組產生的任何 Cuboid 中，這些聯合維度不是一起出現，就是都不出現。每個分組中可以有 0 個或多個聯合，但是不同的聯合之間不應有共用的維度（否則它們將合併成一個聯合）。如果根據這個分組的業務邏輯，多個維度在查詢中總是同時出現，則可以在該分組中把這些維度設定為聯合維度，如圖 8-56 所示。

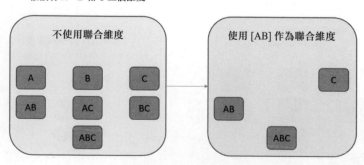

圖 8-56 聯合維度示意

上述操作可以在 Cube Designer 的 Advanced Setting 中的 Aggregation Groups
區域完成，如圖 8-57 所示。

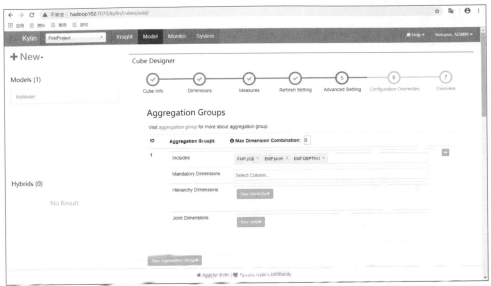

圖 8-57　使用聚合組建置最佳化的示意

聚合組的設計非常靈活，甚至可以用來描述一些極端的設計。假設我們的業
務需求非常單一，只需要某些特定的 Cuboid，那麼可以建立多個聚合組，每
個聚合組代表一個 Cuboid。實際的方法是在聚合組中先包含某個 Cuboid 所需
的所有維度，然後把這些維度都設定為強制維度。這樣目前的聚合組就只能
產生我們想要的那個 Cuboid 了。

有的時候，Cube 中有一些基數非常大的維度，如果不進行特殊處理，它們就
會和其他的維度進行各種組合，進一步產生一大堆包含它們的 Cuboid。包含
高基數維度的 Cuboid 在行數和體積上通常很龐大，這會導致整個 Cube 的膨
脹率變大。如果根據業務需求知道這個高基數的維度只會與許多個維度（而非
所有維度）同時被查詢到，那麼就可以透過聚合組對這個高基數維度進行一
定的「隔離」。我們把這個高基數的維度放入一個單獨的聚合組中，再把所有
可能會與這個高基數維度一起被查詢到的其他維度也放進來。這樣，這個高
基數的維度就被「隔離」在一個聚合組中了，所有不會與它一起被查詢到的

維度都沒有和它一起出現在任何一個分組中,因此也就不會有多餘的 Cuboid 產生。這點也大幅減少了包含該高基數維度的 Cuboid 的數量,可以有效地控制 Cube 的膨脹率。

3. RowKey 最佳化

Kylin 會把所有的維度按照順序組合成一個完整的 RowKey,並且按照這個 RowKey 昇冪排列 Cuboid 中所有的行。

設計良好的 RowKey 可以更有效地完成資料的查詢過濾和定位,減少 I/O 次數,加強查詢速度,RowKey 中的維度次序,對查詢效能有顯著的影響。

RowKey 的設計原則如下。

(1)被用作 where 過濾的維度放在沒有被用作 where 過濾的維度前邊,如圖 8-58 所示。

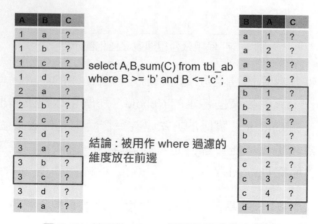

圖 8-58　被用作 where 過濾的維度放在前邊

(2)基數大的維度放在基數小的維度前邊,如圖 8-59 所示。二維 Cuboid, 由 3DCuboid 計算而來,此處,3DCuboid-1110/1101,均可透過計算獲得二維 Cuboid-1100,Kylin 的規則是選擇 Cuboid 小的,即選擇 3DCuboid-1101。我 們應保障 Kylin 所選的 Cuboid-1101 為資料量較小的,而 3DCuboid-1110/1101 的資料量大小實際是由維度 C 和 D 的基數決定的。為保障 Cuboid-1101 的資 料量小於 Cuboid-1110,需保障維度 D 的基數小於維度 C 的基數。

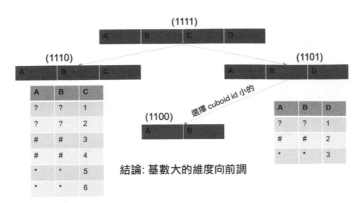

圖 8-59　基數大的維度放在基數小的維度前邊

4. 平行處理粒度最佳化

當 Segment 中某一個 Cuboid 的大小超出一定的設定值時，系統會將該 Cuboid 的資料分配到多個分區中，以實現 Cuboid 資料讀取的平行化，進一步最佳化 Cube 的查詢速度。實際的實現方式如下；建置引擎根據 Segment 估計的大小，以及參數 "kylin.hbase.region.cut" 的設定決定 Segment 在儲存引擎中需要的分區數量，如果儲存引擎是 HBase，那麼分區的數量就對應於 HBase 中的 Region 數量。"kylin.hbase.region.cut" 的預設值是 5.0，單位是 GB，也就是說，對於一個大小是 50GB 的 Segment，建置引擎會給它分配 10 個分區。使用者還可以透過設定 "kylin.hbase.region.count.min"（預設值為 1）和 "kylin.hbase.region.count.max"（預設值為 500）兩個參數來決定每個 Segment 最少或最多被劃分成幾個分區。

由於每個 Cube 的平行處理粒度控制不同，因此建議讀者在 Cube Designer 的 Configuration Overwrites 中進行設定，如圖 8-60 所示，可以通過點擊 "Property" 按鈕，為每個 Cube 量身訂製控制平行處理粒度的參數。假設將目前 Cube 的 "kylin.hbase.region.count.min" 的參數值設定為 2，"kylin.hbase.region.count.max" 的參數值設定為 100。這樣無論 Segment 的大小如何變化，它的分區數量最小都不會小於 2，最大都不會大於 100。相對地，Segment 背後的儲存引擎（HBase）為了儲存 Segment，也不會使用小於 2 個或大於 100 個的分區。我們還調整了 "kylin.hbase.region.cut" 參數的預設值，這樣 50GB

的 Segment 基本上會被分配到 50 個分區中，相比預設設定，Cuboid 最多會獲
得 5 倍的平行處理量。

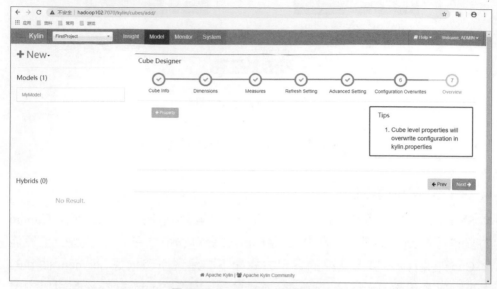

圖 8-60　Kylin Cube 屬性設定

8.3.7　Kylin BI 工具整合

可以與 Kylin 結合使用的視覺化工具有很多，如下所示。

- ODBC：與 Tableau、Excel、Power BI 等工具整合。
- JDBC：與 Saiku、BIRT 等 Java 工具整合。
- REST API：與 JavaScript、Web 網頁整合。

1.　JDBC

（1）新增專案並匯入依賴。

```
<dependencies>
    <dependency>
        <groupId>org.apache.kylin</groupId>
        <artifactId>kylin-jdbc</artifactId>
        <version>2.5.1</version>
```

```
            </dependency>
        </dependencies>
```

（2）編碼，敘述如下。

```java
package com.atguigu;

import java.sql.*;

public class TestKylin {

    public static void main(String[] args) throws Exception {

        // Kylin_JDBC 驅動
        String KYLIN_DRIVER = "org.apache.kylin.jdbc.Driver";

        // Kylin_URL
        String KYLIN_URL = "jdbc:kylin://hadoop102:7070/FirstProject";

        // Kylin的使用者名稱
        String KYLIN_USER = "ADMIN";

        // Kylin的密碼
        String KYLIN_PASSWD = "KYLIN";

        // 增加驅動資訊
        Class.forName(KYLIN_DRIVER);

        // 取得連接
        Connection connection = DriverManager.getConnection(KYLIN_URL,
KYLIN_USER, KYLIN_PASSWD);

        // 預先編譯SQL敘述
        PreparedStatement ps = connection.prepareStatement("SELECT
sum(sal) FROM emp group by deptno");

        // 執行查詢
        ResultSet resultSet = ps.executeQuery();
```

```
        // 檢查列印
        while (resultSet.next()) {
            System.out.println(resultSet.getInt(1));
        }
    }
}
```

（3）結果展示，如圖 8-61 所示。

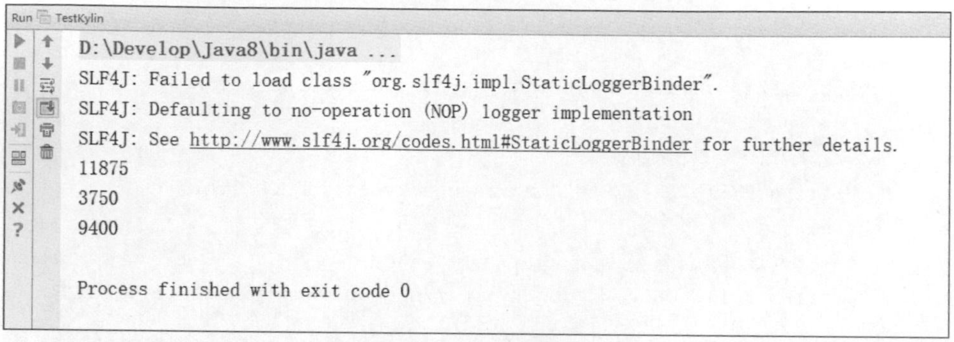

```
Run    TestKylin
       D:\Develop\Java8\bin\java ...
       SLF4J: Failed to load class "org.slf4j.impl.StaticLoggerBinder".
       SLF4J: Defaulting to no-operation (NOP) logger implementation
       SLF4J: See http://www.slf4j.org/codes.html#StaticLoggerBinder for further details.
       11875
       3750
       9400

       Process finished with exit code 0
```

圖 8-61 JDBC 結果展示

2. Zeppelin

1）Zeppelin 安裝與啟動

（1）將 zeppelin-0.8.0-bin-all.tgz 上傳至 Linux。

（2）解壓 zeppelin-0.8.0-bin-all.tgz 到 /opt/module 目錄下。

```
    [atguigu@hadoop102 sorfware]$ tar -zxvf zeppelin-0.8.0-bin-all.tgz -C
/opt/module/
```

（3）修改名稱。

```
    [atguigu@hadoop102 module]$ mv zeppelin-0.8.0-bin-all/ zeppelin
```

（4）啟動 Zeppelin。

```
    [atguigu@hadoop102 zeppelin]$ bin/zeppelin-daemon.sh start
```

讀者可登入 Zeppelin 網頁（http://hadoop102:8080/#/）進行檢視，Web 預設通訊埠為 8080，如圖 8-62 所示。

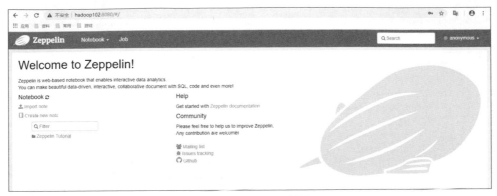

圖 8-62　Zeppelin 網頁

2）設定 Zeppelin 支援 Kylin

（1）選擇 "anonymous" → "Interpreter" 選項，設定解譯器，如圖 8-63 所示。

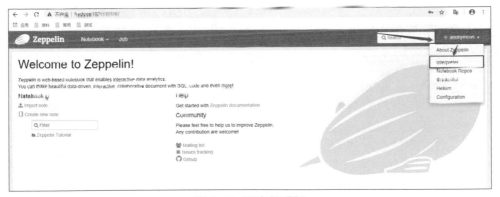

圖 8-63　設定解譯器

（2）搜尋 Kylin 外掛程式並修改對應的設定，如圖 8-64 所示。

圖 8-64　搜尋 Kylin 外掛程式並修改對應的設定

（3）修改完成後，點擊 "Save" 按鈕儲存修改內容，如圖 8-65 所示。

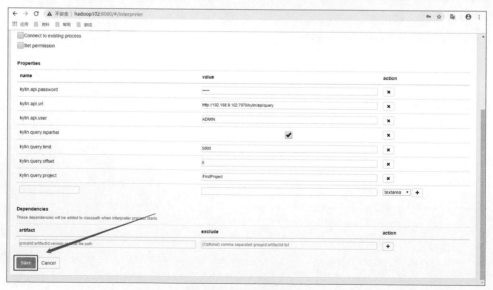

圖 8-65　儲存修改內容

3）案例實操

需求：查詢員工的詳細資訊，並使用各種圖表進行展示。

（1）選擇 "Notebook" → "Creat new note" 選項，建立新的 note，如圖 8-66 所示。

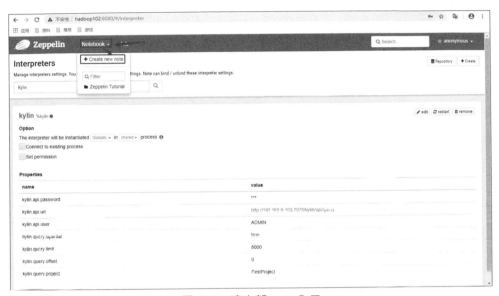

圖 8-66 建立新 note 入口

（2）在 "Note Name" 文字標籤中輸入 "test_kylin" 並點擊 "Create" 按鈕，如圖 8-67 所示，note 建立成功的頁面如圖 8-68 所示。

圖 8-67 建立新 note

可以輸入 SQL 敘述查詢啦，注意：需要以 %kylin 開頭，讓 Zepplin 知道用哪個外掛程式

圖 8-68 note 建立成功的頁面

（3）執行查詢操作，如圖 8-69 所示。

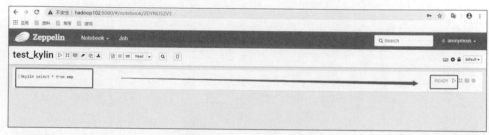

圖 8-69 執行查詢操作

（4）展示結果，如圖 8-70 所示。

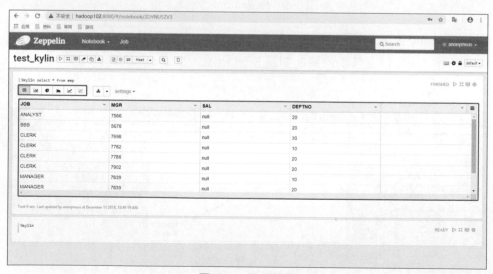

圖 8-70 展示結果

（5）其他圖表格式，如圖 8-71 ～圖 8-75 所示。

圖 8-71　柱狀圖

圖 8-72　圓餅圖

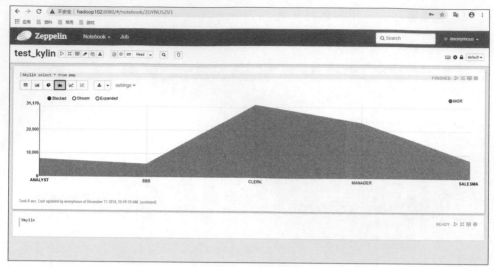

圖 8-73 面積圖

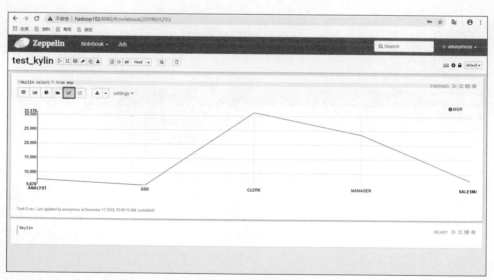

圖 8-74 聚合線圖

圖 8-75　散點圖

8.4　即席查詢架構比較

目前應用比較廣泛的幾種即席查詢架構有 Druid、Kylin、Presto、Impala、Spark SQL 和 Elasticsearch，針對回應時間、資料支援、技術特點等方面的比較如表 8-3 所示。

表 8-3　即席查詢架構比較

比較專案	Druid	Kylin	Presto	Impala	Spark SQL	Elasticsearch
微秒級回應	Y	Y	N	N	N	N
百億資料集	Y	Y	Y	Y	Y	Y
SQL 支援	Y	Y	Y	Y	Y	N
離線	Y	Y	Y	Y	Y	Y
即時	Y	Y	N	N	N	Y
精確去重	N	Y	Y	Y	Y	N
多表 join	N	Y	Y	Y	Y	N
DBC for BI	N	Y	Y	Y	Y	N

針對上表的比較情況，分析整理如下。

- Druid：即時處理時序資料的 OLAP 資料庫，因為它的索引首先按照時間進行分片，查詢的時候也是按照時間線去路由索引的。
- Kylin：核心是 Cube，Cube 是一種預計算技術，基本想法是預先對資料進行多維索引，查詢時只掃描索引而不存取原始資料，進一步加強查詢速度。
- Presto：它沒有使用 MapReduce，大部分場景下比 Hive 快一個數量級，其中的關鍵是所有的處理都在記憶體中完成。
- Impala：基於記憶體運算，速度快，支援的資料來源沒有 Presto 多。
- Spark SQL：基於 Spark 平台上的 OLAP 架構，基本想法是增加機器以實現平行計算，進一步加強查詢速度。
- Elasticsearch：最大的特點是使用倒排索引解決了索引存在的問題。根據研究，Elasticsearch 在資料取得和聚集時用的資源比在 Druid 中高。

架構選型如下。

- 從超巨量資料的查詢效率來看：Druid > Kylin > Presto > Spark SQL。
- 從支援的資料來源種類來看：Presto > Spark SQL > Kylin > Druid。

8.5 本章歸納

本章主要對三個應用比較廣泛的即席查詢架構進行了說明。對資料倉儲系統來說，即席查詢是不可或缺的環節。本章對三個即席查詢架構的特點、安裝部署等方面進行了說明，並對目前比較流行的幾個即席查詢架構進行了比較，在實際應用中，讀者可以根據自己專案的實際情況進行選取。

中繼資料管理模組

企業進行中繼資料管理的方向有三個：一是基於資料平台進行中繼資料管理，由於巨量資料平台的興起，目前企業逐步開始針對 Hadoop 環境進行中繼資料管理；二是基於企業資料整體管理規劃開展對中繼資料的管理，也是企業資料資產管理的基礎；三是將中繼資料作為某個平台的元件進行此平台特有的中繼資料管理，它作為一個仲介或中轉互通平台各元件間的資料。基於資料平台的中繼資料管理相對成熟，也是業界最早進行中繼資料管理的切入點或説是資料平台建設的必備。社區中開放原始碼的中繼資料管理系統方案，常見的有 Hortonworks 主推的 Apache Atlas，本書將以 Atlas 為例對中繼資料管理介紹。

9.1 Atlas 入門

Atlas 為組織提供開放式中繼資料管理和治理功能，用以建置組織的資料資產目錄，對這些資產進行分類和管理，並為資料科學家、資料分析師和資料治理團隊提供圍繞這些資料資產的協作功能。

9.1.1 Atlas 概述

Atlas 的整體設計偏重於資料血緣關係的擷取，以及表格維度的基本資訊和業務屬性資訊的管理。為了達到這個目的，Atlas 設計了一套通用的 Type 系統來描述這些資訊。Type 的主要基礎類型包含 DataSet 和 Process，前者用來描述各種資料來源本身，後者用來描述一個資料處理的流程，舉例來說，一個 ETL 任務。

Atlas 現有的 Bridge 實現，從資料來源的角度來看，主要覆蓋了 Hive、HBase、HDFS 和 Kafka，此外，還有轉換於 Sqoop、Storm 和 Falcon 的 Bridge，不過這三者更多是從 Process 的角度入手的，最後落地的資料來源還是上述四種。

9.1.2 Atlas 架構原理

Atlas 架構如圖 9-1 所示。

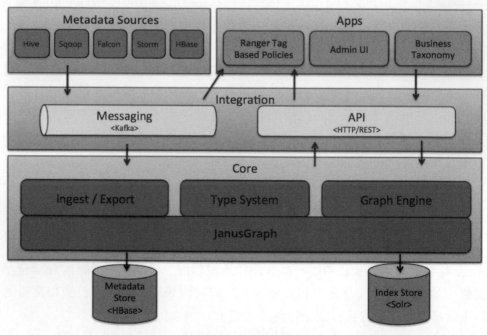

圖 9-1 Atlas 架構

Atlas 架構具有以下關鍵元件。

- Metadata Store<HBase>：採用 HBase 儲存中繼資料。

- Index Store<Solr>：採用 Solr 建索引。
- Ingest/Export：擷取元件允許將中繼資料增加到 Atlas。同樣地，匯出元件將 Atlas 檢測到的中繼資料更改並公開為事件。
- Type System：使用者為他們想要管理的中繼資料物件定義模型。Type System 稱為實體的類型實例，表示受管理的實際中繼資料物件。
- Graph Engine：Atlas 在內部使用 Graph 模型持久儲存它管理的中繼資料物件。
- API<HTTP/REST>：Atlas 的所有功能都透過 REST API 向最後使用者展示，該 API 允許建立、更新和刪除類型和實體。它也是查詢和發現 Atlas 管理的類型和實體的主要機制。
- Messaging<Kafka>：除了 API，使用者還可以選擇使用以 Kafka 為基礎的訊息傳遞介面與 Atlas 整合。
- Metadata Sources：目前，Atlas 支援從以下來源分析和管理中繼資料：HBase、Hive、Sqoop、Storm 和 Falcon。
- Admin UI：該元件是一個以 Web 為基礎的應用程式，允許資料管理員和科學家發現和註釋中繼資料。這裡最重要的是搜尋介面和類似 SQL 的查詢語言，可用於查詢 Atlas 管理的中繼資料類型和物件。
- Ranger Tag Based Policies：許可權管理模組。
- Business Taxonomy：業務分類。

9.2 Atlas 安裝及使用

Atlas 的安裝及使用需要基於 Hadoop、Zookeeper、Kafka、HBase、Solr、Hive、Azkaban 等，所以需要先對以上架構進行安裝部署。

9.2.1 安裝前環境準備

1. 安裝 JDK8、Hadoop 2.7.2

（1）安裝 JDK、Hadoop 叢集，可參考 3.3 節中的相關內容。

（2）啟動 Hadoop 叢集。

```
[atguigu@hadoop102 hadoop-2.7.2]$ sbin/start-dfs.sh
[atguigu@hadoop103 hadoop-2.7.2]$ sbin/start-yarn.sh
```

2. 安裝 Zookeeper 3.4.10

（1）安裝 Zookeeper 叢集，可參考 4.3.1 節中的相關內容。

（2）啟動 Zookeeper 叢集。

```
[atguigu@hadoop102 zookeeper-3.4.10]$ zk.sh start
```

3. 安裝 Kafka 0.11.0.2

（1）安裝 Kafka 叢集，可參考 4.3.3 節中的相關內容。

（2）啟動 Kafka 叢集。

```
[atguigu@hadoop102 kafka]$ kf.sh start
```

4. 安裝 HBase 1.3.1

（1）安裝 HBase 叢集，可參考 8.3.2 節中的相關內容。

（2）啟動 HBase 叢集。

```
[atguigu@hadoop102 hbase]$ bin/start-hbase.sh
```

5. 安裝 Solr 5.2.1

（1）Solr 的版本必須是 5.2.1，參見官方網站。

（2）下載 Solr 安裝套件。

（3）把 solr-5.2.1.tgz 上傳到 hadoop102 的 /opt/software 目錄下。

（4）解壓 solr-5.2.1.tgz 到 /opt/module/ 目錄下。

```
[atguigu@hadoop102 software]$ tar -zxvf solr-5.2.1.tgz -C /opt/module/
```

（5）修改 solr-5.2.1 的名稱為 solr。

```
[atguigu@hadoop102 module]$ mv solr-5.2.1/ solr
```

（6）進入 solr/bin 目錄，修改 solr.in.sh 檔案。

```
[atguigu@hadoop102 solr]$ vim bin/solr.in.sh
# 增加下列指令
ZK_HOST="hadoop102:2181,hadoop103:2181,hadoop104:2181"
SOLR_HOST="hadoop102"
# Sets the port Solr binds to, default is 8983
# 可修改通訊埠
SOLR_PORT=8983
```

（7）分發 Solr，進行 Cloud 模式部署。

```
[atguigu@hadoop102 module]$ xsync solr
```

★ 提示：分發完成後，將每台機器 /opt/module/solr/bin 目錄下的 solr.in.sh 檔案的 "SOLR_HOST=" 修改為對應主機名稱即可。

（8）在三台節點伺服器上分別啟動 Solr，即 Cloud 模式。

```
[atguigu@hadoop102 solr]$ bin/solr start
[atguigu@hadoop103 solr]$ bin/solr start
[atguigu@hadoop104 solr]$ bin/solr start
```

★ 提示：啟動 Solr 前，需要先啟動 Zookeeper 叢集。

（9）透過 Web 網頁存取 8983 通訊埠，可指定三台節點伺服器中任意一台節點伺服器的 IP 位址（http://hadoop102:8983/solr/#/），開啟 Solr UI 介面，如圖 9-2 所示。

★ 提示：只有 UI 介面出現 Cloud 功能表列時，Solr 的 Cloud 模式才算部署成功。

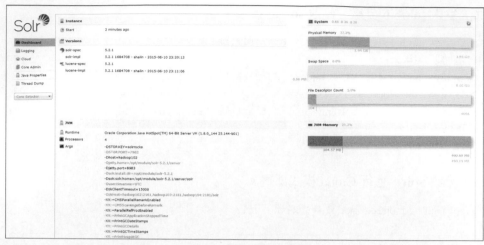

圖 9-2 Solr UI 介面

（10）撰寫 Solr 啟動、停止指令稿。

① 在 hadoop102 的 /home/atguigu/bin 目錄下建立指令稿 s.sh。

```
[atguigu@hadoop102 bin]$ vim s.sh
```

在指令稿中撰寫以下內容。

```
#!/bin/bash

case $1 in
"start"){
    for i in hadoop102 hadoop103 hadoop104
    do
        ssh $i "/opt/module/solr/bin/solr start"
    done
};;
"stop"){
    for i in hadoop102 hadoop103 hadoop104
    do
        ssh $i "/opt/module/solr/bin/solr stop"
    done
};;
esac
```

② 增加指令稿執行許可權。

```
[atguigu@hadoop102 bin]$ chmod 777 s.sh
```

③ Solr 叢集啟動指令稿。

```
[atguigu@hadoop102 module]$ s.sh start
```

④ Solr 叢集停止指令稿。

```
[atguigu@hadoop102 module]$ s.sh stop
```

6. 安裝 Hive 1.2.1

安裝 Hive 1.2.1，可參考 6.2.2 節中的相關內容。

7. 安裝 Azkaban 2.5.0

安裝 Azkaban 2.5.0，可參考 6.9.1 節中的相關內容。

8. 安裝 Atlas 0.8.4

（1）將 apache-atlas-0.8.4-bin.tar.gz 上傳到 hadoop102 的 /opt/software 目錄下。

（2）解壓 apache-atlas-0.8.4-bin.tar.gz 到 /opt/module/ 目錄下。

```
[atguigu@hadoop102 software]$ tar -zxvf apache-atlas-0.8.4-bin.tar.gz
-C /opt/module/
```

（3）修改 apache-atlas-0.8.4 的名稱為 atlas。

```
[atguigu@hadoop102 module]$ mv apache-atlas-0.8.4/ atlas
```

9.2.2 整合外部架構

Atlas 的安裝方式分為整合附帶的 HBase+Solr 和整合外部的 HBase+Solr。通常在企業開發中選擇整合外部的 HBase+Solr 安裝方式，方便專案整體進行整合操作。

1. Atlas 整合 HBase

（1）進入 /opt/module/atlas/conf 目錄，修改設定檔 atlas-application.properties。

```
[atguigu@hadoop102 conf]$ vim atlas-application.properties

# 修改Atlas儲存資料的主機
atlas.graph.storage.hostname=hadoop102:2181,hadoop103:2181,hado
op104:2181
```

（2）進入 /opt/module/atlas/conf/hbase 目錄，增加 HBase 叢集的設定檔到 ${Atlas_Home}。

```
[atguigu@hadoop102 hbase]$ ln -s /opt/module/hbase/conf/ /opt/module/
atlas/conf/hbase/
```

（3）在 /opt/module/atlas/conf 目錄下的 atlas-env.sh 檔案中增加 HBASE_CONF_DIR。

```
[atguigu@hadoop102 conf]$ vim atlas-env.sh

# 增加HBase設定檔路徑
export HBASE_CONF_DIR=/opt/module/atlas/conf/hbase/conf
```

2. Atlas 整合 Solr

（1）進入 /opt/module/atlas/conf 目錄，修改設定檔 atlas-application.properties。

```
[atguigu@hadoop102 conf]$ vim atlas-application.properties

# 修改以下設定
atlas.graph.index.search.solr.zookeeper-url=hadoop102:2181,hadoop103:
2181,hadoop104:2181
```

（2）將 Atlas 設定的 solr 資料夾複製到外部 Solr 叢集的各節點中。

```
[atguigu@hadoop102 conf]$ cp -r /opt/module/atlas/conf/solr/opt/
module/solr/
```

（3）進入 /opt/module/solr 目錄，修改複製過來的 solr 資料夾名稱為 atlas_
conf。

```
[atguigu@hadoop102 solr]$ mv solr atlas_conf
```

（4）在 Cloud 模式下，啟動 Solr（先啟動 Zookeeper 叢集），並建立 collection。

```
[atguigu@hadoop102 solr]$ bin/solr create -c vertex_index -d /opt/
module/solr/atlas_conf -shards 3 -replicationFactor 2

[atguigu@hadoop102 solr]$ bin/solr create -c edge_index -d /opt/
module/solr/atlas_conf -shards 3 -replicationFactor 2

[atguigu@hadoop102 solr]$ bin/solr create -c fulltext_index -d /opt/
module/solr/atlas_conf -shards 3 -replicationFactor 2
```

★ **注意**：如果需要刪除 vertex_index、edge_index、fulltext_index 等 collection，
則可以執行以下敘述。

```
[atguigu@hadoop102 solr]$ bin/solr delete -c ${collection_name}
```

（5）驗證 collection 是否建立成功
登入 Solr Web 主控台（http://hadoop102:8983/solr/#/~cloud），顯示的內容如圖
9-3 所示。

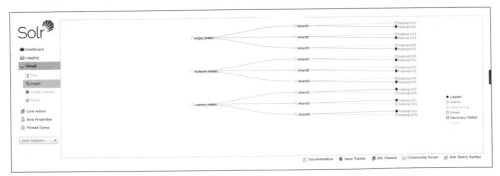

圖 9-3 Solr 主控台顯示的內容

3. Atlas 整合 Kafka

（1）進入 /opt/module/atlas/conf 目錄，修改設定檔 atlas-application.properties。

```
[atguigu@hadoop102 conf]$ vim atlas-application.properties

#########  通知設定  #########
atlas.notification.embedded=false
atlas.kafka.zookeeper.connect=hadoop102:2181,hadoop103:2181,hado
op104:2181
atlas.kafka.bootstrap.servers=hadoop102:9092,hadoop103:9092,hado
op104:9092
atlas.kafka.zookeeper.session.timeout.ms=4000
atlas.kafka.zookeeper.connection.timeout.ms=2000

atlas.kafka.enable.auto.commit=true
```

（2）啟動 Kafka 叢集，並建立 Topic。

```
[atguigu@hadoop102 kafka]$ bin/kafka-topics.sh --zookeeper hadoop102:2181,
hadoop103:2181,hadoop104:2181 --create --replication-factor 3
--partitions 3 --topic _HOATLASOK

[atguigu@hadoop102 kafka]$ bin/kafka-topics.sh --zookeeper hadoop102:2181,
hadoop103:2181,hadoop104:2181 --create --replication-factor 3
--partitions 3 --topic ATLAS_ENTITIES
```

4. Atlas 的其他設定

（1）進入 /opt/module/atlas/conf 目錄，修改設定檔 atlas-application.properties。

```
[atguigu@hadoop102 conf]$ vim atlas-application.properties

#########  伺服器屬性  #########
atlas.rest.address=http://hadoop102:21000
# If enabled and set to true, this will run setup steps when the server
starts atlas.server.run.setup.on.start=false

#########  實體審核設定  #########
atlas.audit.hbase.zookeeper.quorum=hadoop102:2181,hadoop103:2181,hado
op104:2181
```

（2）記錄效能指標，進入 /opt/module/atlas/conf 目錄，修改 atlas-log4j.xml 檔
案。

```
[atguigu@hadoop102 conf]$ vim atlas-log4j.xml

# 去掉以下程式的註釋
<appender name="perf_appender" class="org.apache.log4j.
DailyRollingFileAppender">
    <param name="file" value="${atlas.log.dir}/atlas_perf.log" />
    <param name="datePattern" value="'.'yyyy-MM-dd" />
    <param name="append" value="true" />
    <layout class="org.apache.log4j.PatternLayout">
        <param name="ConversionPattern" value="%d|%t|%m%n" />
    </layout>
</appender>

<logger name="org.apache.atlas.perf" additivity="false">
    <level value="debug" />
    <appender-ref ref="perf_appender" />
</logger>
```

5. Atlas 整合 Hive

（1）進入 /opt/module/atlas/conf 目錄，修改設定檔 atlas-application.properties。

```
[atguigu@hadoop102 conf]$ vim atlas-application.properties

######### Hive Hook Configs #######
atlas.hook.hive.synchronous=false
atlas.hook.hive.numRetries=3
atlas.hook.hive.queueSize=10000
atlas.cluster.name=primary
```

（2）將 atlas-application.properties 設定檔加入 atlas-plugin-classloader-0.8.4.jar
中。

```
[atguigu@hadoop102 hive]$ zip -u /opt/module/atlas/hook/hive/atlas-
plugin-classloader-0.8.4.jar /opt/module/atlas/conf/atlas-application.
properties
```

```
[atguigu@hadoop102 hive]$ cp /opt/module/atlas/conf/atlas-application.
properties /opt/module/hive/conf/
```

> ✦ **注意**：這個設定不能參照官方網站的做法：將設定檔複製到 Hive 的 conf
> 目錄中。參照官方網站的做法，一直讀取不到 atlas-application.properties 設定
> 檔，筆者檢視原始程式發現這個設定檔是在 classpath 中讀取的，所以將它解
> 壓到 jar 套件中。

（3）在 /opt/module/hive/conf 目錄下的 hive-site.xml 檔案中設定 Atlas hook。

```
[atguigu@hadoop102 conf]$ vim hive-site.xml
<property>
     <name>hive.exec.post.hooks</name>
     <value>org.apache.atlas.hive.hook.HiveHook</value>
</property>

[atguigu@hadoop102 conf]$ vim hive-env.sh

#在Tez引擎依賴的jar套件後面追加與Hive外掛程式相關的jar套件
export HIVE_AUX_JARS_PATH=/opt/module/hadoop-2.7.2/share/hadoop/common/
hadoop-lzo-0.4.20.jar$TEZ_JARS,/opt/module/atlas/hook/hive/atlas-plugin-
classloader-0.8.4.jar,/opt/module/atlas/hook/hive/hive-bridge-shim-
0.8.4.jar
```

9.2.3 叢集啟動

（1）啟動 Hadoop 叢集。

```
[atguigu@hadoop102 hadoop-2.7.2]$ sbin/start-dfs.sh
[atguigu@hadoop103 hadoop-2.7.2]$ sbin/start-yarn.sh
```

（2）啟動 Zookeeper 叢集。

```
[atguigu@hadoop102 zookeeper-3.4.10]$ zk.sh start
```

（3）啟動 Kafka 叢集。

```
[atguigu@hadoop102 kafka]$ kf.sh start
```

（4）啟動 HBase 叢集。

```
[atguigu@hadoop102 hbase]$ bin/start-hbase.sh
```

（5）啟動 Solr 叢集。

```
[atguigu@hadoop102 solr]$ bin/solr start
[atguigu@hadoop103 solr]$ bin/solr start
[atguigu@hadoop104 solr]$ bin/solr start
```

（6）進入 /opt/module/atlas 目錄，重新啟動 Atlas 服務。

```
[atguigu@hadoop102 atlas]$ bin/atlas_stop.py

[atguigu@hadoop102 atlas]$ bin/atlas_start.py
```

錯誤訊息檢視路徑為 /opt/module/atlas/logs/*.out 和 /opt/module/atlas/logs/application.log。

造訪網址：http://hadoop102:21000。
帳戶：admin。
密碼：admin。

9.2.4 匯入 Hive 中繼資料到 Atlas

（1）設定 Hive 環境變數。

```
[atguigu@hadoop102 hive]$ sudo vim /etc/profile

# 設定Hive環境變數
export HIVE_HOME=/opt/module/hive
export PATH=$PATH:$HIVE_HOME/bin/

[atguigu@hadoop102 hive]$ source /etc/profile
```

（2）啟動 Hive。

```
[atguigu@hadoop102 hive]$ hive
```

（3）進入 /opt/module/atlas 目錄，將 Hive 中繼資料匯入 Atlas 中。

```
[atguigu@hadoop102 atlas]$ bin/import-hive.sh

Using Hive configuration directory [/opt/module/hive/conf]
Log file for import is /opt/module/atlas/logs/import-hive.log
log4j:WARN No such property [maxFileSize] in org.apache.log4j.PatternLayout.
log4j:WARN No such property [maxBackupIndex] in org.apache.log4j.
PatternLayout.
```

輸入使用者名稱（admin）和密碼（admin）。

```
Enter username for atlas :- admin
Enter password for atlas :-
Hive Meta Data import was successful!!!
```

9.3 Atlas 介面檢視及使用

9.3.1 檢視基本資訊

1. 登入

（1）在瀏覽器網址列中輸入 http://hadoop102:21000/login.jsp，登入 Atlas，如圖 9-4 所示。

圖 9-4 Atlas 登入

（2）帳號、密碼預設都為 admin，登入成功的頁面如圖 9-5 所示。

圖 9-5　Atlas 登入成功的頁面

2. 檢視 Hive 資料庫

檢視對應的 Hive 資料庫，類型選擇 "hive_db"，如圖 9-6 所示。

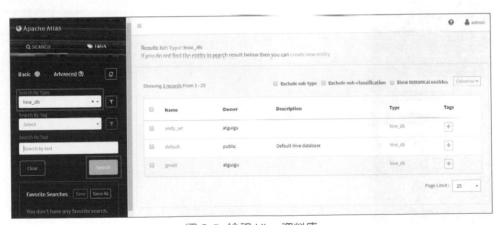

圖 9-6　檢視 Hive 資料庫

3. 檢視 Hive 處理程序

檢視 Hive 處理程序，類型選擇 "hive_process"，如圖 9-7 所示。

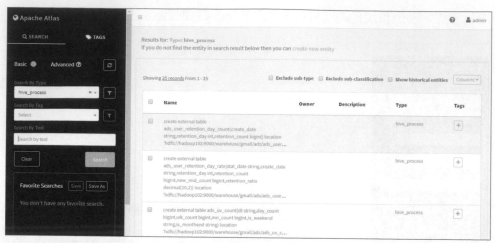

圖 9-7 檢視 Hive 處理程序

4. 檢視 Hive 表

檢視 Hive 表，類型選擇 "hive_table"，如圖 9-8 所示。

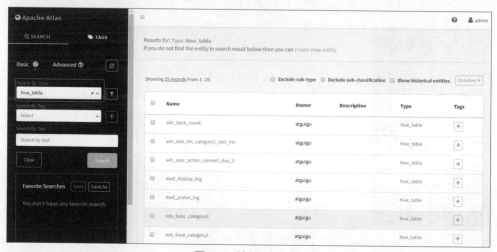

圖 9-8 檢視 Hive 表

5. 檢視 Hive 列

檢視 Hive 列，類型選擇 "hive_column"，如圖 9-9 所示。

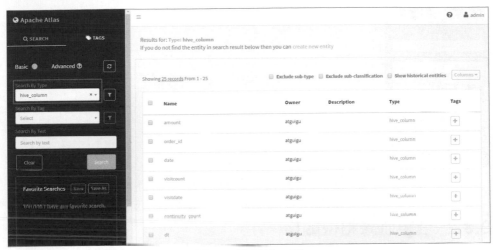

圖 9-9 檢視 Hive 列

6. 篩選查詢準則

舉例來說，要查詢 name 為 id 的列，則在 "Search By Text" 文字標籤中輸入 "where name="id""，其他選項篩選條件的寫法一樣，如圖 9-10 所示。

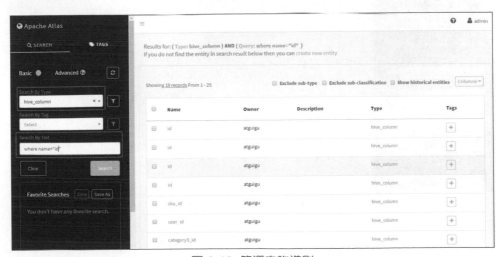

圖 9-10 篩選查詢準則

7. 檢視具有血緣依賴列的資料

檢視類型選擇 "hive_column_lineage"，如圖 9-11 所示。

圖 9-11 檢視具有血緣依賴列的資料

9.3.2 檢視血緣相依關係

1. 第一次檢視表血緣相依關係

（1）先選擇 "hive_db" 類型，然後點擊 "Type" 列的 "hive_db"，檢視對應的資料庫。舉例來説，檢視 gmall 資料庫，如圖 9-12 所示。

圖 9-12 檢視指定資料庫

（2）點擊 "Tables" 按鈕，檢視 gmall 資料庫中的所有表，如圖 9-13 所示。

圖 9-13　檢視 gmall 資料庫中的所有表

（3）選擇 "Propertics" 標籤，檢視表詳情，如圖 9-14 所示。

圖 9-14　檢視表詳情

（4）選擇 "Lineage" 標籤，檢視表血緣相依關係，如圖 9-15 所示。

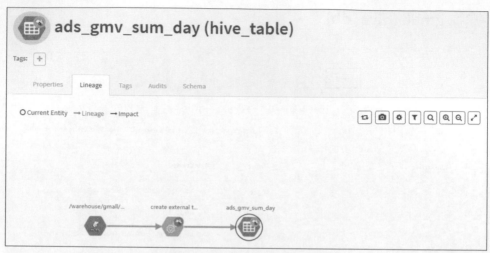

圖 9-15　檢視表血緣相依關係

（5）選擇 "Audits" 標籤，檢視表修改操作對應的時間和詳情，如圖 9-16 所示。

圖 9-16　檢視表修改操作對應的時間和詳情

（6）選擇 "Schema" 標籤，檢視表欄位資訊，如圖 9-17 所示。

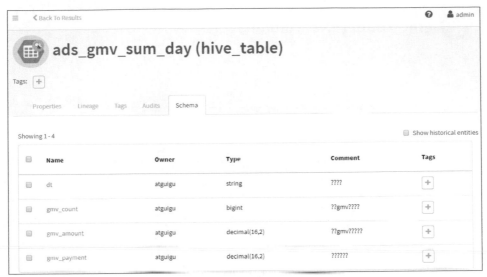

圖 9-17　檢視表欄位資訊

2. 第一次檢視欄位血緣相依關係

（1）點擊 "gmv_count" 欄位檢視對應資訊，如圖 9-18 所示。

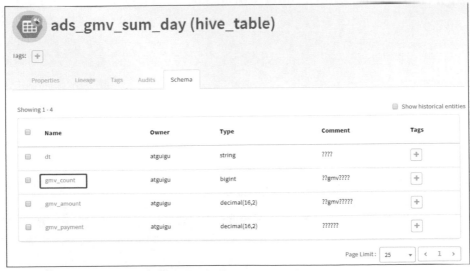

圖 9-18　檢視 gmv_count 欄位

（2）選擇 "Properties" 標籤，顯示欄位詳情，如圖 9-19 所示。

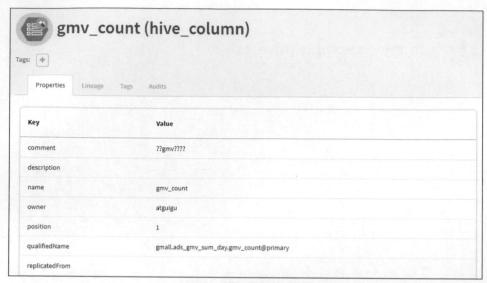

圖 9-19 檢視欄位詳情

（3）選擇 "Lineage" 標籤，檢視欄位血緣關係，如圖 9-20 所示。

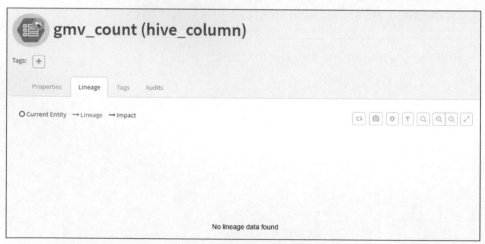

圖 9-20 檢視欄位血緣關係

（4）選擇 "Audits" 標籤，檢視欄位修改操作對應的時間和詳情，如圖 9-21 所示。

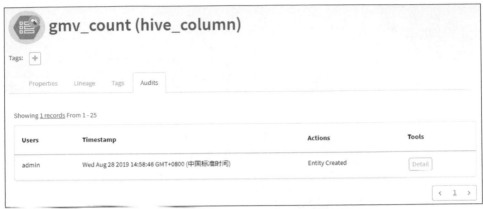

圖 9-21 檢視欄位修改操作對應的時間和詳情

3. 啟動 GMV 全流程任務

（1）啟動 Azkaban。

① 啟動 Executor 伺服器。在 Exccutor 伺服器目錄下執行啟動指令。

```
[atguigu@hadoop102 executor]$ pwd
/opt/module/azkaban/executor
[atguigu@hadoop102 executor]$ bin/azkaban-executor-start.sh
```

② 啟動 Web 伺服器。在 Azkaban Web 伺服器目錄下執行啟動指令。

```
[atguigu@hadoop102 server]$ pwd
/opt/module/azkaban/server
[atguigu@hadoop102 server]$ bin/azkaban-web-start.sh
```

③ 檢視 Web 頁面：https://hadoop102:8443。

（2）上傳任務。參考 6.9.3 節中的相關內容。

（3）等待 Azkaban 執行結束，檢視結果。

（4）檢視 Atlas 表血緣相依關係，如圖 9-22 所示。

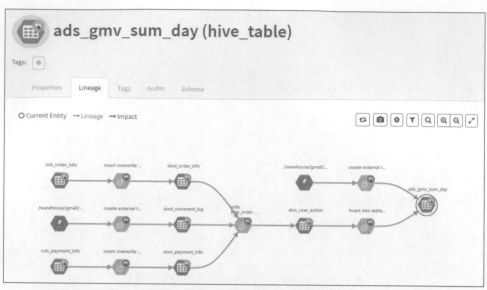

圖 9-22 檢視 Atlas 表血緣相依關係

（5）檢視 Atlas 欄位血緣相依關係，如圖 9-23 所示。

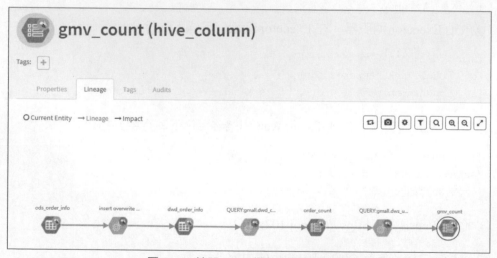

圖 9-23 檢視 Atlas 欄位血緣相依關係

9.4 本章歸納

本章以 Atlas 為例，主要從 Atlas 的概述和架構原理、Atlas 安裝前的環境準備、與外部架構的整合、介面的檢視及使用幾個方面，為讀者說明了巨量資料的中繼資料管理系統。Atlas 只是為巨量資料的中繼資料管理提供了一種解決方案，其本身也存在一定的限制，讀者如果有興趣，可以對 Atlas 進行深入了解或探索學習其他的中繼資料管理架構。

Note

Note

Note